THE NATURE OF UNIDENTIFIED GALACTIC HIGH-ENERGY GAMMA-RAY SOURCES

ASTROPHYSICS AND SPACE SCIENCE LIBRARY

VOLUME 267

THE NATURE OF UNIDENTIFIED GALACTIC HIGH-ENERGY GAMMA-RAY SOURCES

Proceedings of the Workshop held at Tonantzintla, Puebla, México, 9–11 October 2000

edited by

ALBERTO CARRAMIÑANA

Instituto Nacional de Astrofísica,
Óptica y Electrónica,
Tonantzintla, México

OLAF REIMER

and

DAVID J. THOMPSON

NASA/Goddard Space Flight Center,
Maryland, U.S.A.

KLUWER ACADEMIC PUBLISHERS

DORDRECHT / BOSTON / LONDON

A C.I.P. Catalogue record for this book is available from the Library of Congress.

ISBN 1-4020-0010-3

Published by Kluwer Academic Publishers,
P.O. Box 17, 3300 AA Dordrecht, The Netherlands.

Sold and distributed in North, Central and South America
by Kluwer Academic Publishers,
101 Philip Drive, Norwell, MA 02061, U.S.A.

In all other countries, sold and distributed
by Kluwer Academic Publishers,
P.O. Box 322, 3300 AH Dordrecht, The Netherlands.

Printed on acid-free paper

Printed in the Netherlands.

This book is dedicated to Carl Fichtel, Robert Hofstadter, and Klaus Pinkau.

Contents

Preface

The Energetic Gamma-Ray Experiment Telescope (EGRET) instrument on the *Compton Gamma-Ray Observatory* left as a legacy its Third Catalog of High Energy Gamma-Ray Sources, whose detections include a large number of blazars, some pulsars, the Large Magellanic Cloud and a solar flare. Most of the newly discovered objects - a majority of the catalog - are unidentified sources, with a clearly predominant Galactic population. Are all these radio-quiet pulsars, like Geminga, or is there a novel type of celestial object, awaiting identification? In spite of the limited angular resolution provided by EGRET and COMPTEL, there is still much to learn about unidentified γ-ray sources: correlation studies, multiwavelength observations and theoretical work can provide valuable clues, specially if these efforts are carried out in a coordinated manner.

The aim of this workshop, held from October 9 to 11, 2000, at the Instituto Nacional de Astrofísica, Óptica y Electrónica, at Tonantzintla, México, was to gather experts on the subject, including observational astronomers specialized in other regions of the electromagnetic spectrum, in an effort to address the question of the Nature of Galactic high-energy gamma-ray sources, both from the theoretical and observational perspective, and elaborate schemes for future identification studies which can make use of existing and forthcoming facilities.

Our thanks are extended to the sponsoring organization: el Consejo Nacional de Ciencia y Tecnología, México;

- to the Scientific Organizing Committee: Alberto Carramiñana (INAOE), Olaf Reimer (GSFC), Gottfried Kanbach (MPE), Dave Thompson (GSFC), Matthew Baring (GSFC), Michael Catanese (SAO), Karen Brazier (Durham), Gustavo Romero (IAR/UNLP), Roger Romani (Stanford), and Dale Frail (NRAO).

- and Local Organizing Committee: Alberto Carramiñana, Divakara Mayya, Daniel Rosa, Gabriela López, and Germán de Santos,

and to the staffs of the Coordinación de Astrofísica, Dirección de Investigación y Posgrado, and Dirección Administrativa of INAOE.

This volume represents the results of this workshop, a look both backward and forward that we hope will provide a useful guide for the next generations of high-energy gamma-ray telescopes.

Alberto Carramiñana, Olaf Reimer & David J. Thompson

Participants to the Workshop

Alvarez, César	INAOE, Tonantzintla	calvarez@inaoep.mx
Baring, Matthew†	NASA/GSFC/USRA	baring@twinkie.gsfc.nasa.gov
Barrera, Pablo	INAOE, Tonantzintla	pbarrera@inaoep.mx
Brazier, Karen	University of Durham	Karen.Brazier@durham.ac.uk
Carramiñana, Alberto	INAOE, Tonantzintla	alberto@inaoep.mx
Combi, Jorge Ariel	Instituto Argentino de Radioastronomía	combi@irma.iar.unlp.edu.ar
Chavushyan, Vahram	INAOE, Tonantzintla	vahram@inaoep.mx
Digel, Seth	NASA/GSFC/USRA	digel@gsfc.nasa.gov
Ellison, Don	North Carolina State University	don_ellison@ncsu.edu
Fegan, Steve	CfA, Harvard & University of Arizona	sfegan@egret.sao.arizona.edu
Gallardo, Agustín	INAOE, Tonantzintla	delfin-x@inaoep.mx
Gaztañaga, Enrique	INAOE, Tonantzintla	gazta@inaoep.mx
Gehrels, Neil	NASA/GSFC	Gehrels@lheapop.gsfc.nasa.gov
Gómez, Yolanda	UNAM	gocy@astrosmo.unam.mx
Grenier, Isabelle	Université de Paris & CEA, Saclay	isabelle.grenier@cea.fr
Kaaret, Philip	CfA, Harvard	pkaaret@cfa.harvard.edu
Kanbach, Gottfried	MPE	gok@mpe.mpg.de
Luna, Abraham	INAOE, Tonantzintla	aluna@inaoep.mx
Martí, Josep	University Jaen	jmarti@ujaen.es
Mayya, Divakara	INAOE, Tonantzintla	ydm@inaoep.mx
Mereghetti, Sandro	IFC CNR -MILANO	sandro@ifctr.mi.cnr.it
Mukherjee, Reshmi	Columbia University	muk@astro.columbia.edu
Ortíz, Martha	INAOE, Tonantzintla	ahtramop@inaoep.mx
Paredes, Josep	Universitat de Barcelona	josepmp@am.ub.es
Pedraza, Rocio	UNAM	navegante@messagez.com
Petry, Dirk	Iowa State University	petry@iastate.edu
Punsly, Brian		brian.punsly@hsc.com
Recillas, Elsa	INAOE, Tonantzintla	elsare@inaoep.mx
Reimer, Olaf	NASA/GSFC/NRC	olr@egret.gsfc.nasa.gov
Roberts, Mallory	McGill University, Montréal	roberts@physics.mcgill.ca
Rodríguez Hoy, Luis	INAOE, Tonantzintla	hoy@inaoep.mx
Rodríguez, Ary	INAOE, Tonantzintla	ary@inaoep.mx
Rodríguez, Luis Felipe	IA-UNAM, Morelia	l.rodriguez@astrosmo.unam.mx
Romani, Roger	Stanford University	rwr@astro.stanford.edu
Romero, Gustavo	Instituto Argentino de Radioastronomía	romero@irma.iar.unlp.edu.ar
Rosa, Daniel	INAOE, Tonantzintla	danrosa@inaoep.mx
Slane, Patrick	CfA, Harvard	slane@cfa.harvard.edu

Participants to the Workshop *continued*

Tenorio-Tagle, Guillermo	INAOE, Tonantzintla	gtt@inaoep.mx
Thompson, David J.	NASA/GSFC	djt@bartok.gsfc.nasa.gov
Torres, Diego	Instituto Argentino de Radioastronomía	dtorres@venus.iar.unlp.edu.ar
Villicaña, Ilhuiyolitzin	UNAM	ilhui@starmedia.com
Zharikov, Sergey	IA-UNAM, Ensenada	zhar@astrosen.unam.mx

†: Matthew Baring now at Rice University, baring@spacsun.rice.edu

I

UNIDENTIFIED CELESTIAL SOURCES AT GAMMA-RAY ENERGIES

UNIDENTIFIED GAMMA-RAY SOURCES: AN INTRODUCTION

David J. Thompson
Laboratory for High Energy Astrophysics
NASA Goddard Space Flight Center
Greenbelt, Maryland
USA
djt@egret.gsfc.nasa.gov

Keywords: Gamma-ray sources, unidentified

Abstract

What makes a gamma-ray source "unidentified"? Typical gamma-ray source error boxes are large by astronomical standards; therefore a positional correlation with a known object is usually not enough to identify a source. Historically, identifications have come in several ways: (1) finding a time-variable signal seen at another wavelength; (2) repeatedly seeing within gamma-ray error boxes members of a small, well-defined class of energetic sources; (3) seeing spatially-resolved gamma-ray emission from an extended source; and (4) finding a positionally-correlated, highly-unusual object with special parameters that might be expected to produce gamma rays.

1. INTRODUCTION

The goal of this meeting was to examine current work and develop better ways to approach the problem of the unidentified sources in the future. By way of introduction, it is worth noting that this is not a brand-new field. The first satellite-detected unidentified gamma-ray source was found over 25 years ago, and the first *COS-B* catalog dates back over 20 years. Extensive efforts were made to identify these sources, extending over many years. We can learn from these experiences, particularly in light of what EGRET found.

This paper will examine the pre-EGRET and EGRET eras and ask, "Which methods of identifying gamma-ray sources were successful and which were not?" Topics covered are: a study of the first two unidentified

A. Carramiñana et al. (eds.),
The Nature of Unidentified Galactic High-energy Gamma-ray Sources, 3–16.

sources, a review of some possible source classes, and a look at some early population studies. The conclusion presents a list of identification techniques.

2. THE EARLIEST UNIDENTIFIED SOURCES

2.1. THE SOURCE IN VELA

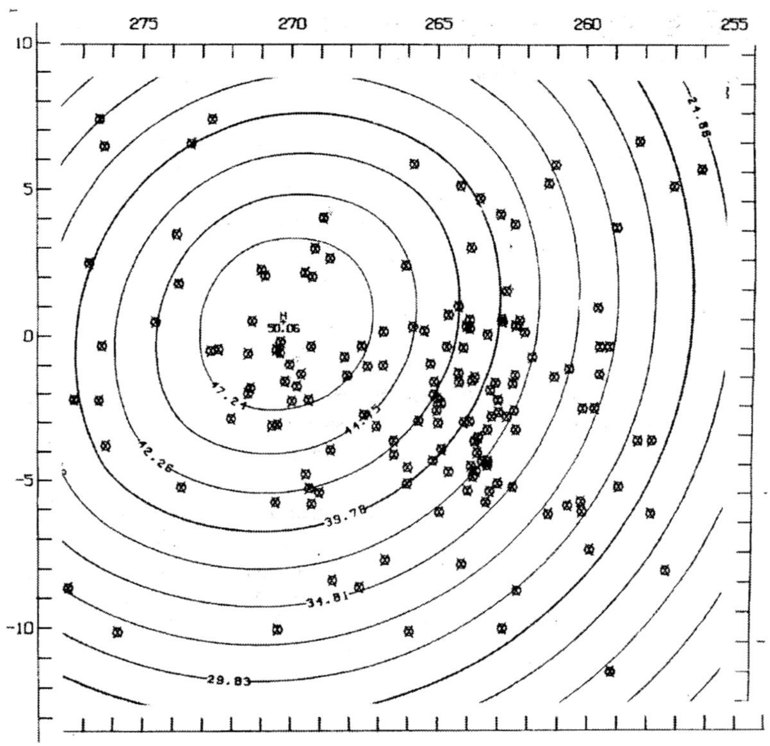

Figure 1 *SAS-2* map of the Vela region, in Galactic coordinates.

Figure 1 is a map of the first clearly-recognized unidentified source. This map, from *SAS-2*, shows the arrival directions of individual $E > 100$ MeV gamma rays collected for a week in 1973. The contours show the exposure function for the observation. There is an obvious concentration around Galactic coordinates $l = 263$, $b = -3$, but two things stand out:

- It is not easy to determine exactly where that source is. It is a statistical problem with varying exposure, a broad point spread function, and a non-zero (and probably non-uniform) background.

The inherent limitations of astronomy in the pair-production energy regime make source localization difficult.

- Unless a source is very large (quite a few degrees), it is extremely difficult to determine whether it is extended. Large error boxes and uncertain sizes have always characterized gamma-ray sources. EGRET did better than *SAS-2* and *COS-B*, especially at the higher energies, but the problem remains. Such systematic effects are discussed in the following paper (Reimer 2001).

The first paper on this source (Thompson et al. 1974) gave a tentative identification as the Vela supernova remnant, based on position, the possibility that it might be extended, and some calculations of the energetics of possible particle acceleration. The possibility that it was the Vela pulsar was largely neglected, because the models of the time expected the Crab to be the only gamma-ray pulsar (Pacini 1971), and the X-ray emission showed little evidence of pulsation (Harnden and Gorenstein 1973).

That paper represented one of the first publications to identify a gamma-ray source incorrectly. A year later, a check of the pulsar timing (Thompson et al. 1975) showed that the pulsar was indeed the source. The lesson learned was that a positional coincidence by itself does not give much confidence for gamma-ray source identification, even when accompanied by plausible energetic arguments.

2.2. THE SECOND SOURCE IN THE GALACTIC ANTICENTER - GEMINGA

The second unidentified gamma-ray source, Fig. 2 (Kniffen et al. 1975), was a much more difficult problem. Unlike the Crab and Vela, there was no obvious candidate object in the large error box. Here are some early efforts to identify what we now know as Geminga:

- The original Kniffen et al. paper suggested an OB association, I-Geminorium, based on position. That particular OB association is too far away.

- Fichtel et al. (1975) suggested a SNR, IC443, based on position and the possibility that the source was extended.

- Lamb, in Thompson et al. (1977) came very close to getting the answer. He used the ratio of gamma ray to radio intensity to suggest it resembled the Crab and Vela and might be an undiscovered radio pulsar. He also found a hint of a 59 s period, which we now know was incorrect.

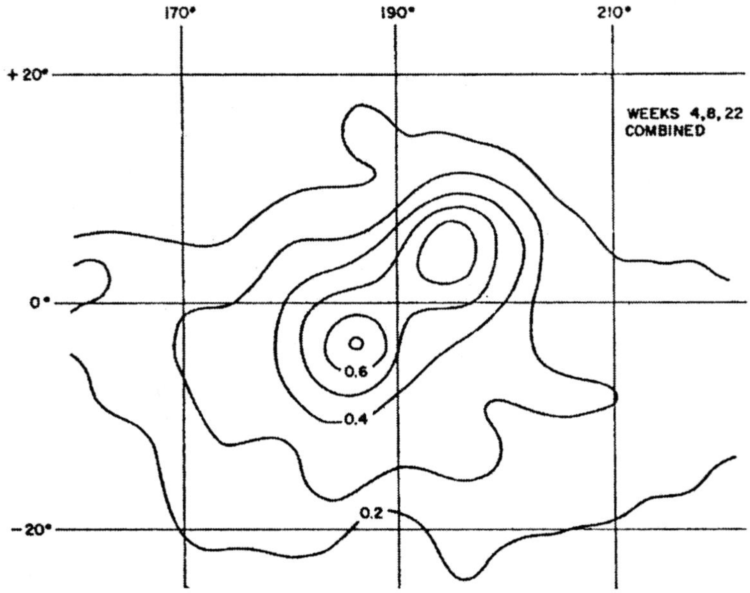

Figure 2 *SAS-2* intensity contour map of the Crab/Geminga region, in Galactic coordinates. Note the reversal of longitude from the standard astronomical convention. Geminga is the source to the upper right.

- The 59 s periodicity suggested to Maraschi and Treves (1977) a shrouded accretion source model and a fast pulsar with free precession model.
- Davies, Fabian, and Pringle (1978) also devised a binary model.
- After the discovery of 3C273 by *COS-B*, Moffat et al. (1983) suggested that it was a quasar, QSO 0630+180, that fell in the error box.

With the exception of Lamb's work, all of these early ideas turned out to be incorrect. This example illustrates how hard it is to deal with sources that have no obvious counterparts.

Here is another list, showing the path that finally led to the correct identification.

- A key observation was finding an unusual X-ray source in the Geminga error box. Bignami, Caraveo, and Lamb (1983) then used the gamma ray/X-ray and X-ray/optical values to suggest that it was an isolated neutron star, i.e. a radio pulsar.

- Having the good X-ray position allowed a deep search for an optical counterpart. Bignami et al. (1987) found the G" star, but by that time they had abandoned the pulsar idea due to the lack of a radio counterpart. They had the right star, but the wrong interpretation.

- It was Halpern and Tytler (1988) who finally put all the data together. They independently found the G" star, but they then constructed a full spectral energy distribution, including the gamma-ray, X-ray, optical, and radio - and from this concluded that it must be a radio-quiet pulsar.

- The fact that it was a pulsar was confirmed with the discovery of the period in the ROSAT data by Halpern and Holt (1992).

- Bertsch et al. (1992) finished the identification by showing the period in the gamma rays and finding the period derivative, confirming that it is a radio-quiet, spin-down pulsar. Seventeen years had passed from the first discovery to the final identification, with many unsuccessful efforts along the way.

These early sources illustrate two powerful identification techniques: periodicity and unusual broad-band spectral energy distributions. In the case of Geminga, the value of working downward in energy to wavelengths where better positioning is possible is also clear.

3. SOURCE CLASSES

3.1. EXTENDED SOURCES

The first recognized cosmic gamma-ray source was an extended one: our Galaxy, seen with *OSO-3* (Kraushaar et al. 1972). The *COS-B* group pioneered the use of extended emission on smaller scales as a source identification technique. Their detection of the Orion cloud complex (Caraveo et al. 1980, Bloemen et al. 1984) was based on correlation of gamma-ray intensity with gas column density on a map size of over 200 square degrees, large enough that the broad point spread function of a gamma-ray telescope does not affect the result. The EGRET group applied a similar technique to confirm the identification of the Large Magellanic Cloud (Sreekumar et al. 1992). Digel (2001) describes some of the other work on extended Galactic emission in the EGRET data. Although this identification technique has not been applicable to most gamma-ray sources seen with previous telescopes, it may be valuable for future observations with *GLAST*.

3.2. ACTIVE GALACTIC NUCLEI (AGN)

Another individual source that deserves mention, even though it is not Galactic, is 3C273. The identification (Swanenburg et al. 1978) was based on position, which is more feasible at high latitudes where the sky is not so crowded and the diffuse radiation is more smooth than along the Galactic Plane. 3C273 was almost surely the source that was seen, but it shows the pitfalls of "one-of-a-kind" identifications. Although we now recognize that blazars are the relevant gamma-ray source class, it just happened that 3C273 was the one bright blazar seen by *COS-B*. Because 3C273 is also a nearby quasar, however, the blazar connection was missed. The *COS-B* papers (Swanenburg et al. 1978; Bignami et al. 1981) had all the pieces of the puzzle. The double-peaked Spectral Energy Distribution (SED) was visible in the first paper, showing the now-familiar synchrotron and Compton components. They also recognized the importance of beaming and the problem of photon-photon absorption. If it had been any other blazar, the *COS-B* team might very well have determined that blazars are the class, not nearby quasars.

The interpretation of the 3C273 result by nearly everyone focused on the idea of nearby AGN rather than beamed ones. As a result, nearby AGN were invoked as explanations in cases that have not been confirmed by later observations:

- The suggestion of a QSO identification for Geminga, mentioned above, was based on similarity with 3C273.
- Apparao et al. (1978) associated 2CG135+01 with the nearby QSO 2S0241+622. The EGRET position for this source is inconsistent with the QSO.
- Strong and Bignami (1983) identified an excess in the *COS-B* map with NGC1275, a Seyfert Galaxy in the Perseus Cluster. Neither the source nor the identification were confirmed by EGRET.

The recognition of blazars as the gamma-ray AGN class did not come until EGRET used several techniques to identify the population:

- The first blazar detection, 3C279 (Hartman et al. 1992), was based on position, using the fact that 3C279 was a powerful source in a relatively small error box. It had previously been noted as a candidate gamma-ray source (Kanbach et al. 1988) based on its X-ray luminosity.
- As other high-latitude sources were seen by EGRET, it was realized that they were systematically associated with bright, flat-spectrum

radio sources of the blazar class. A summary of this correlation analysis is given by Mattox et al. (1997).

- The blazar association was confirmed by the detection of simultaneous variability seen in gamma-rays and longer wavelengths (e.g. Wagner et al. 1995; Wehrle et al. 1998).

3.3. PULSARS

The first "class" of gamma-ray sources was pulsars, and extensive searches for pulsar periodicity were made of both the *SAS-2* and *COS-B* data bases. Table 1 shows the results of those searches. The significance listed is the probability of finding the resulting distribution of gamma-ray arrival times by chance.

Table 1 Early Searches for Gamma-Ray Pulsars

Reference	*Identification*	*Significance*
Ögelman et al. 1976	PSR B1818−04	4×10^{-5}
Thompson et al. 1976	PSR B1747−46	8×10^{-5}
Buccheri et al. 1979 [a]	PSR B1822−09	2×10^{-4}
Buccheri et al. 1979 [a]	PSR B0740−28	1×10^{-5}

[a] Unpublished preprint.

Four candidate pulsars were found. The light curves appear fairly convincing, with the two *COS-B* candidates showing the same double-peaked light curves as seen for the Crab and Vela at gamma-ray energies. Not one of these four has been confirmed. In fact, three of the four turned out to be derived using incorrect radio timing information. The *COS-B* team learned that fact before their paper was published (Buccheri et al. 1983), and so the paper was withdrawn.

Other periodicity studies have also given indications of positive results that remain unverified; in particular, the *SAS-2* claim of the 4.8 hour Cygnus X-3 period (Lamb et al. 1977) was not confirmed either by *COS-B* (Hermsen et al. 1987) or EGRET (Mori et al. 1997). There have also been a number of claims of pulsed radiation at TeV energies, none of which have been verified.

As strong an identification tool as periodicity can be, it is only convincing if the statistics are good. The six pulsars seen by EGRET have probabilities in the 10^{-9} range or better (Thompson 2000). Several candidate pulsars have been claimed in the EGRET data with 10^{-4} probabilities. Experience suggests they should remain candidates until better statistics are available.

4. POPULATION STUDIES

4.1. THE SECOND *COS-B* CATALOG

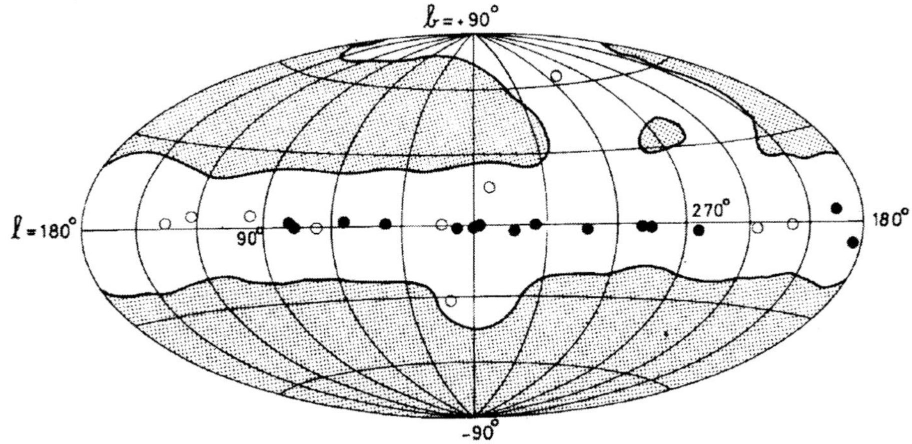

Figure 3 The Second *COS-B* catalog. The shaded region was not included in the analysis. The filled circles are the brighter sources (Swanenburg et al. 1981).

Population studies gained a real boost with *COS-B*, especially the first (Hermsen et al. 1977) and second *COS-B* catalog (Swanenburg et al. 1981), which were systematic analyses of the data along the Galactic Plane. Figure 2 shows the second catalog, in Galactic coordinates. Table 2 presents a summary of the derived properties of these sources based on their spatial and energy distributions, from Hermsen (1980).

These catalogs were the real beginning of the search for the unidentified Galactic gamma-ray sources. These derived properties did not match any known population, especially the lack of bright X-ray or radio counterparts, suggesting one or more unknown classes of objects.

4.2. SUPERNOVA REMNANT OB ASSOCIATIONS

Even before the *COS-B* catalog analysis was complete, there were searches for source classes that might explain the results. Table 3 is a summary of one of the best of these, Montmerle's search for SNOBs:

Table 2 Properties of Sources in the Second *COS-B* Catalog

Characteristic	*Range of Values*
Angular size	$< (1^o - 2^o)$
Intensity > 100 MeV	$(1 - 5) \times 10^{-6}$ ph cm^{-2} s^{-1}
Energy flux > 100 MeV	$(4 - 20) \times 10^{-10}$ erg cm^{-2} s^{-1}
Energy spectrum	Diverse, ave. consistent with E^{-2}
Time variability	Not excluded
Distance range	(2 - 7) kpc
Luminosity > 100 MeV	$(0.4 - 5) \times 10^{36}$ erg s^{-1}
From lack of identification	$L_X \leq 0.1 L_\gamma$
From lack of identification	$L_{\rm radio} \ll L_\gamma$

supernova remnant OB associations (Montmerle 1979). His concept was that the young stars injected cosmic rays into the regions where they could be accelerated to high energies by the SN shocks and then collide with ambient matter to produce gamma rays. The correlation was examined based on the positions in Galactic coordinates of sources or excesses that appeared in the *COS-B* maps.

Table 3 Possible Supernova Remnant/OB Association Identifications

COS-B *Source/location*	*Identification*	*Later Results*
2CG006−00	W28	3EG J1800−2338
25 0	K39A/B	No *COS-B* source
2CG036+01	W44	3EG J1856+0114
2CG078+01	DR3+DR4 (γ Cyg)	3EG J2020+4017
189 +3	IC443	No *COS-B* source 3EG J0617+2238
2CG311−01?	RCW86	Position mismatch
327 0	Kes 27	No *COS-B* source
2CG333+01	RCW103	No EG source
335 0	G335.2+0.1	No *COS-B* source
2CG 342−02	Kes 45	PSR B1706−44

This table shows what became of this analysis, which was based almost entirely on positional association. A few of the suggested sources remain in the EGRET data (Hartman et al. 1999). On further analysis, some did not even appear in the *COS-B* catalog, and one turned out to be a pulsar. Judging from the relatively low rate of confirmation, it seems unlikely that this is a class of gamma-ray sources, although later analyses have shown a general correlation of EGRET sources with young star-forming regions such as these (Kaaret and Cottam, 1996; Yadigaroglu and Romani 1997). Nevertheless, Montmere's work established a clear methodology, because it started with a reasonable physical hypothesis and then made specific, testable predictions for source locations and spectra.

4.3. WOLF-RAYET STARS

Pollock (1987) carried out another population study, with Wolf-Rayet stars as the hypothesized gamma-ray sources. The testable hypothesis was that the WR stars with nonthermal emission seen in radio must be accelerating particles, and these have the potential to upscatter local photons to gamma-ray energies. In particular, the best candidates were thought to be those WR stars with variability and nonthermal X-ray emission.

As shown in Table 4, Pollock found four candidates in the *COS-B* data. All four of the *COS-B* sources have counterparts in the third EGRET catalog, but all turn out to lie outside the corresponding EGRET error boxes; therefore this hypothesis seems unlikely to be valid.

Table 4 Possible Wolf-Rayet Star Identifications

COS-B *Source*	*Identification*	*Later Results*
083+03	HD193793 (WR140)	Outside 3EG error box
2CG054+01	MR93 (WR125)	Outside 3EG error box
2CG006−00	AS268 (WR105)	Outside 3EG error box
2CG356+00	HDE318016 (WR98)	Outside 3EG error box

5. DISCUSSION

If the individual analyses and these population studies did not solve the mystery of the *COS-B* sources, then what happened to them? Table 5 shows that many of them disappeared when the underlying gas model was improved (Mayer-Hasselwander and Simpson 1988; Grenier, Hermsen, and Pollock 1990). The original analysis had not taken the structure of the interstellar gas into account, and in many cases that structure was comparable in size to the point spread function. Roughly

Table 5 *COS-B* Sources Explained as Structure in the Interstellar Gas

COS-B *Source*	*Later Results*
2CG006−00	3EG J1800−2338
2CG013+00	No source
2CG036+01	3EG J1856+0114 nearby
2CG054+01	3EG J1928+1733 nearby
2CG065+00	3EG J1958+2909
2CG095+04	No source
2CG121+04	No source
2CG218−00	No source
2CG235−01	No source
2CG288−00	Two 3EG sources nearby
2CG311−01	3EG J1410−6147 nearby
2CG333+01	No source

half of the original source catalog could be accounted for by an improved gas model. A few of these turned out to be EGRET sources, or have EGRET sources nearby, but most of them have vanished completely. I cannot emphasize too strongly that a gamma-ray source catalog is only as good as the gas model used to generate it. The model used for EGRET (Hunter et al. 1997) represents a tremendous amount of effort and matches the large-scale observations well, but it is not perfect, and so a few of the EGRET sources could be (and probably are) artifacts. We just do not know which ones.

In the end, about half the cataloged *COS-B* sources are still seen, and for the most part they have still not been identified. Table 6 summarizes the status of the 2CG sources that have plausible counterparts in the 3EG catalog. Only two of the 12 have positive identifications (Geminga and PSR B1706−44). There is one other candidate pulsar, and the possibility remains of a correlation with the Galactic Center, although that is not a physical model. EGRET did not solve the mystery, but it did provide a vastly-expanded catalog of source positions, energy spectra, and time variability information on which to base future studies.

6. CONCLUSIONS

What have we learned from this "ancient history?" Based on the successes and failures of the past, I propose this ordered list of what

Table 6 *COS-B* Sources - An Unsolved Mystery

COS-B *Source*	EGRET Source(s)	Identification
2CG006−01	3EG J1800−2338	?
2CG065+00	3EG J1958+2909	?
2CG075+00	3EG J2016+3657	?
	3EG J2021+3716	?
2CG078+01	3EG J2020+4017	?
2CG135+01	3EG J0229+6151	?
	3EG J0241+6103	?
2CG195+04	3EG J0633+1751	Geminga Pulsar
2CG284−00	3EG J1027−5817	?
2CG288−00	3EG J1048−5840	PSR B1046−58?
2CG311−01	3EG J1410−6147	?
2CG342−02	3EG J1710−4439	PSR B1706−44
2CG356+00	3EG J1734−3232	?
2CG359−00	3EG J1744−3011	?
	3EG J1746−2851	Galactic Center?

seems to work in identifying gamma-ray sources, ranging from highest confidence to lowest confidence:

1 Periodicity - high significance (pulsars).

2 Time variability correlated with other wavelengths (blazars).

3 Spatially-extended source (molecular clouds).

4 Large sample of well-defined, energetic object class spatially correlated with sources (blazars).

5 Spectral Energy Distribution with an “unusual” object at multiple wavelengths (Geminga, Cen A?).

6 Plausible class of objects with testable model predictions (radio-quiet pulsars).

7 Periodicity - lower significance (pulsars).

8 Sample of plausible objects spatially correlated with sources and having testable predictions (SNR).

9 One plausible object in a gamma-ray error box.

10 Hypothesized class of objects without testable predictions.

The first five at the top are ones that have given relatively solid results. I would be inclined to believe results based on these types of identification. The next four below that give lower confidence, because they are techniques that have not always been successful. Low-significance periodicity, for example, has produced some false identifications, and various types of SNR have been invoked unsuccessfully. Radio-quiet gamma-ray pulsars certainly seem likely, but Geminga remains the only proven case. Finally, to round out the list to ten, I include the obvious "no confidence" case of a hypothesis that cannot be tested.

7. SUMMARY

Results from two generations of high-energy gamma-ray telescopes provide a baseline set of tests for identifying sources. Some techniques have been more successful than others, and those comparisons provide useful lessons for ongoing work and as suggestions for the future. In many respects, however, the past is just a starting point. Continuing efforts such as those at this meeting can be thought of as a series of questions:

Are there better ways to identify gamma-ray sources?

What are the source classes we should be looking for?

How much identification can be done without more gamma ray data?

These are the subjects of other contributions to this volume.

References

Bertsch, D.L., et al. (1992) *Nature*, **357**, 306

Apparao, K.M.V. (1978) *Nature*, **273**, 450

Bignami, G.F., Caraveo, P.A., Lamb, R.C. (1983) *ApJ*, **272**, 9

Bignami, G.F., et al. (1983) *A&A*, **93**, 71

Bignami, G.F., Caraveo, P.A., Paul, J.A., Salotti, L., Vigroux, L. (1987) *ApJ*, **319**, 358

Bloemen, J.B.G.M. et al. (1984) *A&A*, **139**, 37

Buccheri, L. et al. (1981) unpublished preprint

Buccheri, L. et al. (1983) *A&A*, **128**, 245

Caraveo, P.A. et al. (1980) *A&A*, **91**, 3

Davies, R.E., Fabian, A.C., Pringle, R.E. (1978) *Nature*, **271**, 634

Digel, S.W. (2001), these proceedings

Fichtel, C.E., Hartman, R.C., Kniffen, D.A., Thompson, D.J., Bignami, G.F., Ögelman, H.B., Tümer, T. (1975) *ApJ*, **198**, 163

Harnden, F.R. Jr., and Gorenstein, P. (1973) *Nature*, **241**, 107

Halpern, J.P., and Tytler, D. (1988) *ApJ*, **330**, 201

Halpern, J.P., and Holt, S.S. (1992) *Nature*, **357**, 222
Hartman, R.C. et al. (1992) *ApJ*, **385**, L1
Hartman, R.C. et al. (1999) *ApJS*, **123**, 79
Hermsen, W. (1980), PhD thesis, Leiden University
Hermsen, W. et al. (1977) *Nature*, **269**, 494
Hermsen, W. et al. (1987) *A&A*, **175**, 141
Kaaret, P., and Cottam, J. (1996) *ApJ*, **462**, L35
Kanbach, G. et al. (1988) *Sp. Sci. Rev.*, **49**, 69
Kniffen, D.A., Hartman, R.C., Thompson, D.J., Bignami, G.F., Fichtel, C.E., Ögelman, H.B., Tümer, T. (1975) *Proc. 14th Internat. Cosmic Ray Conf.*, **1**, 100
Kraushaar, W. et al. (1972) *ApJ*, **177**, 341
Lamb, R.C., Fichtel, C.E., Hartman, R.C., Kniffen, D.A., Thompson, D.J. (1977) *ApJ*, **212**, L63
Maraschi, L. and Treves, A. (1977) *A&A*, **61**, L11
Mattox, J.R., Schachter, J., Molnar, L., Hartman, R.C., Patnaik, A.R. (1997) *ApJ*, **481**, 95
Moffat, A.F.J., Schlickeiser, R., Shara, M.M., Sieer, W., Tuffs, R., Kuhr, H. (1983) *ApJ*, **271**, L45
Montmerle, T. (1979) *ApJ*, **231**, 95
Mori, M. et al. (1997) *ApJ*, **476**, 842 Ögelman, H.B., Fichtel, C.E., Kniffen, D.A., Thompson, D.J. (1976) *ApJ*, **209**, 584
Pacini, F. (1971), *ApJ*, **163**, L17
Pollock, A.M.T. (1987), *A&A*, **171**, 135
Reimer, O. (2001) these proceedings
Strong, A.F., and Bignami, G.F. (1983) *ApJ*, **274**, 549
Sreekumar, P. et al. (1992) *ApJ*, **400**, L67
Swanenburg, et al. (1978) *Nature*, **275**, 298
Swanenburg, et al. (1981) *ApJ*, **243**, L69
Thompson, D.J. (2000) *Proc. Heidelberg Conference on High-Energy Gamma-Ray Astronomy*, in press.
Thompson, D.J., Bignami, G.F., Fichtel, C.E., Kniffen, D.A. (1974) *ApJ*, **190**, L51
Thompson, D.J., Fichtel, C.E., Kniffen, D.A., Lamb, R.C. (1976) *Astrophys. Lett.*, **17**, 173
Thompson, D.J., Fichtel, C.E., Kniffen, D.A., Ögelman, H.B. (1975) *ApJ*, **200**, L79
Thompson, D.J., Fichtel, C.E., Hartman, R.C., Kniffen, D.A., Lamb, R.C. (1977) *ApJ*, **13**, 252.
Wagner, S. J. et al. (1995) *ApJ*, **454**, L97.
Wehrle, A.E. et al. (1998) *ApJ*, **497**, 178.
Yadigaroglu, I.A. and Romani, R.W. (1997) *ApJ*, **476**, 347.

THE GAMMA-RAY PROPERTIES OF UNIDENTIFIED EGRET SOURCES

Olaf Reimer
Laboratory for High Energy Astrophysics
NASA Goddard Space Flight Center
Greenbelt, MD 20771
USA
olr@egret.gsfc.nasa.gov

Keywords: High–energy gamma–ray sources, unidentified EGRET sources, population studies

Abstract Although the majority of gamma-ray sources still remain unidentified, we have various kinds of information to characterize the observational properties of unidentified EGRET sources. Despite astronomical properties like locations of individual sources or the collective arrangement of the class as such, the nine years of *CGRO* observations provide the ability to investigate flux variability at different timescales, enable us to perform periodicity searches, determine gamma-ray source spectra between 30 MeV and 10 GeV and even investigate spectral variability. The basic observational properties of unidentified high-energy gamma-ray sources discovered by EGRET are reviewed. Various instrumental and observational peculiarities affecting the interpretation of the EGRET data are pointed out, also describing the way such biases might affect scientific conclusions drawn from the EGRET data.

1. INTRODUCTION

With NASAs *Compton Gamma-Ray Observatory* mission terminated in June 2000, the EGRET data base will remain a unique and extremely important source of scientific information. Although being an archival data base from now on, it has to be considered as state-of-the-art for several more years, because presently no other instrument covering the high-energy gamma-ray wavelengths is in orbit. Even when instruments like *AGILE* and *GLAST* will become operational, the EGRET data will be the reference for new observational results. *CGRO*s coverage of a long time period between 1991 and 2000 will be used in determining the

A. Carramiñana et al. (eds.),
The Nature of Unidentified Galactic High-energy Gamma-ray Sources, 17–34.

long-term behaviour of gamma-ray sources and very likely for subsequent archival research. To a lesser degree, this has been already demonstrated by comparing EGRET data with results from previous missions, in particular with *COS-B*. Not only were earlier reported source detections considered for positional comparison, in cases like Geminga archival data were analyzed in conjunction with the EGRET data for tracing its long term periodicity behavior. When going into the subject of discussing unidentified EGRET sources, *COS-B* source findings are still an interesting aspect of reference, since predictions from *COS-B* population studies could be supported or rejected using the EGRET data or to establish the long-term coverage of individual sources (i.e. 2CG 135+1 or 2CG 075+0). Here, I will review how EGRET data were used to construct the source catalogs, flux histories and source spectra, and discuss the pecularities of the existing point-source catalogs, their positional accuracy and underlying systematics. The importance of understanding in which way such biases might affect scientific conclusions will be addressed. Also, different approaches to deal with variability are compared. Quantitatively, the flux determination needs to be related to EGRETs instrumental response in orbit over time and energy. The determination of photon spectra will be described and a view beyond simple single power-law fits needs to be given. In several cases with exceptional observational coverage, also spectral variability could be adressed. Having accumulated detailed knowledge of the spacial, temporal and spectral properties of individual unidentified EGRET sources, the quest for finding signatures in the collective could be challenged. Conclusions are drawn on the validity of assumptions, selections and cuts in population studies, mainly under aspects of known instrumental biases or pecularities with the gamma-ray point source catalogs and questioning barely justified speculations.

2. EGRET SOURCE CATALOGS AND GAMMA–RAY SOURCE LOCATIONS

Omitting all low level EGRET data products and therefore the complete process of event reconstuction and event quality classes, the Third EGRET catalog of high-energy gamma-ray sources (Hartman et al. 1999) has been constructed on the basis of individual viewing periods, meaningfully selected combinations of individual viewing periods (i.e. subsequent short observations or sequences with nearly identical pointings) and annual superpositions of all observations synchronized with *CGRO* observation cycles. For each of those selections appropriate count, exposure and intensity maps have been constructed, usually by applying a standard $< 30°$ field–of–view cut. These maps have been analysed by means of a maximum-likelihood procedure (Mattox et al. 1996). Although fluxes are consistently given for $E > 100$ MeV, similar analyses have been performed for the energy intervals 300-1000 MeV and above 1000 MeV. The different likelihood test statistics (TS) maps were compared and, as long as $(TS)^{1/2} > 4$, the one which produced the smallest error contours was chosen to represent the actual source location. Figure 1 shows the EGRET all-sky TS map, the result of the maximum–likelihood analysis procedure carried out from all viewing periods between April 1991 and September 1995 at > 100 MeV. The intense and

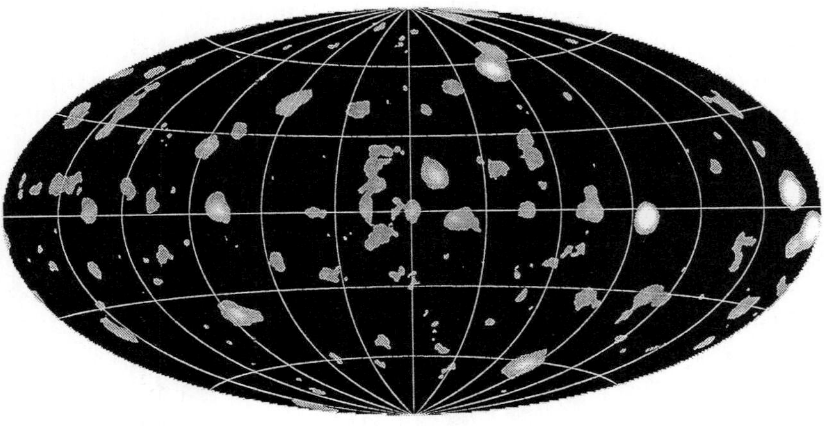

Figure 1 Test statistics (to say detection significance) map as result of the maximum–likelihood analysis of the EGRET data from cycle 1 to 4 observations

highly structured diffuse emission along the Galactic plane makes the determination of sources in the Galactic plane more subject to systematic uncertainties than those at higher latitudes. Therefore, the detection criterion for a gamma-ray source is also different for its location: in at least one of the derived maps $(TS)^{1/2} > 4$ for sources at $|b| > 10°$ or $(TS)^{1/2} > 5$ at $|b| < 10°$ must be fulfilled. This peculiar step in the acceptance criterion is somewhat arbitrary, however translates directly into a bias taken care of in source population studies. Additionally, the acceptance criterion could be fulfilled in any of the derived maps, resulting in detections on the basis of very uneven exposure times. One-time flaring sources will be included as well as sources which build up the detection significance in a purely statistical way originating from more and more observations, i.e. exposure. The resulting point source catalog (Fig. 2) therefore represents a rather uneven sample for population study applications.

Also, the underlying diffuse emission model (Hunter et al. 1997) is known to be imperfect on smaller scales. The likelihood source detection algorithm will translate such inaccuracies of the diffuse model into the detection significance and, very likely, into acceptance issues for sources near the catalog thresholds. The procedure of independently scaling of the nominal values of the diffuse emission model within the radius–of–analysis (usually 15°) in order to account for small-scale structures might not be in each case the most accurate one in order to discriminate point-like excesses against features in the diffuse emission. In confused regions with sources of significantly overlapping point spread functions (PSF), the order in which sources are optimized becomes important, too. These systematics are generally not easy to quantify, but obviously need consideration in 3EG catalog-based population studies. Figure 2 shows the high-energy gamma-ray sources fulfilling the 3EG catalog acceptance criteria. The size of the symbols represents the maximum intensity seen for this source.

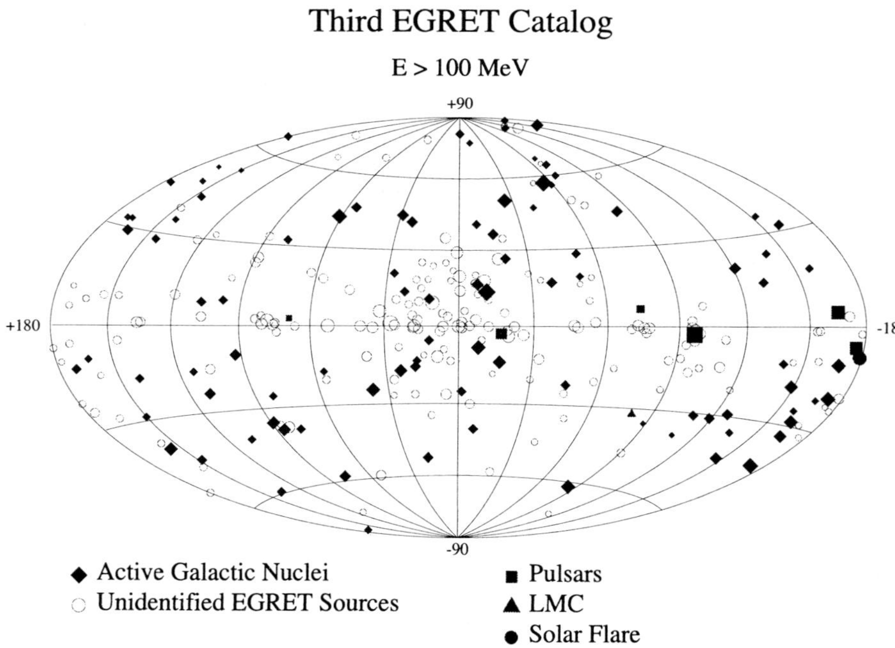

Figure 2 Gamma-ray point sources in the Third EGRET catalog

With the pecularities of the construction of the Third EGRET catalog in mind, nevertheless the longitudinal and latitudinal characteristics of the sample can be sketched. Fig. 3 shows the complete 3EG catalog sample (outlined) and a selection made of sources fulfilling a common

$(TS)^{1/2} > 5$ acceptance criterion on the basis of the summed map from *CGRO* observation cycles 1 to 4 only (shaded). As can easily be seen, the distribution of unidentified sources represents a distinct Galactic population, but also shows some spreading at mid-latitudes on top of a rather flat component present at all latitudes.

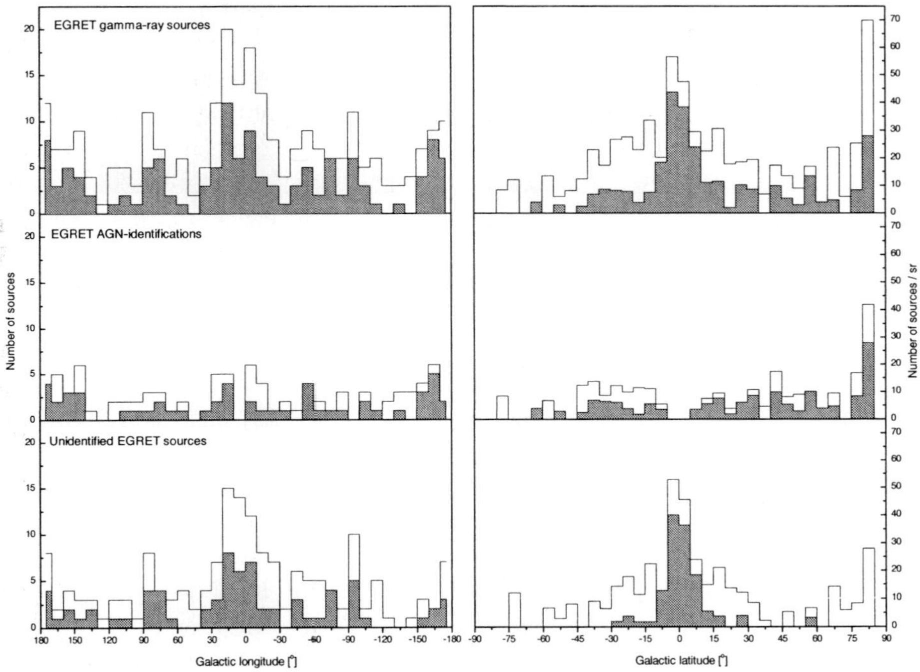

Figure 3 Longitudinal and latitudinal distributions of the 3EG catalog sources. See text for details.

Additional compilations of high-energy sources have been made, i.e. for GeV sources (Lamb & Macomb 1997; Reimer et al. 1997). The obvious reason to compile a catalog of a higher energy threshold is the significantly reduced Galactic diffuse emission component (and therefore suppression of systematic uncertainties originating from it during source determination procedures) in conjunction with a narrower instrumental PSF. These advantages are offset by the reduced photon flux at higher energies and therefore a loss in photon statistics. However, in cases of bright sources, hard photon spectral indices or regions suffering from source confusion, the trade between limited statistics and better angular resolution often leads to significantly narrower error contours. Figure 4 compares the error contours of the unidentified low-latitude source 3EG J2020+4017 (γ Cyg) at energies above 100 MeV and above 1 GeV, respectively. A proper way to treat error contours in a mathematical-

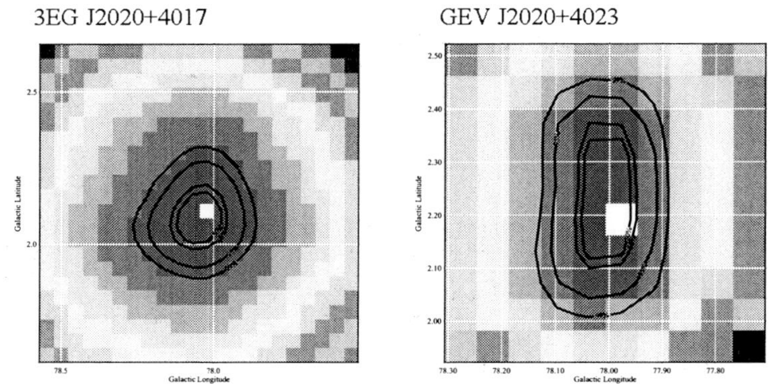

Figure 4 Gamma-ray emission from γ Cygni at $E > 100$ MeV and $E > 1$ GeV, respectively. Note the different scaling of the figures. The visible pixelation of 0.05° by 0.05° is the same in both images. The best position is similar, but the uncertainties are smaller at higher energies, dedicating the GeV-image for counterpart studies.

statistical way (i.e. for counterpart propability tests) has been found in elliptical fits to source contours. Narrower error contours to a gamma-ray source do not automatically imply a more exact point source location in the coordinates. Comparing the catalog source positions determined at different energy thresholds with precise coordinates of astronomical objects could only be performed if an identification has been established, i.e. from observations at other wavelengths. For the high-energy gamma-ray sources this can be accomplished using pulsars (PSRs) and active galactic nuclei (AGN). Figure 5 compares the gamma-ray source locations from the 3EG catalog (Hartman et al. 1999) and the GeV-catalog (Lamb & Macomb 1997) with the radio positions of these objects, providing estimates of the precision with which EGRET typically determines gamma-ray point source coordinates.

3. EGRET SOURCE DETECTABILITY AND CONSEQUENCES

The significance s for a detection of an isolated point source with EGRET ($E > 100$ MeV) is adequately represented by

$$s \propto f \sqrt{\frac{e}{b_g}}, \tag{2.1}$$

where f is the flux, e the exposure and b_g the intensity of the diffuse gamma-ray emission at the region of the source (Mattox et al. 1996).

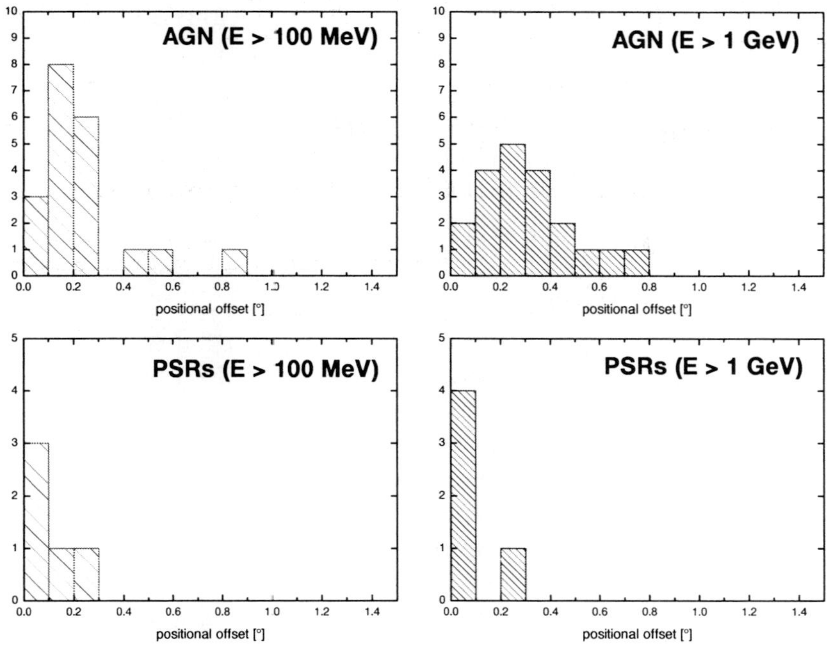

Figure 5 Source location offsets comparing the gamma-ray coordinates of identified objects with precise coordinates from optical or radio counterparts. The GeV positions are not significantly more precise than those determined for $E > 100$ MeV.

In order to simulate a corresponding picture of the gamma-ray sky as given in the 3EG catalog, all three observables needs to be looked at in detail.

- Exposure: the observational history of the EGRET instrument is highly non–uniform and so is the exposure. EGRET pointings have a typical field–of–view of 40°. However for most applications a 30° cut is recommended as applied, for instance, in the 3EG catalog. If population studies compare with the EGRET source catalog, the exposure needs to be determined for each individual source. As mentioned above, catalog sources are not consistently included on the basis of equal exposure time. Therefore, exposure time as well as the corresponding number of sources matching the detection criterion at any considered time interval should be looked at.

- Diffuse gamma-ray emission: the detectability for EGRET sources also depends on the diffuse gamma-ray background in the source region. To compensate for these non-uniformities, several ways are applicable. The diffuse gamma-ray emission model constructed from HI and CO distributions (Hunter et al. 1997) is available in the standard EGRET energy intervals. Alternatively, a point–source–removed intensity map could be used as has been done by Strong, Moskalenko & Reimer (2000). Although both approaches are different in the scientific content, they provide a good measure of the characteristics of the diffuse gamma-ray emission for the purpose of use in population studies.

- Flux: Monte Carlo based population simulations are required to reproduce the actually observed $\log N - \log S$ distribution of gamma-ray sources, globally as well as locally. As noticed by Gehrels et al. (2000), the $\log N - \log S$ distribution for unidentified sources close to the Galactic plane differs from the one obtained at high Galactic latitudes, although partly as a result of the non-uniform detectability function of EGRET itself. Nevertheless, the $\log N - \log S$ of identified gamma-ray sources is distinctly different, i.e. for AGN (Özel & Thompson 1996). Estimates of the fraction of unresolved point sources will come to play when concluding on the basis of simulations involving large numbers of sources as expected in the *GLAST* era.

A rather simplified approch to account for the EGRET detectability function can be made by determining upper limits for a grid on the sky. These upper limits need to be determined at comparable statistical significance. An example is given in the 3EG catalog, using the summed exposures of *CGRO* cycles 1, 2, 3 and 4. Note that the catalog acceptance criterion and 95% confidence upper limits leave room for excesses in the test statistics in between, which needs to be accounted for in simulations. The full composition of the 3EG catalog might be investigated by following this scheme through each of the individual viewing periods in order to account for transient sources. Lastly, upper limits near bright catalog sources are expected to be underestimated due to the width of the EGRET point spread function.

Figure 6 shows a detectability map for equally bright sources, determined for EGRET observations from *CGRO* cycles 1, 2, 3, and 4 ($E > 100$ MeV). Note the features close to the Galactic plane, where the low detectability directly in the plane increasingly becomes compensated from the exposure. However, the highly non-uniform character of EGRETs detectability function is easy to recognize. On a 0.5° x 0.5° grid, the value of the detectability function for sources of comparable flux differs by more than a factor of ten!

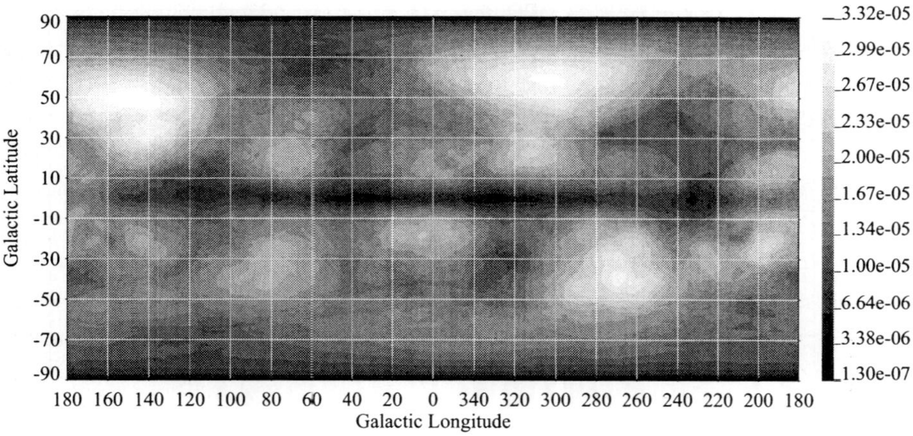

Figure 6 Two-dimensional detectability function for γ-ray sources above 100 MeV and comparable flux. This map corresponds only to the analysis of the summed EGRET cycle 1 to 4 observations, listed as P1234 in the 3EG catalog (units of cm^2 s).

4. VARIABILITY OF GAMMA-RAY SOURCES

Gamma-ray source variability is even more difficult to quantify than test statistics excesses, non-uniform detection thresholds and absolute coordinates for gamma-ray sources. At first glance, one needs to look at the spark chamber efficiency of the EGRET instrument. As described in Esposito et al. (1999), the spark chamber efficiency is strongly time-dependent. For any meaningful determination of gamma-ray source fluxes, the response has to be normalized. Scale-factors have been constructed for a given energy and time interval, primarily by comparing the level of the ever present diffuse emission component. Neither the determination of the inital response nor the scaling to a nominal value is perfect. Although we think that the response of the spark chamber should underly a rather smooth degradation between major events impacting its performance (gas refills), some individual viewing periods appear to be slightly off the generally smooth normalization scheme. However, the individual normalization factors are accessible for each observation period and the individual case could be well investigated beyond the flux value listed in a source catalog. Figure 7 gives an impression of the spark chamber efficiency versus time, before and after the normalization has been applied.

So far, variablity studies have been performed largely on the basis of entries in EGRET source catalogs. The first systematic study by McLaughlin et al. (1996) made use of individual viewing periods from

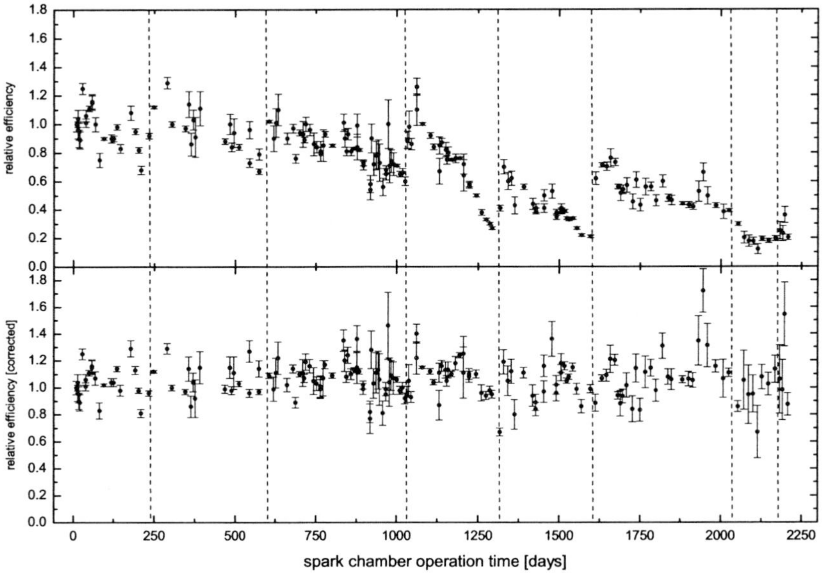

Figure 7 EGRET efficiency as function of observation time, Upper panel: before normalization, lower panel: after normalization has been applied.

the 2EG catalog (Thompson et al. 1995), and has been expanded online to include *CGRO* cycle 3 observations. It has been used also by Wallace et al. (2000) to study short-term time variability on the basis of sub-viewing periods. The method is effectively a measure of inconsistency of the gamma-ray data with the assumption of a constant source flux (for details see McLaughlin et al. 1996).

A different approach to properly quantify flux variability has been carried out by Tompkins (1999), consistently computed for the 3EG catalog sources and source fluxes. A method has been introduced to obtain a fractional variability measure by calculating the likelihood for obtaining any source flux and compare to the actual observation (for details see Tompkins 1999).

Following a method successfully applied to radio data (Romero, Combi & Colomb 1994), Zhang et al. (2000) and Torres et al. (2001) determined a gamma-ray flux variability measure by means of a weighted fluctuation index with pulsars as "standard candle". Unfortunately, both teams use different subsets of gamma-ray sources in the variability study which introduces problems to directly compare the results of the different methods.

Although being a selection of steady sources by comparing detection significances at different observational timescales only (individual and combined viewing periods), Gehrels et al. (2000) effectively obtained a qualitative measure of source variability. This source ensemble is selected against one-time flaring sources (transients) and variable but dim sources close to the detection threshold of the EGRET instrument.

Each method attempted to distinguish gamma-ray sources by means of quantifying the degree of flux variability. At present, different classes of gamma-ray emitters are most clearly distinguishable in terms of variability by applying the method developed by Tompkins. Figure 8 gives the sketch of the separation achieved on the basis of the variability criterion τ (as the inverse fraction of the average and the standard deviation). Among the unidentified sources, there is a tendency that sources at low Galactic latitudes are less variable than at high Galactic latitudes. It is striking that the variability distribution of unidentified sources as a whole is distincly different than the one of pulsars and active galactic nuclei.

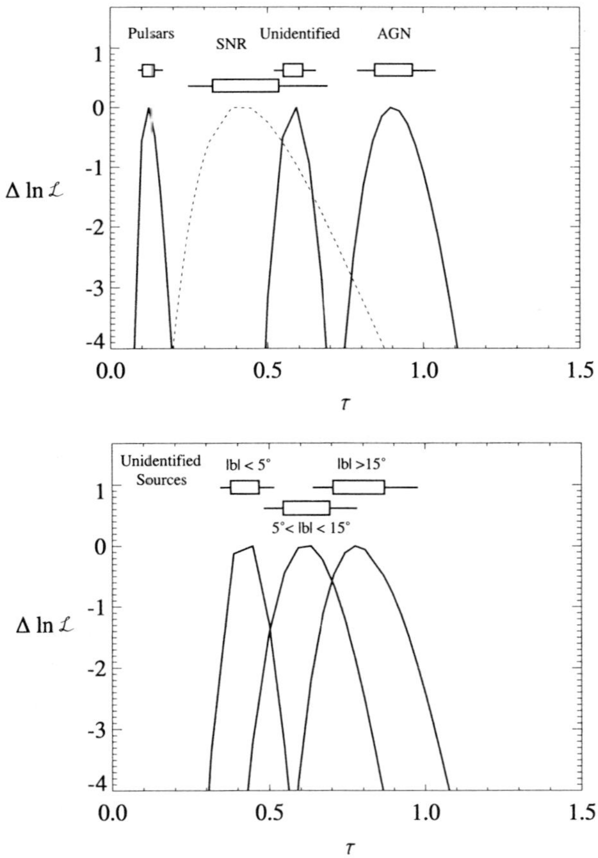

Figure 8 Variability index for different source populations. From: [Tompkins 1999]

When comparing the results of the different methods against each other, the level of consistency among the results of the variability studies is incredibly low. Unfortunately, each study has been performed using different source ensembles, hampering efforts to trace the origin of such discrepancies. Figure 9 compares, for instance, results from the variability study by Tompkins with the one by McLaughlin, using sources common to two samples. Note that the difference of the number of viewing periods included in the study by Tompkins and by McLaughlin could not account for the spread in the variability results.

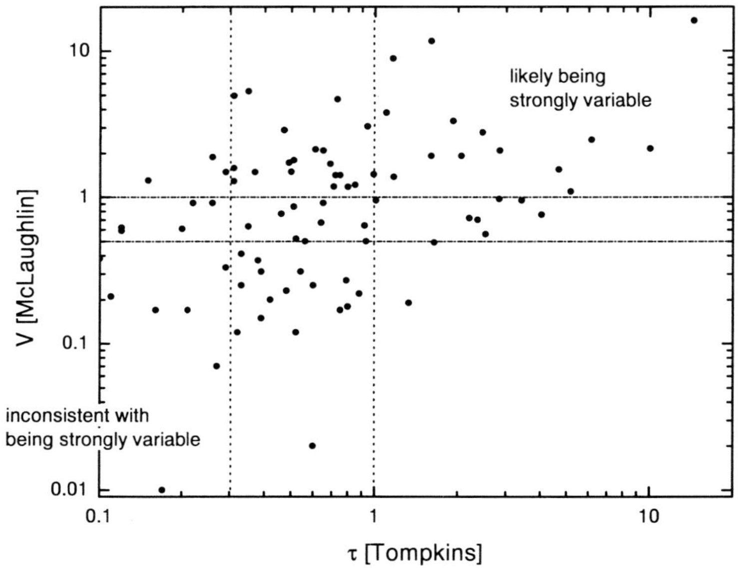

Figure 9 Comparison of the results from different variability studies, here given for McLaughlin et al. (1996) and Tompkins (1999). The vertical and horizontal grid represent the suggested boundaries by the autors for low variability, uncertain, and high variability characteristics.

In consequence, not only the flux uncertainty of a particular source in a given time interval needs to be accounted for, also the uncertainties of the normalization procedure. Studies aimed to compute a common measure of gamma-ray flux variation should be aware of the level of underlying systematics before drawing conclusions. Variability examined up to the very detail for an individual source might stand here as an example, as recently revised for the high-latitude unidentified source 3EG J1835+5918 (Reimer et al. 2001). For many years believed to be a variable source of gamma-ray emission, only the combination of its expanded observational history, appropriate cuts to establish comparable quality in the data to be compared and an understanding of the systematics revealed that this source actually is compatible being a non-variable source, in this case with direct implications of its likely identification.

5. SPECTRAL CHARACTERISTICS OF GAMMA-RAY SOURCES

The efficency of the EGRET spark chamber is not only a function of time, but also of the energy. Figure 10 shows the spark chamber efficiency as function of time for the ten energy intervals typically chosen for determining the spectrum of an EGRET detected gamma-ray source. For clarification, the normalization functions obtained by fitting the data (compare with upper panel Fig. 7) are given here. Consequences beyond the systematics as already mentioned for gamma-ray source fluxes are expected.

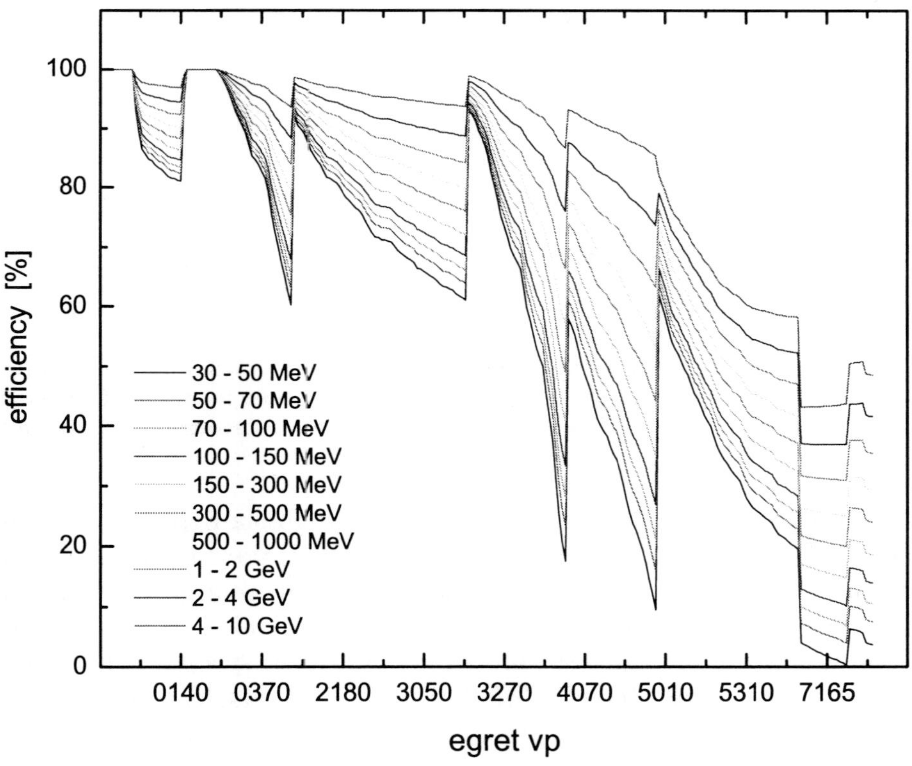

Figure 10 EGRET efficiency as function of observation time and energy. The ten subsequent energy bands are sketched which are used to determine source spectra. For clearer view, here the correction functions are plotted only.

The Third EGRET catalog lists for the majority of the sources the photon spectral index in $F(E) \propto E^{-\gamma}$, consistently determined for the combined observations from *CGRO* observation cycles 1 to 4. In various cases the P1234 sum is not the most significant detection and a meaningful or even better spectrum could be determined in an individual viewing period. For the EGRET detected AGN a spectral study on the level of individual viewing periods has been performed (if sufficient counts have been recorded) by Mukherjee et al. (1997). Fierro et al. (1997) also published phase-resolved spectra for the brightest gamma-ray pulsars. At present, individual EGRET source spectra are investigated beyond a single power-law model fit (Bertsch et al. 2000). Additionally, the spectral index could be used in order to conclude on spectral variability. Such information goes beyond flux variability and an individual power-law spectral index only, but has not often be used in studies of individual as well as population studies so far.

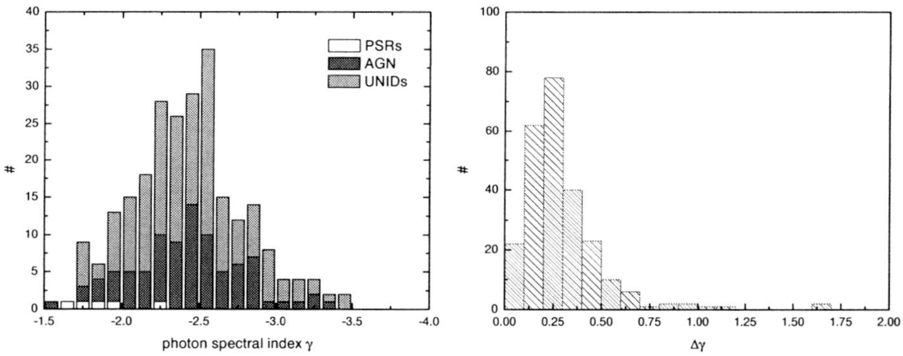

Figure 11 Distribution of the power–law spectral index for AGN, PSRs and unidentified EGRET sources and its uncertainty. It gives a vague impression about the difficulty to distinguish unidentified gamma-ray sources by its spectral characteristics only.

The spectral characteristics of individual gamma-ray sources have been used in a similar way to attempt a distinction between source classes in population studies as variability, exclusively on the basis of the photon spectral indices given in the 3EG catalog. Generally ignored by only picking up the numerical value of the spectral index when drawing conclusions, the significant uncertainties in the spectral index, especially for dim sources, put conclusions of distinct spectral characteristics rather into perspective. At present, conservative conclusions could only be drawn from the hardness of the spectrum of gamma-ray pulsars and the indication of a cut-off at GeV-energies. AGN seem to have a different spectrum when observed at high activity states (outbursts/flares) compared to their average spectra. However this needs to be investigated further. The vast majority of the unidentified EGRET sources are currently not distinguishable from the identified gamma-ray source populations by spectral characteristics only. Due to the wide spread found among the spectal index, and accounting for the uncertainties in the power-law spectral fits as well as the systematic bias towards finding hard spectrum gamma-ray sources significantly easier to discriminate against the diffuse emission component than soft spectra sources, it appears that viable conclusions might arise rather from the detailed spectrum of an individual source than from populations studies. Merck et al. (1996) conducted a survey for pulsar-like characteristics among the sources near the Galactic plane. At present, three of the sources suggested therein support the validity of this approach (PSR B1046–58/3EG J1048–5840, RX J2020.2+4026/3EG J2020+4017, SAX J0635+0533/3EG J0634+0521).

6. SIGNATURES FROM SPATIAL, TEMPORAL AND SPECTRAL PROPERTIES

Seeking signatures from the spatial, temporal and spectral characteristics of the unidentified EGRET sources is the subject of population studies. At present, the ensemble of persistent (Grenier 1999) or steady sources (Gehrels et al. 2000) appears to be the most promising in order to give hints on the nature of unidentified sources. Here I only sketch the general associations between spatial, temporal, and spectral properties for the unidentified EGRET sources and address the most prominent features or lack of features, respectively.

In Fig. 12a, the tendency to find hard spectrum sources predominantely close to the Galactic plane is apparent. However, this is at least partly due to the inability to discriminate soft spectrum sources near the plane against the dominant diffuse gamma-ray emission component. Fig. 12b gives the impression that variability is a common phenomenon for unidentified sources at all Galactic latitudes, although a clustering of more steady sources towards the Galactic plane is indicated, at least for the sources with extremely low probability to exhibit strong flux variability. Fig. 12c shows the correlation between spectral and variability index. Only a rough tendency is indicated, that the softer sources are also characterized by higher variability. Counterexamples could be found for both extremes: hard spectrum sources exhibiting a high degree of flux variability and soft sources consistent with being non-variable. Fig. 12d shows the flux distribution of unidentified sources. The few extremely

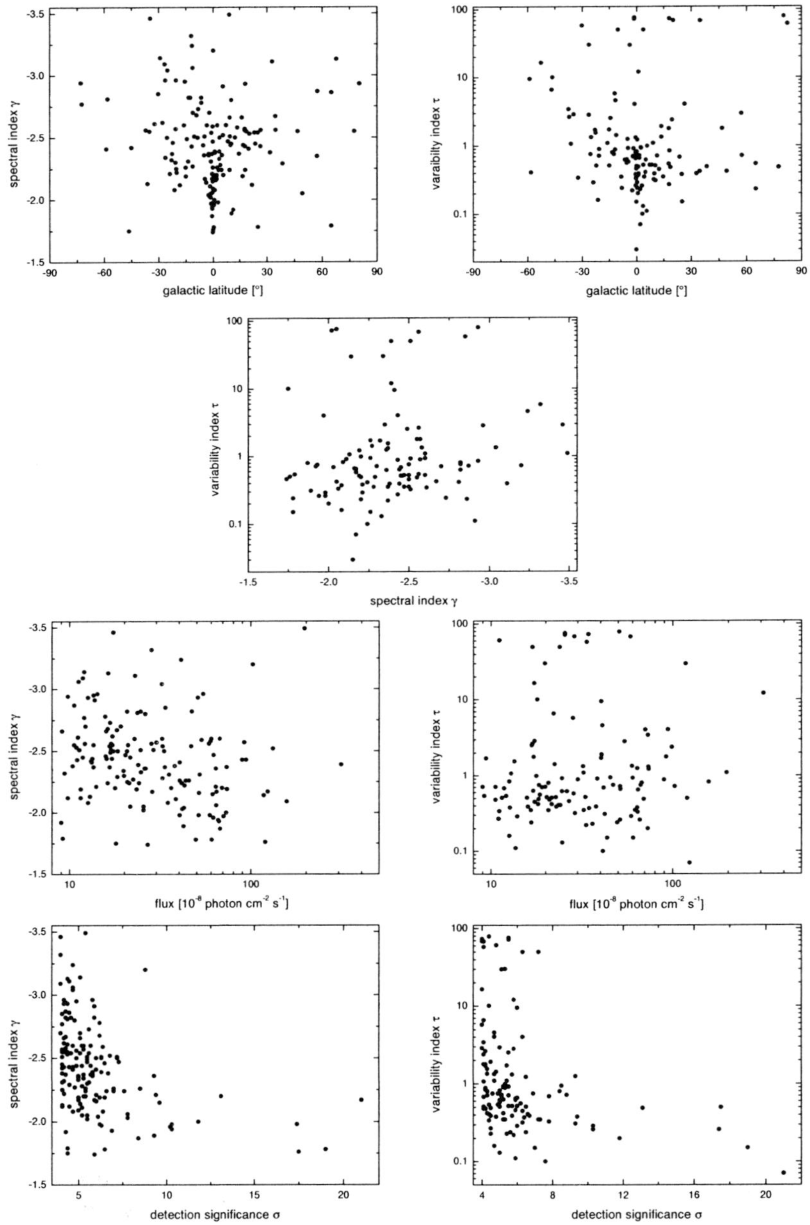

Figure 12 Associations between the observables determined to characterize the gamma-ray properties of unidentified EGRET sources. See text for details.

high fluxes are the signatures of transient sources, arising from hard as well as soft spectrum emitters. Fig. 12e shows the same flux distribution against the variability index. The flaring or transient behaviour of unidentified sources is not in direct relation to either extreme flux values or strong source variability. This seems rather curious, but becomes obvious when comparing with a similar arrangement of these quantities for the EGRET detected AGN. Fig. 12f and 12g show spectral and variability indices as function of the source detection significance. High detection significances could be found predominantly for hard spectrum and low variable sources, giving the combination of both the rather distinct feature as apparent in the selections called "steady" (Gehrels et al. 2000) or "persistent" (Grenier et al. 1999).

However, nearly all signatures have to be put into perspective when supplemented by the appropriate uncertainties. The less apparent correlations are basically at the 1σ level in the uncertainties of the observables themselves. Only carefully chosen combinations between the observables still reveal higher degrees of confidence in correlations among the observable parameter of the unidentified EGRET sources.

7. CONCLUSIONS

At present, we still fail to identify the nature of the majority of gamma-ray sources on the sky. However, the various observables at hand help to characterize individual sources rather well. For applications beyond individual sources (population studies, selected source ensembles) the degree of systematic biases and individual uncertainties needs to accounted for. Above all, in most cases gamma-ray astronomy still suffers from statistical limitations. Applying cuts is a valid procedure only if the particular cut is well understood in all its consequences for the data space. Conclusions drawn from subsets have to address its implication for the residual sources also. Generally, selections have to be made to avoid additional non–uniformities as already present in the 3EG source catalog. It needs to be investigated, at which level the known systematics and biases might put existing population studies in perspective. Efforts should be made to obtain more unbiased subsets instead of picking up any and each of the catalog listed gamma-ray sources. A compensation against the various biases by performing appropriate corrections is a requirement for comprehensive population studies.

The EGRET data will remain unique in gamma-ray astronomy until follow-up missions, in particular *GLAST*, will clarify a lot concerning the identity of individal gamma-ray sources and hypotheses of the composition of unidentified EGRET sources in the collective. Until then we have to work out the open questions on the basis of already acquired data. The tremendous potential offered by the nine years of EGRET data should not be neglected.

Acknowledgments

The author wish to thank Alberto Carramiñana and the staff of INAOE for the kind hospitality to make this workshop happened in a way as successfully and memorable as it apparently became. Also, I like to thank NRC for travel support.

References

Fierro, J.M. et al. 1997, ApJ 494, 734
Esposito, J.A. et al. 1999, ApJS 123, 203
Gehrels, N. et al. 2000, Nature 404, 363
Grenier, I., 1999, AIP Conf. Proc. 515, 261
Hartman, R.C. et al. 1999, ApJS 123, 79
Hunter, S.D. et al. 1997, ApJ 481, 205
Lamb, R.C. & Macomb, D.J 1997, ApJ 488, 872
Mattox, J.R. et al. 1996, ApJ 461, 396
McLauchlin, M. et al. 1996, ApJ 473, 763
Merck, M. et al. 1996, A&A Suppl. Ser. 120, 465
Mukherjee, R. et al. 1997, ApJ 490, 116
Özel, M.E. & Thompson, D.J. 1996, ApJ 463, 105
Reimer, O. et al. 1997, Proc. 25th ICRC, Vol.3, 97
Reimer, O. et al. 2001, MNRAS, in press (astro-ph/0102150)
Romero, G.E. et al. 1994, A&A 288, 731
Strong, A.W., Moskalenko, I.V. & Reimer, O. 2000, ApJ 537, 763
Thompson, D.J. et al. 1995, ApJS 101, 259
Torres, D.F. et al. 2001, A&A, 370, 468
Wallace, P. et al. 2000, ApJ 540, 184
Zhang, L., Zhang, Y.J. & Cheng, K.S. 2000, A&A 357, 957

MULTIWAVELENGTH SEARCHES AND SPECTRAL ASPECTS OF UNIDENTIFIED HIGH ENERGY GAMMA-RAY SOURCES

Reshmi Mukherjee
Barnard College, Columbia University
Department of Physics & Astronomy
New York, NY 10027
muk@astro.columbia.edu

Jules Halpern
Columbia Astrophysics Laboratory
Columbia University
New York, NY 10027
jules@astro.columbia.edu

Keywords: High energy gamma-ray sources, unidentified EGRET sources

Abstract More than half the sources in the Third EGRET (3EG) catalog are unidentified, having no firmly established counterparts at other wavelengths. Some of the EGRET unidentified sources have remained a mystery since the first surveys of the γ-ray sky with the *COS-B* satellite. The unidentified sources generally have large error circles, and finding counterparts has often been a challenging job. A multiwavelength approach, using X-ray, optical, and radio data, is often needed to understand the nature of these sources. This technique has been used recently to find likely identifications for several EGRET sources, such as 3EG J1835+5918, 3EG J2227+6122, and the *COS-B* field 2CG 075+00, which overlaps with two EGRET sources 3EG J2016+3657 and 3EG J2021+3716. This article reviews multiwavelength studies of unidentified EGRET sources and their spectral energy distributions, using the above-mentioned sources as examples.

1. INTRODUCTION

One of the first surveys of the γ-ray sky was carried out by the *COS-B* satellite in the 1970s, which detected several point sources of high energy

A. Carramiñana et al. (eds.),
The Nature of Unidentified Galactic High-energy Gamma-ray Sources, 35–48.

γ-rays along the Galactic plane (Swanenburg et al. 1981). The majority of the *COS-B* sources did not have firmly established counterparts at any other waveband. With the launch of the *Compton Gamma Ray Observatory (CGRO)* in 1991, the on-board EGRET (Energetic Gamma-ray Experiment Telescope) instrument carried out improved surveys of the γ-ray sky above 100 MeV, at relatively better angular resolution. The third EGRET (3EG) catalog lists 271 point sources of γ-rays (Hartman et al. 1999), the detection of which has indeed been one of the major advances in high energy astrophysics in the past decade. However, of these 170 sources remain unidentified, some of them originally detected by the *COS-B* satellite. The nature of these persistent γ-ray sources in the Galaxy, at $>$ 100 MeV, is an outstanding mystery in high energy astrophysics, almost three decades after their discovery.

Figure 3.2 shows the sky map in Galactic coordinates of the high energy γ-ray sources in the 3EG catalog. The majority of the identified sources are blazars at high latitudes and pulsars at low latitudes (Table 3.2). In addition, EGRET has detected one radio Galaxy, Cen A, and the LMC. The principle method of identification of 3EG sources relies on finding positional coincidences between EGRET sources and FSRQ/mm sources (e.g. Thompson et al. 1995) or is based on the statistical evidence that blazars are the dominant population. (e.g. Mattox et al. 1997; Bloom et al. 1997).

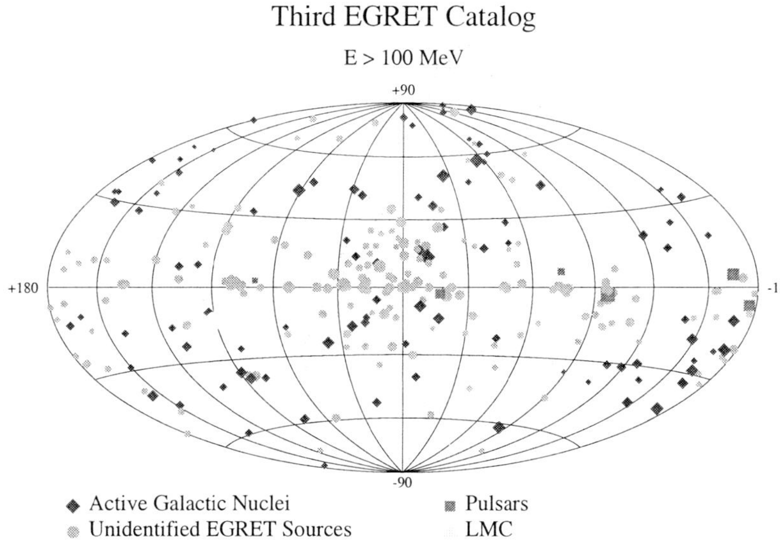

Figure 1 Point sources (from the Third EGRET Catalog) detected by EGRET at $>$ 100 MeV (Hartman et al. 1999).

Table 1 Sources in the Third EGRET (3EG) Catalog

Source Class	Number
Blazars	67 (94)*
Pulsars	6
Unidentified	170
Normal Galaxy	1 (LMC)
Radio Galaxy	1 (Cen A)
X-ray Binary	1 (Cen X-3)

* 27 AGN have been marginally identified.

The identification of the EGRET sources, particularly those close to the Galactic plane has proved to be challenging. About 40% of these sources lie within $|b| \leq 10^\circ$ of the Galactic plane. Identification on the basis of position alone has been difficult because the size of the EGRET error contours is typically large, $\sim 0.5^\circ - 1^\circ$. In addition, the presence of strong Galactic diffuse emission along the plane, and a lack of tight correlation between the γ-ray flux and other properties, like X-ray flux, core radio flux, etc., allows only the strongest sources to be identified on the basis of position alone.

Comprehensive surveys of the fields associated with the EGRET unidentified sources have met with limited success (see Mukherjee, Thompson & Grenier 1997 for a review). Several researchers have considered the possibility that rotation-powered pulsars are likely identifications for the unidentified 3EG sources (e.g. Halpern & Ruderman 1993; Helfand 1994). It has been noted previously that the unidentified EGRET sources in the Galactic plane lie in proximity to star formation sites and supernova remnants (Yadigaroglu & Romani 1997, Sturner & Dermer 1995, Esposito et al. 1996), or are correlated with OB associations and massive stars (Montmerle 1979; Kaaret & Cottam 1996; Kaul & Mitra 1997; Romero et al. 1999).

Studying multiwavelength data in the fields of EGRET unidentified sources could be an approach that might help in the identification of some of the high energy 3EG sources. Efforts to identify the γ-ray sources at other wavelengths include systematic multifrequency radio observations (e.g. Özel et al. 1988) and X-ray imaging studies (e.g., Brazier et al. 1996; 1998; Roberts & Romani 1998; Mirabal et al. 2000; Mukherjee et al. 2000; Reimer et al. 2000; Roberts et al. 2001; Halpern et al. 2001a). In this article we selectively review the recent efforts to identify EGRET sources using a multiwavelength approach by considering a few specific examples: 3EG J2016+3657 & 3EG J2021+3719, 3EG J1835+5918, & 3EG J2227+6122. It is impossible to comprehensively describe the exciting recent developments in the field in just a few pages, and the reader is therefore referred to the specific references given in the following sections.

2. MULTIWAVELENGTH STUDY OF 3EG J2016+3657 & 3EG J2021+3719

3EG J2016+3657 & 3EG J2021+3719 are two sources in the Cygnus region consistent with the unidentified *COS-B* source 2CG 075+00 (Pollack et al. 1985). The 3EG catalog gives their Galactic coordinates as $l = 74.^{\circ}76$, $b = 0.^{\circ}98$ (3EG J2016+3657) and $l = 75.^{\circ}58$, $b = 0.^{\circ}33$ (3EG J2021+3716). In the cumulative Phase 1 through Cycle 4 EGRET observations, 3EG J2016+3657 and 3EG J2021+3716 were detected at significances of 6.4σ and 10.3σ, respectively. Recently, Mukherjee et al. (2000) carried out a multiwavelength study of this region, and concluded that the weaker of the two sources, 3EG J2016+3657, is most likely the blazar-like radio source B2013+370 (G74.87+1.22). This identification was later confirmed by Halpern et al. (2001b) who found optical spectroscopic identifications of all soft and hard X-ray sources in the error circle of the EGRET source, leaving B2013+370 as the most likely counterpart of 3EG J2016+3657. This section summarizes the highlights of the quest for finding an identification for 3EG J2016+3657.

The error circles of both 3EG J2016+3657 & 3EG J2021+3716 were covered by archival X-ray imaging observations with *ROSAT* (PSPC and HRI) and *ASCA*, as well as *Einstein* IPC (Wilson 1980). Figure 2 shows the *ROSAT* soft X-ray (0.2–2.0 keV) image of this region. The minimum detectable intrinsic flux for the *ROSAT* image was 6.5×10^{-13} erg cm^{-2} s^{-1}. Several of the *ROSAT* point sources were visible in the *ASCA* image of this region. The positions derived from the *ROSAT* analysis were used to search for counterparts to the X-ray sources. The X-ray observations of 3EG J2016+3657 and 3EG J2021+3719 are described in more detail in Mukherjee et al. (2000).

A complete set of optical identifications of all X-ray point sources within the error circles of the two EGRET sources was recently obtained by Halpern et al. (2001b) using the MDM 2.4 m and the KPNO 2.1 m telescopes. Other than source # 1 and # 3 in figure 2, the other sources in the EGRET fields are either cataclysmic variables (CVs), or Wolf-Rayet stars or binary O stars. CVs are not known to be γ-ray emitters, and it remains to be demonstrated that either Wolf-Rayet stars or binary

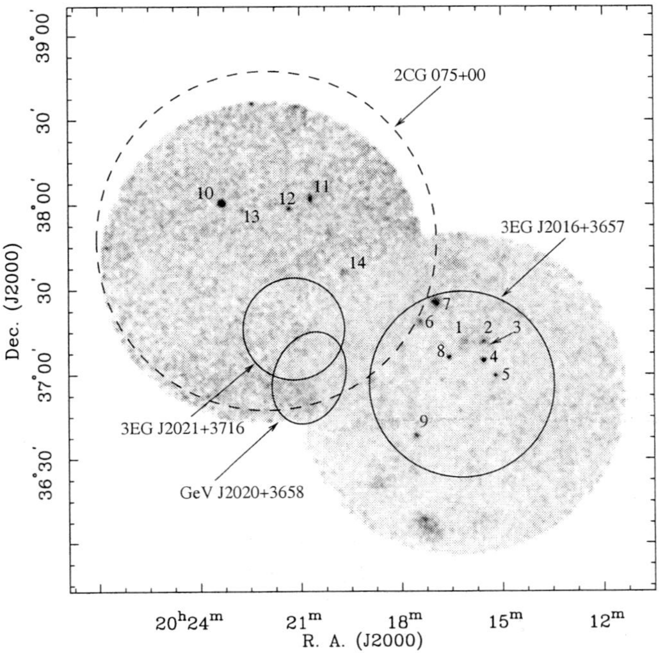

Figure 2 *ROSAT* soft X-ray image of 3EG J2016+3657 and 3EG J2021+3716. The circles for the two 3EG sources correspond to the $\sim$ 95 % confidence contours. The dashed circle corresponds to the *COS-B* source 2CG 075+00. The GeV Catalog source (Lamb & Macomb 1997) is also shown (Mukherjee et al. 2000).

O stars contribute at all to the EGRET source population. The two sources of interest in this field are the source numbers 1 and 3. Of these, # 1 is the supernova remnant (SNR) CTB 87 and # 3 is the blazar-like radio source B2013+370. Source #3, which is barely resolved in figure 2, is clearly seen in the *ROSAT* HRI image shown in figure 3. Although SNRs have previously been associated with EGRET sources, in this case CTB 87 is disfavored because it is simply too weak and too far away to be the likely candidate (see Halpern et al. 2001b; Mukherjee et al 2000 for detailed arguments). Instead, the most likely counterpart for 3EG J201657 is the compact, flat-spectrum radio source B2013+370.

B2013+370 has all the standard characteristics of a compact, extragalactic, non-thermal radio source, similar to the others detected by EGRET. It has a 5 GHz flux of $\sim$ 2 Jy, is detected at 90 GHz and 142 GHz with IRAM 30 m telescope, and exhibits variability at mm wavelengths. Figure 4 shows two images of B2013+370 taken with the MDM 2.4 m telescope in April and July, 2000. The blazar is seen to be clearly variable. The images illustrate its change from $R = 21.40 \pm 0.04$ in April

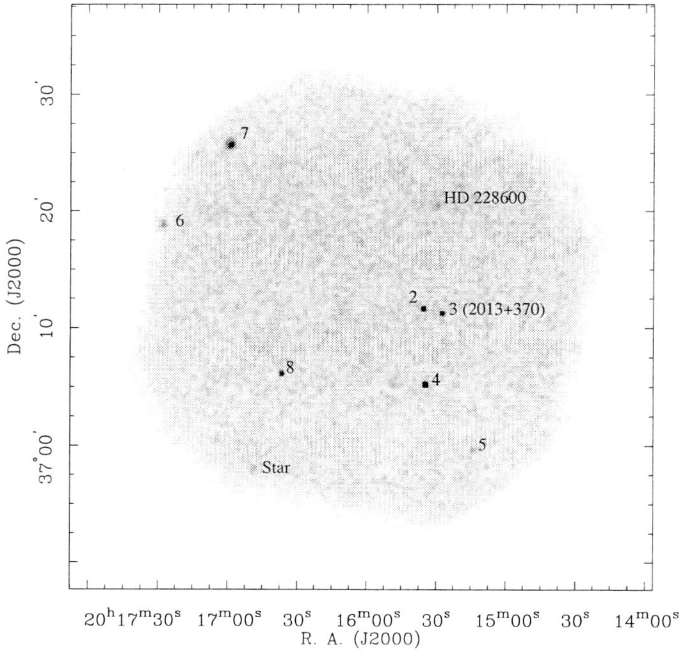

Figure 3 *ROSAT* HRI X-ray image of the field around 3EG J2016+3657. The image shows the sources 2 and 3 (B2013+370) as clearly resolved point sources (Mukherjee et al. 2000).

to $R = 21.81 \pm 0.05$ in July. Detection of optical variability bolsters the arguments that B2013+370 is very likely to be a FSRQ or BL Lac object.

Fig. 5 shows the spectral energy distribution (SED) of 3EG J2016+3657, assuming that it is the blazar B2013+370. The SED is characterized by a synchrotron peak at lower energies, a Compton peak at higher energies, and is dominated by power output in γ-rays. The SED of 3EG J2016+3657 is similar to that of other blazars detected by EGRET. In summary, there appears to be very little doubt that 3EG J2016+3657 is the blazar B2013+370. 3EG J2021+2716, on the other hand, remains unidentified. While it is not surprising to find one blazar only 1° from the Galactic equator, the expectation of finding more blazars in the Galactic plane is low. (Using the number of well-identified blazars in the 3EG catalog, one expects just one blazar within the zone $-1° < b < +1°$.) It is unlikely, therefore, that blazars will make a significant contribution to the low-latitude unidentified EGRET sources.

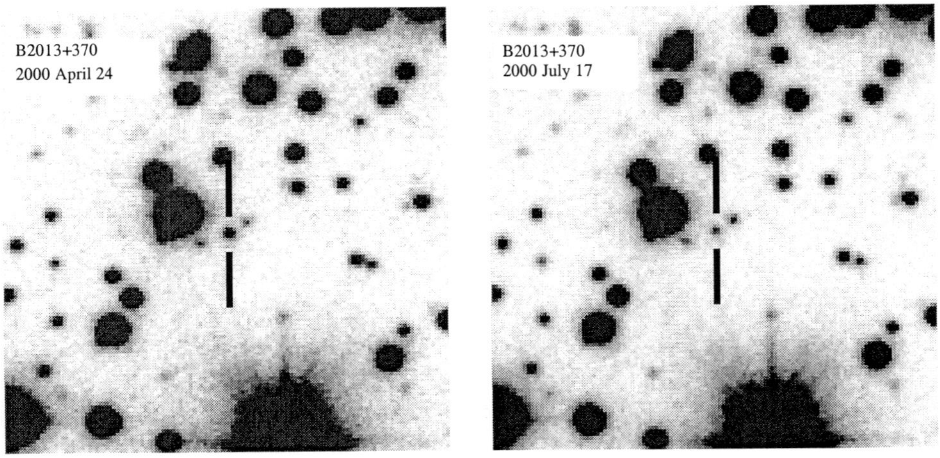

Figure 4 R-band CCD images of B2013+370 (Source # 3 in figure 3) illustrating change from $R = 21.40 \pm 0.04$ in April to $R = 21.81 \pm 0.05$ in July. The two images are each $35'' \times 35''$ and were taken with the MDM 2.4 m telescope. (Halpern et al. 2001b).

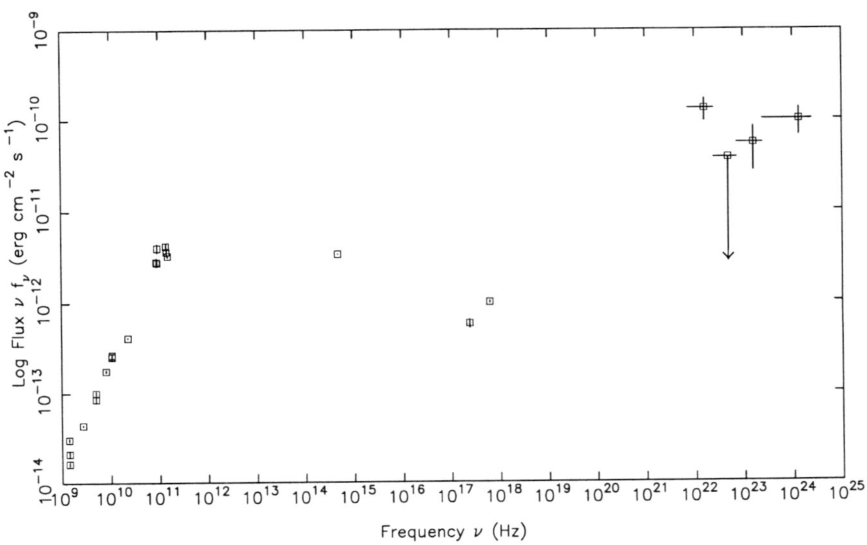

Figure 5 Broad band spectrum of B2013+370, assuming that it is the γ-ray source 3EG J2016+3657 (Mukherjee et al. 2000).

3. THE CASE OF 3EG J2227+6122

3EG J2227+6122 is another source in the Galactic plane that was the subject of recent multiwavelength study (Halpern et al. 2001a). It is located at Galactic coordinates $l = 106.^\circ5$, $b = 3.^\circ2$ and has been persistently detected by EGRET (Hartman et al. 1999). COMPTEL observations of the source were described in the past by Iyudin et al. (1997). X-ray and radio observations of the EGRET field strongly point to the possibility that 3EG J2227+6122 is a young pulsar (Halpern et al. 2001c). Rotation-powered pulsars have long been believed to dominate the Galactic γ-ray source population, and many authors have considered this hypothesis from both statistical and theoretical viewpoints (Halpern & Ruderman 1993; Helfand 1994; Mukherjee et al. 1995; Kaaret & Cottam 1996). This section reviews the arguments made in favor of 3EG J2227+6122 being a young pulsar.

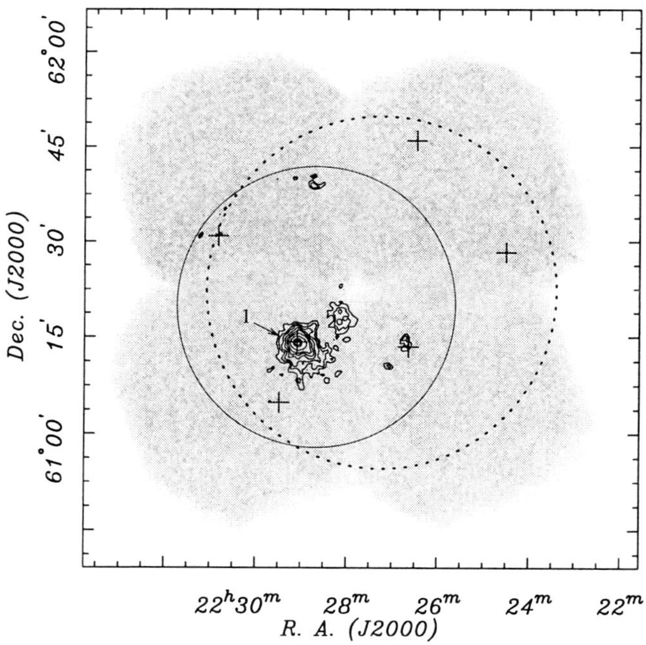

Figure 6 Composite *ROSAT* HRI image of the 3EG J2227+6122 field. The dashed circle corresponds to the 95% error contour of the EGRET source. Except for # 1, all the X-ray point sources (plus signs) are bright stars. # 1 is the only unidentified HRI source, and is coincident with a bright, hard source seen in the *ASCA* GIS image (contours). The solid circle corresponds to the *ASCA* GIS field (Halpern et al. (2001a).

Figure 6 shows a composite *ROSAT* HRI image of the error circle of 3EG J2227+6122, showing 6 point-like X-ray sources within the EGRET 95% contour (Halpern et al. 2001a). All sources, except # 1 have optical spectroscopic identifications using the KPNO 2.1 m telescope and Goldcam spectrograph, and are either bright K and M type stars, or emission-line stars. None of these could possibly be responsible for the high energy γ-ray emission. The interesting source in figure 6 is # 1, RX J2229.0+6114, which has no optical counterpart. It has been detected in the *ASCA* image of the region, and the contours in figure 6 correspond to the *ASCA* GIS image of the source AX J2229.0+6114.

Halpern et al. (2001a) obtained 20 cm (NVSS) and 6 cm (VLA) images of the error circle of the EGRET source. Interestingly, they found that there was only one radio source that was coincident with an X-ray source in the field, and that was source # 1, RX J2229.0+6114 in figure 6. The two radio maps are shown in figure 7. The radio source VLA J2229.0+6114 has an incomplete circular shell-like structure, with a high degree of linear polarization evident throughout the shell. Halpern et al. (2001a) have presented convincing arguments that VLA J2229.0+6114 and RX/AX J2229.0+6114 are associated with each other. These arguments are further supported by the recent detection of radio and X-ray pulsations from RX/AX J2229.0+6114 in follow-up observations of the region (Halpern et al. 2001c). The radio observations were carried out at

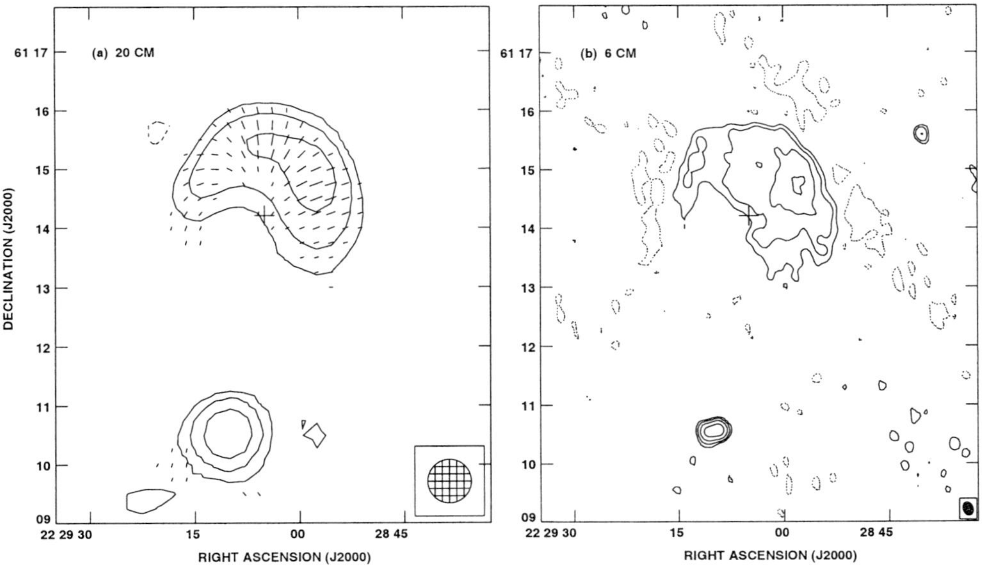

Figure 7 (Right) 20 cm NVSS map and (left) 6 cm VLA map, showing the shell-like radio source. The position of the X-ray RX J2229.0+6114 source is shown with a '+'. Notice the polarization vectors in the 20 cm NVSS map (Halpern et al. (2001a).

Jodrell Bank in 2001 February. The pulsations measured were found to have a period of 51.6 ms, and a $\dot{E}$ of 2.2×10^{37} erg s^{-1}. An image from the *Chandra* X-ray Observatory shows a point source surrounded by diffuse emission, contained within the incomplete radio shell, at the location of RX/AX J2229.0+6114. Halpern et al. (2001c) suggest that this is a new supernova remnant, G106.6+2.9, with the radio shell likely to be a shock between the pulsar wind nebula and the surrounding medium. Given these new results, the identification of 3EG J2227+6122 with the pulsar PSR J2229+6114 is quite compelling.

4. THE NATURE OF 3EG J1835+5918

3EG 1835+5918 is the brightest and most accurately positioned unidentified EGRET source that has been persistently detected at high energy γ-rays (Nolan et al. 1996). 3EG J1835+5918 is located at high Galactic latitude at $l = 88.74°$, $b = 25.07°$, well away from the confusing diffuse emission. The source shows no strong evidence of variability (Reimer et al. 2000), and has a spectral index in the 70 MeV to 4 GeV range of -1.7 (Hartman et al. 1999). Despite its small error circle, 3EG J1835+5918 has remained a mystery, and has been the subject of several multifrequency studies (Reimer et al. 2000; Carramiñana et al. 2000; Mirabal et al. 2000; Reimer et al. 2001; Mirabal et al. 2001). No known flat-spectrum radio source has been found in its error circle (Mattox et al. 1997). Its temporal and spectral variability indicate that it is more similar to pulsars than blazars.

ROSAT HRI and PSPC as well as *ASCA* observations of the EGRET field yield several point-like X-ray sources within the error circle of 3EG J1835+5918 (Mirabal et al. 2000; Reimer et al. 2001). Optical identifications of the X-ray sources were carried out independently by Mirabal et al. (2000) and Carramiñana et al. (2000). Most of the sources were found to be either radio-quiet QSOs or coronal emitting stars or a galaxy cluster. In addition, analysis of archival radio data (VLA, NRAO and WENSS) revealed only three sources within the 99% error contour of 3EG J1835+5918, all of which were fainter than 4 mJy at 1.4 GHz. The positions of the quasars and radio sources in the vicinity of the EGRET source are shown in figure 8 (Mirabal et al. 2000). Three of the interesting sources are individually marked in the figure: RX J1834.1+5913 is the brightest quasar in the EGRET error ellipse, VLA J1834.7+5918 is the brightest of the three weak radio sources within the EGRET error circle, and RX J1836.2+5925 is an object that does not seem to have an optical counterpart. No blazar-like radio sources were found in the vicinity of the EGRET source. The brightest neighbouring radio sources were

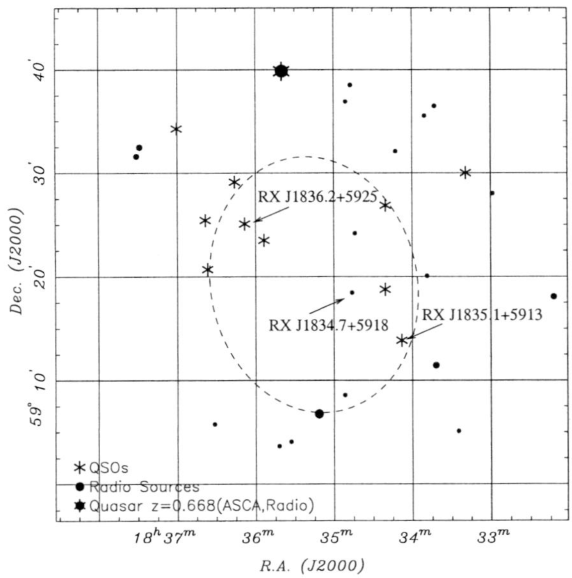

Figure 8 Quasars and radio sources within the error circle (shown with dashed lines) of 3EG J1835+5918. The positions of the three interesting objects, described in the text, are individually marked (Mirabal et al. 2000).

steep-spectrum radio galaxies or quasars. In fact, the broadband characteristics of 3EG J1835+5918 were examined to see if they fall within the multiwavelength parameters of the blazar class of sources seen by EGRET. Figure 9 (Mirabal et al. 2000) shows the radio, optical, X-ray and γ-ray fluxes of the sample of well-identified blazars in Mattox et al. (1997). For comparison, the fluxes of the brightest possible QSO counterpart, RX J1834.1+5913, and the most likely radio counterpart VLA J1834.7+5918 are also shown. A low energy synchrotron component and a high energy inverse Compton component are assumed. Note that both the candidates are found to lie at the faint end of the distribution, making it unlikely that 3EG J1835+5918 is a blazar.

RX J1836.2+5925, indicated in figure 8, is the most intriguing object within the error circle of 3EG J1835+5918. This object has no optical counterpart to a limit of $V > 25$ (Mirabal et al. 2001), and has been suggested as a radio quiet pulsar, and the most promising counterpart to the enigmatic γ-ray source 3EG J1835+5918 (Reimer et al. 2001; Mirabal et al. 2001). The ratio of the γ-ray flux above 100 MeV of 3EG J1835+5918 to the X-ray flux (0.12 - 2.4 keV) of RX J1836.2+5925 is similar to that of other similar candidates considered to be of pulsar origin (Reimer et al. 2001). The lack of an optical counterpart, and

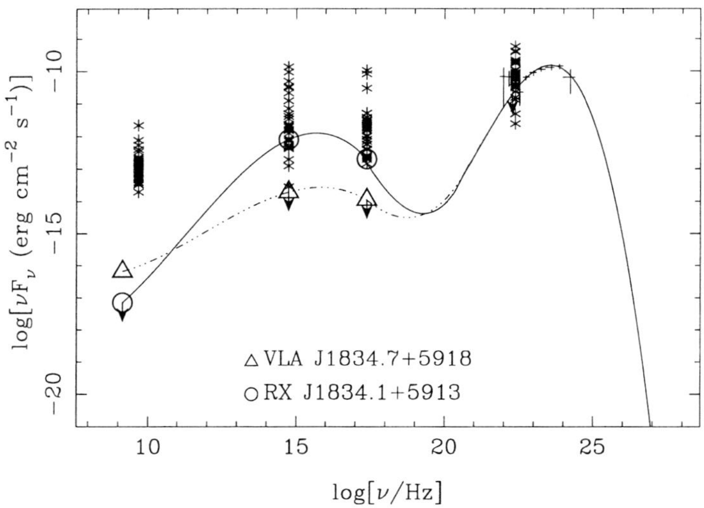

Figure 9 The broadband fluxes of EGRET blazars, as compiled from the literature, are shown as asterisks. The X-ray and radio data of the two most likely counterparts, RX J1834.1+5913 and VLA J1834.7+5918, if 3EG J1835+5918 were a blazar, are also shown. Note that the two candidates lie at the faint end of the distribution, making it unlikely that 3EG J1835+5918 is a blazar, at least with properties similar to other EGRET-detected blazars (Mirabal et al. 2000).

the non-variability of the γ-ray source are all characteristic signatures for a radio-quiet pulsar. Recently, Mirabal et al. (2001) have presented arguments that RX J1836.2+5925 is indeed a neutron star, and could be a nearby, rotation-powered radio-quiet γ-ray pulsar. Although its X-ray flux is at least 10 times fainter than that of Geminga, RX J1836.2+5925 is possibly older or more distant than Geminga, and the most likely counterpart of 3EG J1835+5918.

5. SUMMARY

We have presented a review of some recent work done towards understanding the nature of the unidentified EGRET sources using a multiwavelength approach. This review is by no means comprehensive, and just gives the reader a flavor of the field by using three specific examples. Other examples of studying the unidentified 3EG sources through a broadband approach may be found elsewhere in these proceedings. For a comprehensive search in the *ASCA* data for potential X-ray counterparts of GeV sources, the reader is referred to "The *ASCA* Catalog of Potential X-ray Counterparts of GeV Sources" (Roberts et al. 2001).

Resolving the nature of the unidentified EGRET sources remains a daunting task. Although progress has been made in the effort to identify individual EGRET sources, the majority still remain unidentified. It is possible that there may be a new class of γ-ray emitters, yet to be identified. Future observations with *GLAST* or *AGILE* will enable us to determine γ-ray source positions more accurately, and perhaps search for pulsations directly in the γ-ray data. In the interim before *GLAST* or *AGILE*, several of the unidentified EGRET sources will continue to be observed above 250 GeV by ground-based instruments like Whipple (e.g. Fegan et al. 2001, these proceedings), as well as by the new generation low-threshold ground-based Cherenkov detectors like STACEE (Williams et al. 2000) and CELESTE (de Naurois et al. 2000), sensitive to energies as low as 50 GeV. Both STACEE and CELESTE have recently reported the detection of the Crab nebula (Oser et al. 2000; de Naurois et al. 2000) and will soon be observing the unidentified EGRET sources along with blazars and SNRs. In the future, unidentified 3EG sources are likely to be studied not only by satellite-based experiments like *GLAST*, but also by next generation ground-based detectors like VERITAS (Bradbury et al. 1999), MAGIC (Martinez et al. 1999) and HESS (Kohnle 1999). Detection of very high energy γ-ray emission from unidentified 3EG sources with ground-based atmospheric Cherenkov telescopes is likely to open an exciting new chapter in the study of these sources.

This research was supported in part by the National Science Foundation and the Research Corporation.

References

Bloom, S. D., et al. 1997, ApJ, 488, L23.

Bradbury, S. M., et al. 1999, Proceedings of the 26th ICRC, Salt Lake City, Utah, astro-ph/9907248.

Brazier, K. T. S., et al. 1996, M. N. R. A. S., 281, 1033.

Brazier, K. T. S., et al. 1998, M. N. R. A. S., 295, 819.

Carramiñana, A. et al. 2000, Proceedings of the Fifth Compton Symposium, M. L. McConnell & J. M. Ryan Eds., AIP Vol. 510, p. 494.

de Naurois 2000, To appear in the Proceedings of the Heidelberg Symposium on High-Energy Gamma-Ray Astronomy, astro-ph/0010265.

Esposito, J. A., et al. 1996, ApJ, 461, 820.

Fegan, S. 2001, in these proceedings.

Halpern, J. & Ruderman 1993, ApJ, 415, 286.

Halpern, J., Gotthelf, E. V., Leighly, K. M., Helfand, D. J. 2001a, ApJ, 547, 323.

Halpern, J., Eracleous, M., Mukherjee, R., E. V. Gotthelf 2001b, ApJ, 551, 1016.

Halpern, J., et al. 2001c, ApJL, 552, L125.
Hartman, R. C., et al. 1999, ApJS, 123, 79.
Helfand, D. J. 1994, MNRAS, 267, 490.
Iyudin, A., et al. 1997, Proc. 25th ICRC, O. G. 3.1.3, Durban, South Africa.
Kaaret, P. & Cottam, J. 1996, ApJ, 462, L35.
Kaul, R. K. & Mitra, A. K. 1997, Proceedings of the Fourth Compton Symposium, C. D. Dermer, M. S. Strickman,
Kohnle, A. 1999, Proceedings of the 26th ICRC, Salt Lake City, D. Kieda, M. Salomon, B. Dingus (Eds.), OG 4.3.24, Vol. 5, p. 271. & J. D. Kurfess Eds., AIP, New York, p. 1271.
Lamb, R. C., & Macomb, D. J. 1997, ApJ, 488, 872.
Martinez, M., et al. 1999, Proceedings of the 26th ICRC, Salt Lake City, Utah.
Mattox, J. R., et al. 1997, ApJ, 481, 95.
Mirabal, N., et al. 2000, ApJ, 541, 180.
Mirabal, N., et al. 2001, ApJL, 547, L137.
Montmerle, T. 1979, ApJ, 231, 95.
Mukherjee, R., et al. 1995, ApJ, 441, L61.
Mukherjee, R., Grenier, I. A., & Thompson, D. J. 1997, Proceedings of the Fourth Compton Symposium, C. D. Dermer, M. S. Strickman, & J. D. Kurfess Eds., AIP, New York, p. 384.
Mukherjee, R., Gotthelf, E. V., Halpern, J., & Tavani, M. 2000, ApJ, 542, 740.
Nolan, P., et al. 1996, ApJ, 459, 100.
Oser, S., et al. 2000, ApJ, in press, astro-ph/0006304.
Ozël, M. E., et al. 1988, A&A, 200, 195.
Pollock, A. M. T., et al., 1985, A&A, 146, 352.
Reimer, O. 2000, Proceedings of the Fifth Compton Symposium, M. L. McConnell & J. M. Ryan Eds., AIP Vol. 510, p. 489.
Reimer, O., et al. 2001, MNRAS, in press.
Roberts, M. S. E. & Romani. R. W. 1998, ApJ, 496, 827.
Roberts, M. S. E., Romani, R. W., & Kawai, N. 2001, ApJS, 133, 451.
Romero, G. E., Benaglia, P., & Torres, D. F. 1999, A&A, 348, 868.
Sturner, S. J. & Dermer, C. D. 1995, A&A, 281, L17.
Swanenburg, B. N., et al. 1981, ApJ, 243, L69.
Thompson, D. J., et al. 1995, ApJS, 102, 259.
Williams, D. A., et al. 2000, To appear in the Proceedings of the Heidelberg Symposium on High-Energy Gamma-Ray Astronomy, astro-ph/0010341.
Wilson, A. S. 1980, ApJ, 241, L19.
Yadigaroglu, I.-A. & Romani, R. W. 1997, ApJ, 476, 356.

II

CORRELATIONS STUDIES OF UNIDENTIFIED GAMMA-RAY SOURCES

γ-RAY SOURCES OFF THE GALACTIC PLANE: HOW OLD AND FAR AWAY?

Isabelle A. Grenier
Université Paris VII & Service d'Astrophysique, CEA Saclay, 91191 Gif/Yvette, France
isabelle.grenier@cea.fr

Keywords: Gamma-ray source, pulsar, micro-quasar, Gould Belt

Abstract Whereas a distinct population of unidentified EGRET sources at low latitude unquestionably lies at kpc distances within the thin Galactic plane near active star-forming regions, it is increasingly apparent that the source sample off the plane is heterogeneous and involves different parts of the Galaxy, such as the nearby Gould Belt and the thick Galactic disc or the halo. The source spatial, spectral, and temporal characteristics are reviewed below and different options as to their age and nature in these diverse environments are discussed.

1. SOURCES OFF THE GALACTIC PLANE

Apart from active galactic nuclei that power bright, variable radio sources, identification of the EGRET sources off the Galactic plane has been hindered so far by the lack of counterparts in the hard X-ray range due to the limited sensitivity of the instruments or the lack of exposure far from the Galactic plane. The presence, statistically, of 1 to 3 ROSAT X-ray sources in the large error box has not provided interesting clues besides a large average luminosity L_γ/L_X ratio of typically a few hundred. Despite the reduced source confusion at high latitude, no conspicuous radio sources have triggered identification searches. As tradition dictates, clues can be gained by looking for subsets with distinct temporal, spatial, spectral, and luminosity characteristics and by correlating their spatial distribution with other classes of objects at medium and high latitude.

The main difficulty resides in modelling (or keeping in mind!) the numerous and severe observational biases across the sky such as: 1) the non-uniform detection sensitivity (which depends on exposure, the underlying interstellar intensity, and the parent source luminosity func-

A. Carramiñana et al. (eds.),
The Nature of Unidentified Galactic High-energy Gamma-ray Sources, 51–63.

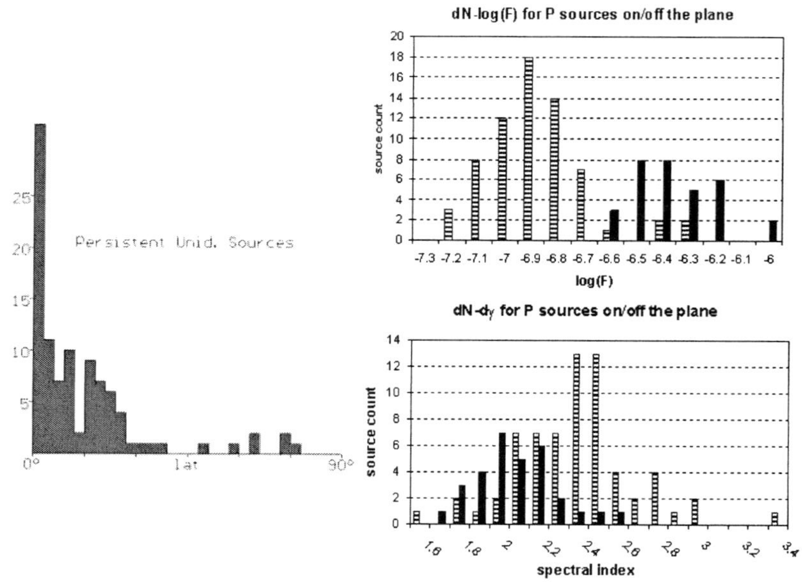

Figure 1 latitude, flux and spectral index distributions of the persistent sources at $|b| < 3°$ and $|b| > 3°$

tion); 2) the non-uniform response of variability indicators (which depend on the source strength with respect to the background, source spectrum and the stability of adjacent sources); 3) the varying precision of spectral measurements (which depend on the source strength and background spectrum, and on gas aging in the spark chamber). Various variability indicators have been introduced in the literature. Those based on the fractional variability, i.e. the ratio of the standard deviation to the average of the set of fluxes, allow to separate steady sources from those with poor flux measurements. This is the case for the τ or the μ indices (Tompkins, 1999, Romero et al., 2000, resp.). The τ indicator was adopted below because its derivation took into account possible fluctuations from the background and adjacent sources and it made use of upper limits as well as detected fluxes. A simple "persistence" criterion has also proved relevant for the 3rd EGRET catalogue sources (Grenier, 2000a): a source is labelled persistent (P) when significantly detected in the cumulative data from April 1991 to October 1995 (P1234) whereas a non-persistent one ($\overline{P}$) is seen in one or more individual observations, but not in the P1234 data. The detection threshold is set at 4σ at $|b| > 10°$ and 5σ elsewhere (Hartman et al., 1999). Most of the persistent sources indeed show no or little variability and the flaring activity of the $\overline{P}$ sources is confirmed by large τ values (Tompkins, 1999).

The persistent sources on and off the plane clearly differ in many aspects, as illustrated in Figure 1. The sharp break near 3° in the latitude profile corresponds to obvious changes in flux and spectrum as well. The sources near the plane are harder: average spectral indices of -2.18 $\pm$ 0.04 and -2.49 $\pm$ 0.04 have been measured at $|b| < 5°$ and $5° < |b| < 30°$, respectively (Gehrels et al., 2000). Different log(N)-log(>S) distributions have been found in the two latitude ranges (Gehrels et al., 2000). The lack of faint sources near the plane is caused by the intense Galactic background. Fainter sources are not resolved, but their integrated flux should not exceed 10% of the interstellar emission (Hunter et al., 1997). The subsequent limit on the number of faint sources indicates that the low-latitude population has a distinctly flatter log(N)-log(>S) distribution than at mid-latitude. The lack of bright sources off the plane is also remarkable. Most of the sources at $|b| < 3°$ show no sign of variability (Tompkins, 1999, Zhang L. et al., 2000), but the sensitivity of the various indicators is reduced above the intense background.

Is the source sample at $|b| > 3°$ homogeneous? P and $\overline{P}$ sources are found in equivalent numbers (67 and 59, resp.). The log-likelihood averages of the spectral and variability indices, which take individual uncertainties into account, indicate that P and $\overline{P}$ sources have distinct spectral and temporal properties. Average spectral indices $< \gamma_P > =$ 2.25 $\pm$ 0.03 and $< \gamma_{\overline{P}} > =$ 2.52 $\pm$ 0.06 were obtained, with a chance probability of 2×10^{-7} of equal index. No systematic difference is expected from the background since both groups are scattered at medium and high latitude and largely overlap. Similarly, averages $< \tau_P > =$ 0.38 $\pm$ 0.06 and $< \tau_{\overline{P}} > =$ 0.95 $\pm$ 0.18 were obtained, with a chance probability of 1.3×10^{-4} of equal τ. The $< \tau_{\overline{P}} >$ average is equivalent to that of 0.90 $\pm$ 0.07 found for the EGRET active galactic nuclei that are well known for their flaring activity. Finally, the two groups exhibit different distributions across the sky (Grenier, 2000a, Grenier, 2000b) and the difference cannot be attributed to systematic biases due to the survey exposure and the intense Galactic background.

To quantify this difference and to locate the sources, their apparent spatial distributions have been confronted with modelled distributions stemming from various parent populations throughout the Galaxy. The model takes into account the non-uniform detection sensitivity across the sky as a function of the source luminosity function which, being unknown, has been left as a free parameter. The method has been checked against the identified active galactic nuclei for which no anisotropic distribution was detected (Grenier, 2000b). Various sites are likely to display sources off the plane:

- a Galactic disc with a gaussian radial scale length of 9 kpc and an exponential scale height z_H varying from 0.1 to several kpc. Values of 0.1 and 0.2 kpc correspond to the interstellar H_2 and HI layers; a value of 0.4 kpc is typical of old radio pulsars wandering at large speed away from their birth place in the H_2 disc (Lyne et al., 1985); a value of 0.8 kpc corresponds to the thick stellar disc component of the Galaxy (Ojha et al., 1996).

- a globular-cluster like distribution, i.e. a spherical halo about the Galactic center with a density profile that is nearly flat up to a radius $R \sim 2.5$ kpc and falls as $R^{-3.5}$ further out (Zinn, 1985).

- a homogeneous Galactic halo with a radius of 20 kpc.

- the structure of the nearby Gould Belt, a 30 Myr-old expanding disc of young massive stars, inclined by $\sim 20°$ onto the Galactic plane. The Belt is traced by the column-density of nearby OB stars of spectral type $<$ B4 (Grenier, 2000b).

As illustrated in Figure 2a, the persistent sources closely follow the curved lane of the Gould Belt at medium latitudes (Grenier & Perrot, 1999, Gehrels et al., 2000, Grenier, 2000b). Their distribution is quite significantly and better correlated with the Gould Belt than with other Galactic structures (those cited above were tested as well as the very local Galactic disc with any scale height - Grenier, 2000b). Their excess at mid-latitude is very significant and clearly points to a nearby Galactic origin. The further significant improvement in the fit from Galactic disc models to the inclined Belt geometry gives strong support to their origin in the Belt system (with a chance probability of only 3×10^{-6} that the improvement be due to a random fluctuation from a Galactic disc distribution). As many as 40 ± 5 sources can be statistically associated with the Belt at $5 < |b| < 30°$. Over 80% of them show no sign of time variability (Tompkins, 1999).

The noteworthy asymmetry of the P sources north of the Galactic center and south of the anticenter vanishes for the $\overline{P}$ sources which are spherically distributed around the Galactic center (see Figure 2b) . For them, the maximum likelihood peaks for a very large scale height of $2^{+1.2}_{-0.6}$ kpc. Equally good fits are obtained for the z_H= 2kpc disc (alias 2kdisc) and the globular-cluster-halo (alias "Glob") fits, as expected from the similitude of the two parent distributions in space. The fit gets moderately worse to larger scales, the chance probability of the sources being

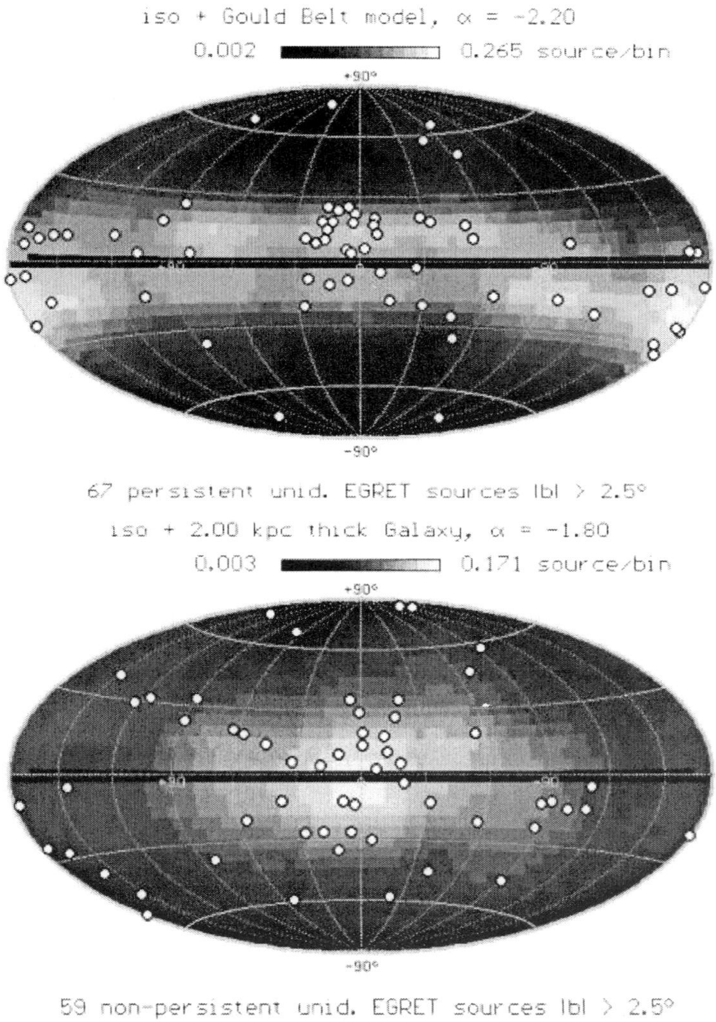

Figure 2 (*l*,*b*) maps of unidentified EGRET sources: harder P sources in (a), softer $\overline{P}$ sources in (b). Modelled source counts following the non-uniform sensitivity of the EGRET survey were generated from a combination of an isotropic distribution and (a) a Gould Belt population as traced by its young massive stars, (b) a Galactic disc population with a 2 kpc scale height.

in the homogeneous halo being 4×10^{-2}. However, the source distribution is not consistent with an extragalactic origin (chance probability of 3×10^{-6}). To smaller scale heights, the fit worsens markedly with decreasing z_H. Chance probabilities are 1.4×10^{-2}, 1.3×10^{-3}, 2.7×10^{-4} that the sources be part respectively of the radio pulsar population or of the HI and H_2 layers. All quoted probabilities allow for the extra degree of freedom in the parent luminosity function.

Three-fold combinations of an isotropic plus two Galactic components have been fitted against the whole P and $\overline{P}$ sample, without distinction. The best fit is obtained for the "iso+Belt+Glob" combination, attributing $N_{iso} = 35 \pm 17$, $N_{Belt} = 46 \pm 9$, and $N_{Glob} = 45 \pm 9$ sources to the three respective components. The second best fit (2.5 σ worse) is obtained for the "iso+Belt+2kdisc" combination. The improvement in the three-fold fit over the two-fold "iso+Belt", "iso+Glob", or "iso+2kdisc" combinations are quite significant (with chance probabilities from 1 to 4×10^{-8}). So, the presence of two distinct spatial components, besides a handful of extragalactic sources, appears to be genuine. An upper limit of 35 to the number of extragalactic sources present in the whole sample is derived from the various tests.

In conclusion, on the basis of their distinct spatial as well as temporal and spectral characteristics, it is proposed that the sources away from the Galactic plane consist of:

- 45 ± 6 relatively hard and stable sources associated with the starburst Gould Belt. They clearly differ (mostly in brightness) from the stable sources found in the Galactic plane. The Belt rim extends to ~150 pc north of the Galactic center and to ~450 pc south of the anticenter, so typical luminosities at 300 pc of 0.3 to 6×10^{33} erg/s over 4π sr can be derived for the Belt sources for E^{-2} spectra above 100 MeV. The Belt origin 30 to 40 Myr ago and the stellar lifetimes of the neutron star and black hole progenitors imply typical source ages of the order of a few Myr.

- 45 ± 9 softer and more variable sources that gather mostly within 60° around the Galactic centre. Their spatial distribution implies an origin in a very thick Galactic disc with a 2 kpc scale height or as the globular clusters in the halo. Yet, no spatial coincidence is found between individual globular clusters and the sources. Both distributions yield a mean distance to the sources ranging from 2 to 8 kpc according to their direction in the sky. It implies luminosities over 4π sr of 10^{35-36} erg/s above 100 MeV. Either born in the thin Galactic disk and having moved out as neutron stars do, or produced high above the plane by the first generation of stars in the still collapsing Galaxy, both scenarii suggest sources of a great age, maybe as old as a few Gyr.

- at most 35 potentially extragalactic sources.

Star-forming regions in the thin Galactic disc shelter another 38 sources described in these proceedings. As in the Gould Belt, they are at most a few Myr of age. These main populations do not preclude the presence of other rarer species, as suggested by the few flaring sources detected along the Galactic plane (Tavani et al., 1997, Romero et al., 2000).

2. SOURCES IN THE GOULD BELT

Strikingly, the stable unidentified EGRET sources appear to be associated with active star-forming regions, both locally in the starburst Gould Belt and throughout the Galactic disc.

The Belt is well delineated by its clouds, so the large number of Belt sources may cast doubts on the reliability of the interstellar background subtraction. It can be shown, however, that they are much too bright to be gas clumps overlooked in the HI and CO surveys (Grenier, 2000b). An origin of the γ-ray emission in the highly supersonic winds expelled by the numerous O stars in the Belt may be envisaged. Their kinetic powers of 10^{35-36} erg/s are large enough. Nuclei and/or electron acceleration is expected at the terminal shock between the wind and the ambient medium (Cassé & Paul, 1980, White, 1985), or along the turbulent expanding wind (White & Chen, 1992). Both sites may be involved to overcome the adiabatic losses suffered by the stellar flare particles on their way to the rim of the wind cavity (Cesarsky & Montmerle, 1983). Synchrotron radiation has indeed been detected from several massive stars. Hadronic interactions in the swept-up gas or inverse-Compton scattering of the stellar UV radiation field could plausibly produce γ rays. The Gould Belt contains 61 ± 7 O-stars with masses > 20 $M_\odot$ that develop powerful winds. Yet, none has been detected with EGRET. At a typical distance of 300 pc, the lack of detection sets an upper limit in luminosity of 10^{33} erg/s above 100 MeV for $E^{-2.5}$ or E^{-2} spectra, if any. As a consequence, the low-latitude sources coincident with OB associations (Romero et al., 1999) cannot be powered by a single active star (see Fig. 3). They may collectively contribute to the observed flux if the angular extent of the association is small enough to produce a point-like source for EGRET (in 2/3 of the cases).

Most of the low-latitude sources can be accounted for, in numbers and longitude distribution, by the population of pulsars born in the Galactic disc in the past million year, within a few kpc, whether the outer gap or polar cap geometries are used (1995, Sturner & Dermer, 1996, Zhang L. et al., 2000). In fact, most of these pulsars must be bright, therefore young, to overcome the intense Galactic background. This is why 90% of the detectable pulsars turn out to be younger than 0.3 Myr. No older

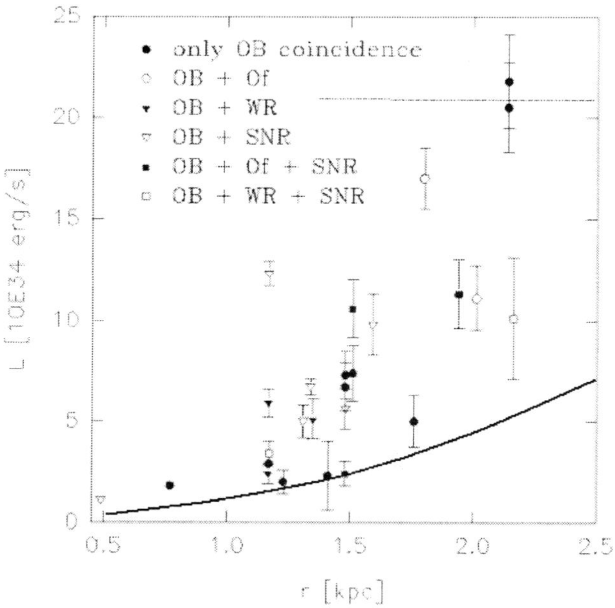

Figure 3 luminosity vs. distance for unidentified sources coincident with OB associations. The curve indicates the maximum luminosity expected from the supersonic wind from a single 0-star as estimated in the nearby Gould Belt.

pulsar is visible beyond 2.5 kpc. A limit of ~10% can be placed on the contribution of fainter pulsars to the diffuse Galactic emission (Sturner & Dermer, 1996; Pohl et al., 1997). Their rapid migration off the plane results in a scale height of ~200 pc after 1 Myr. Yet, both the outer gap and the polar gap models predict 10 times too few pulsars off the plane with respect to the number of Belt sources (1995, Sturner & Dermer, 1996, Zhang L. et al., 2000, Grenier, 2000b). This discrepancy cannot be solved by increasing the pulsar beam width, γ-ray efficiency or birth rate for fear of overproducing sources near the plane. For instance, an increase in birth rate beyond 30% is not supported by the γ-ray data (1995). Larger kick velocities or a larger scale height at birth are not supported by the radio data.

The starburst activity of the Belt has produced supernovae in the last few Myr at a rate that is 3 to 5 times that of the local Galactic disc (Grenier, 2000b). So, the Belt provides candidate pulsars in our neighbourhood, both in excess of the Galactic rate and naturally at mid-latitudes. The source stability is consistent with a pulsar origin. The first example of a Belt pulsar is quite famous (Geminga!) and, notably, the intrinsically faintest pulsar observed so far. Anyhow, it could have been detected up to 1 kpc, hence anywhere inside or around the Belt. Scaling the γ-ray luminosity, L_γ, with the square root of the spin-down power, $\dot{E}$, (Thompson et al. 1995), and extrapolating from Geminga for only half a decade, to $\dot{E} = 10^{33}$ erg/s $\Longrightarrow L_\gamma \sim 6 \times 10^{32}$ erg/s over 1 sr, suggests that a pulsar 10 times as old as Geminga, i.e. 3-Myr old, remains easily visible out to 500 pc. So, there is no strong bias to high-luminosity and youth in the Belt as in the plane. Simulations show that the Belt signature across the sky is preserved over 5 Myr despite the rapid pulsar migration and the expansion and precession of the Belt (Perrot & Grenier, in preparation). For 20 to 27 supernovae produced per Myr in the Belt, the detection of $\sim$40 γ-ray sources requires the product of the beaming fraction $\Delta\Omega/4\pi$ and pulsar age be of order 1.5. The wide outer-gap beam evolves from $\Delta\Omega/4\pi = 0.8$ to 0 with age up to $\sim$2 Myr when pair production in the gap is quenched by the lack of thermal X rays from the star (Romani, 1996), unless the stellar surface is heated by the primary particles flowing down to it (Zhang L. et al., 2000). The polar-cap beam is narrow ($\Delta\Omega/4\pi$ $\sim$0.1, decreasing with time as P^{-1}, Thompson et al. 1995), but emission can be sustained for a few Myr at least. One should also keep in mind that Belt pulsars, being close, may remain visible at large angles, on the side of the main beams. In the recorded lightcurves, side emission is detected off the main peaks over large phase intervals and is 10 times fainter than peak emission. Side emission from a 3 Myr-old pulsar would remain detectable by EGRET up to 350 pc, thus largely increasing the detection probability $\Delta\Omega/4\pi$. Polar cap simulations show that 4-5 times as many off-beam sources as on-beam ones would be detectable above 100 MeV at a given distance (Harding & Zhang, 2001), therefore up to 350 pc for side emission. In this case, the product $\Delta\Omega/4\pi \times 3$ Myr is 1.2-1.5 and the population of neutron stars born in the Belt in the last 3 Myr can plausibly explain the EGRET sources. The softness of side emission ($|\gamma| = 1.8$ to 2.5) may also explain the soft average index $\overline{\gamma} = 2.25 \pm 0.03$ of the Belt sources compared to that of the identified pulsars. So, one would expect a large dispersion of spectra and lightcurves in a population of nearby pulsars, with a harmonic power depending on brightness. Because of the narrow radio beam, radio-quiet γ-ray pulsars should also dominate the sample.

Young (< 0.1 Myr) Belt supernova remnants would be too rare and older, subtending a few degrees, would appear extended, even to EGRET. For mass progenitors > 25 $M_\odot$ (Timmes et al., 1996) and initial stellar-mass-function slopes of -1.1 and -2.0, black holes are 3 to 9 times fewer than neutron stars, respectively. They would be too rare in the young Gould Belt. In the absence of more compelling candidates, $\sim$3 Myr old pulsars, born in the Gould Belt, appear as the most promising counterparts for the Belt sources. They would provide a unique opportunity to constrain pulsar models to longer ages and at a variety of aspect angles.

3. SOURCES AT LARGE SCALE HEIGHT

Likely candidates for such old γ-ray sources include ms pulsars, red micro-blazars accreting from a low-mass star, and old isolated neutron stars accreting from the interstellar medium.

The latter are numerous, $\sim 10^8$ in the Galaxy, but improbable because of their low energetics. The Bondi-Hoyle accretion rate rapidly falls as the 3/2 power of the stellar speed into the ambient medium, so only slow (<20 km/s) objects in a dense cloud ($\sim$100 cm^{-3}) will accrete mass at an appreciable rate. Only slow enough rotators must be retained for the accretion flow to overcome the dipole radiation pressure and centrifugal barrier of the neutron star (Harding & Leventhal, 1992). Finally, the gas is heated by the UV radiation from the neutron star and the formation of a surrounding HII region severely limits the accretion power to a maximum of $\sim 2 \times 10^{32}$ erg/s, only (Blaes et al., 1995). On the other hand, highly magnetized (10^{12} G), fast (>200-400 km/s) neutron stars with long periods can accrete interstellar gas at a much larger rate than the Bondi-Hoyle one through Kelvin-Helmholtz instabilities in the layer of shocked gas in front of the magnetopause (Harding & Leventhal, 1992). The resulting power is, however, still many orders of magnitude too low to match the EGRET luminosities. Moreover, the bow shock would have a typical radius $R_{SH} = 1.8 \times 10^{11}$ cm.$(R_*/10 \text{ km}).(B_*/10^{12} \text{ G})^{1/3}.(n/1\text{cm}^{-3})^{-1/6}.(v/400 \text{ km/s})^{-1/3}$, which would limit particle energies, acquired by diffusive shock acceleration, to $E < 77MeV\,(R_{SH}/1.8 \times 10^{11}cm).(B[R_{SH}]/3.7 \times 10^{-4}G)$, i.e. too low to produce γ rays. The 6 isolated neutron stars discovered by ROSAT share common features, such as soft X-ray emission, no bright radio emission, and no variability. The purely thermal spectra recorded in X rays show, however, no sign of accretion-induced activity (Motch, 1999). Finally, the expected scale height of the isolated old neutron stars, similar to that of the old radio pulsars (i.e. 400 pc) is too small.

Reactivated neutron stars in the form of ms pulsars are quite attractive counterparts since one, PSR J0218+4232, has already been convincingly associated with an EGRET source (3EG J0222+4253) thanks to the timing signature of the pulsar (Kuiper et al., 2000). Its distance of 5.7 kpc, height of 1.6 kpc above the Galactic plane, spectral index of -2.6 in the EGRET band, and luminosity of 2 10^{35} erg/s over 4π sr above 100 MeV are strikingly similar to the characteristics of the halo sources. They share another trait: they do not belong to a globular cluster. The remarkable νF_ν spectrum of PSR J0218+4232 sharply peaks in the MeV range. Aligned pulse profiles have been recorded in X rays, γ rays, and in the radio (Kuiper et al., 2000, Kuiper & Hermsen, in preparation). Despite the low magnetic field (10^{8-9} G), both polar cap and outer gap models predict that a large fraction of the spin-down power may be converted into γ rays (Zhang B. & Harding, 2000, Zhang L. & Cheng, 1998). PSR J0218+4232 does exhibit a 7% efficiency if one sets the unknown beam width to 1 sr (Kuiper et al., 2000). Yet, both models fail to reproduce the spectra: the predicted emission is much too soft in the keV-0.1 MeV band compared with the data from PSR J0218+4232 and two other pulsars; it is much too hard in the EGRET band compared with the PSR J0218+4232 data (Zhang B. & Harding, 2000, Wei et al., 1996, Rudak & Dyks, 1999, Bulik et al., 2000).

Interstellar accretion onto a black hole does not provide the necessary power. For a 10 $M_\odot$ black hole, copious γ rays may be produced from π^0 decay in p-p interactions in the accretion flow if the accretion rate does not exceed 0.3 to 1×10^{18} g/s to prevent the protons from cooling and to limit γ-ray absorption by pair-production on the electron X-ray emission (Colpi et al., 1986). The corresponding maximum luminosity of $\sim 10^{33}$ erg/s above 30 MeV is too low for the halo sources. More massive (35 $M_\odot$) black holes have been proposed (Dermer, 1997). The L_γ/L_X luminosity ratio of 10^{-3} expected from non advection-dominated accretion flows (Colpi et al., 1986) is clearly at variance with the observations where values $\gg 50$, up to a few thousands, are found. γ rays can also be produced in ADAFs via pion decay (Mahadevan et al., 1997), but the modelling of SagA* fails to achieve large L_γ/L_X ratios. A maximum of ~ 10 is reached in the most emissive case using non-thermal proton spectra (Narayan et al., 1998).

To avoid this low ratio, non-thermal acceleration of protons is definitely required, preferably in superluminal jets of μ-quasars. Those have bulk Lorentz factors of 3 to 5. Intense radio flare episodes from the 12.5 kpc distant system GRS 1915+105 could be marginally detected in γ rays through the synchrotron-self-Compton emission of the rapidly expanding radio plasmoids, typically during a few hours after the burst onset ((Atoyan & Aharonian, 1999). The flaring activity of such systems is consistent with the variability of the halo sources and their luminosity of 10^{36-39} erg/s is large enough. Interestingly, μ-quasars can be found up to very high Galactic latitude and height above the plane. For instance, XTE1118+480 at b = -62° lies at z= 1.6 $\pm$ 0.5 kpc above the plane (McClintock et al., 2001, Dwanan et al., 2000). As counterparts to the old EGRET sources they would accrete from a low-mass star companion. One case of a spatial coincidence between a μ-quasar, LS

5039, and an EGRET error box, 3EG J1824-1514, has recently been raised (2000), although the chance coincidence probability is quite high so close to the Galactic plane and the EGRET source is not detected as variable (Tompkins, 1999). There is, however, no present theoretical limitation to the production of γ radiation by such systems, especially in the case of μ-blazars where the jet points at us and we take advantage of the Doppler shift in energy and boosting of the intensity, as in blazars. Their νF_ν spectrum is characterized by two broad peaks, which are often interpreted as collimated synchrotron and inverse Compton emission from the jet. The relative importance of the peaks and their position in energy is related to the total blazar power. The synchrotron and inverse Compton peaks of the powerful "red" blazars peak in the mm-far-IR and MeV bands, respectively. In that case, the inverse Compton peak largely dominates the energy distribution and declines as $E^{-2.6\, to\, -2.5}$ in the EGRET range. So, the halo sources might be red μ-blazars.

In this context, identifying halo sources with ms pulsars or μ-blazars would effectively open a new window either on these old, reactivated neutron stars or on the long-sought high-energy activity of stellar black holes. The halo sources would also shed light on the early generations of compact objects produced by our Galaxy.

References

Atoyan A.M., Aharonian F., 1999, MNRAS 302, 253
Bulik T., Rudak B., Dyks J., 2000, MNRAS 317, 97
Blaes O., Warren O., Madau P., 1995, ApJ 454, 370
Cassé M., Paul J.A., 1980, ApJ, 237, 236
Cesarsky C.J., Montmerle T., 1983, Space Science Rev. 36, 173
Colpi M., Maraschi L., Treves A., 1986, ApJ, 311, 150
Dhawan V., et al., 2000, IAUC 7395
Dermer C.D., 1997, Proc. 4th Compton Symp., AIP 410, 1275
Gehrels N., et al., 2000, Nature 404, 363
Grenier I.A., Perrot C., 1999, Proc. 26th Int. Cosmic Ray Conf. 3, 476
Grenier I.A., 2000, Proc. 6th Toward a major atmospheric Cherenkov detector Symp., AIP 515, 261
Grenier I.A., 2000, A&A Letters, 364, L93
Harding A.K., Zhang B., 2001, ApJ 548, L37
Harding A.K., Leventhal M., 1992, Nature 357, 388
Hartman R.C., et al., 1999, ApJS 123, 79
Hunter S.D., et al., 1997, ApJ 481, 20
Kuiper L. et al., 2000, A&A 359, 615
Lyne A.G., Manchester R.N., Taylor J.H., 1985, MNRAS 213, 613

Mahadevan R., Narayan R., Krolik J., 1997, ApJ 486, 268
McClintock J.E., et al., 2001, astroph-0101421
Motch C., 1999, Proc. Bologna X-ray astronomy 99, astroph 0008485
Narayan R., Mahadevan R., Grindlay J., Popham R., Gammie C., 1998, ApJ, 492, 554
Ojha D.K. et al., 1996, A&A 311, 456
Paredes J.M., et al., 2000, Science 288, 2340
Pohl M., et al., 1997, ApJ 491, 159
Romani R.W., 1996, ApJ 470, 469
Romero G.E., Benaglia P., Torres D.F., 1999, A&A 348, 868
Romero G.E., et al., 2000, Proc. IV INTEGRAL Workshop, Alicante, in press, astroph 0101108
Rudak B., Dyks J., 1999, MNRAS 303, 477
Sturner S.J., Dermer C.D., 1996, ApJ 461, 872
Tavani M. et al., 1997, ApJ 109, 479
Thompson D.J., et al., 1997, Proc. 4th Compton Symp., AIP 410, 39
Timmes F.X., Woosley S.E., Weaver T.A., 1996, ApJ 457, 834
Tompkins B., 1999, PhD thesis, Stanford University
Wei D.M., Cheng K.S., Lu T., 1996, ApJ 468, 207
White R.L., 1985, ApJ 289, 698
White R.L., Chen W., 1992, ApJ 387, L81
Yadigaroglu I.A., Romani R.W., 1997, ApJ 476, 347
Zhang L. & Cheng K.S., 1998, MNRAS 294, 177
Zhang L., Zhang Y.J., Cheng K.S., 2000, A&A 357, 957
Zhang B., Harding A.K., 2000, ApJ 532, 1150
Zinn R., 1985, ApJ 293, 424

LOW-LATITUDE GAMMA-RAY SOURCES: CORRELATIONS AND VARIABILITY

Gustavo E. Romero
Instituto Argentino de Radioastronomía (IAR)
Casilla de Correos No. 5, 1894 Villa Elisa
Buenos Aires, Argentina
romero@irma.iar.unlp.edu.ar

Keywords: Gamma-ray sources, early-type stars, supernova remnants, compact objects

Abstract A review of the main characteristics of low-latitude sources in the third EGRET catalog is presented. There are 75 unidentified gamma-ray sources detected by EGRET at $|b| < 10°$. About a half of these sources are spatially correlated with potential galactic gamma-ray emitters such as supernova remnants, OB associations and early-type stars with very strong stellar winds. The other half is formed by sources without positional correlation with known galactic objects capable to generate a gamma ray flux significant enough as to be detected by EGRET. A variability analysis shows that this second group of sources contains several objects with high levels of gamma-ray variability. These variable sources resemble very much the AGNs detected by EGRET, but without their typical strong radio emission. To establish the nature of these sources is one of the most urgent problems of high-energy astrophysics.

1. INTRODUCTION

The existence of a galactic population of gamma-ray sources is a well-known fact since the early days of the *COS-B* experiment (Bignami & Hermsen 1983). The ESA *COS-B* satellite was launched on August 9, 1975, and was operational until April 25, 1982. The second *COS-B* catalog contains 25 sources, most of which are located very close to the galactic plane (Swanenburg et al. 1981). Although two low-latitude sources were soon identified with pulsars (Crab and Vela), the nature of the remaining, presumably galactic sources, stood uncertain. Montmerle (1979) presented the first correlation study for these high-energy sources. He found that about 50 % of the unidentified *COS-B* detections

A. Carramiñana et al. (eds.),
The Nature of Unidentified Galactic High-energy Gamma-ray Sources, 65–80.

lie in regions containing young objects like massive stars and supernova remnants. He estimated that the chance probability of these associations was as low as $\sim 10^{-4}$ and suggested that the gamma-rays could be the result of π^0-decays originated in hadronic interactions between locally accelerated cosmic rays and ambient gas. The cosmic rays would be produced by a two-step process: low-energy protons or nuclei are firstly accelerated by OB stars and injected at supernova shock fronts where they are subsequently re-accelerated up to high-energies by Fermi mechanism. This scenario predicted by first time the positional correlation between low-latitude gamma-ray sources and star forming regions, a correlation that would be tested by several authors, using increasingly improved data, in the years to come.

A major breakthrough in the study of galactic gamma-ray sources was achieved with the advent of NASA's *Compton Gamma-Ray Observatory (CGRO)* in 1991. During its lifetime, the Energetic Gamma-Ray Experiment Telescope (EGRET) detected 271 point sources, 170 of which have not been clearly identified yet (Hartman et al. 1999). About a half of these unidentified sources are located at low galactic latitudes and many studies looking for correlations with known galactic populations have been performed in recent years. For instance, Sturner & Dermer (1995) and Sturner et al. (1996) have investigated the correlation between gamma-ray sources in the first two EGRET catalogs (Thompson et al. 1995, 1996) and supernova remnants, finding statistical support for the idea that some remnants could be gamma-ray emitters. This idea is also supported by several particular cases studied in detail by Esposito et al. (1996) and Combi et al. (1998, 1999, 2001), where the gamma-ray emission seems to come from molecular clouds overtaken by the expanding shell of nearby remnants.

The correlation between EGRET sources and star forming regions was confirmed by Kaaret & Cottam (1996) and Yadigaroglu & Romani (1997), using data from the second EGRET catalog. Contrary to the original hypothesis of Montmerle, these authors suggested that most of the low-latitude sources could be pulsars. This suggestion is supported by the discovery of several new gamma-ray pulsars since the *COS-B* original identifications (there are at least seven pulsars detected so far, see Thompson 1996) as well as by population studies (e.g. Yadigaroglu & Romani 1995, Zhang et al. 2000, McLaughlin & Cordes 2000). The high level of variability and peculiar spectral features presented by some low-latitude unidentified sources, however, is quite at odds with the hypothesis of a unique population of galactic gamma-ray emitters and seem to open the possibility that our Galaxy could contain yet unknown types of high-energy sources (McLaughlin et al. 1996, Merck et al. 1996, Tavani et al. 1998, Tompkins 1999, Romero et al. 2000, Punsly et al. 2000, Torres et al. 2001).

In this paper we shall review the main correlational properties of the sample of low-latitude unidentified gamma-ray sources in the third and final EGRET catalog. We shall discuss which objects in our Galaxy are expected to generate strong enough gamma-ray emission as to have been detected by EGRET and what are the prospects for future space missions like *INTEGRAL* and *GLAST* regarding the identification of galactic gamma-ray sources.

2. LOW-LATITUDE SOURCES AND THE SPIRAL STRUCTURE OF THE GALAXY

There are 81 unidentified gamma-ray sources in the third EGRET catalog located at less than 10° from the galactic plane. Six of these sources are thought to be artifacts associated with the proximity of the very bright Vela pulsar. These sources do not show up in a map which excludes the Vela pulsation intervals and will not be taken into account in the present discussion. We have, consequently, a sample of 75 unidentified low-latitude sources, whose distribution with galactic latitude is shown in Figure 1. A strong concentration around zero degrees can be clearly seen in this histogram, indicating that most of the sources belong to our Galaxy.

One of the first things that we could ask about the group of low-latitude gamma-ray sources is whether they are correlated with the spiral arms of the Galaxy. The arms are where most stars are formed and the regions where the galactic gas storage is concentrated. In order to test the correlation, we need a tracer for the spiral structure of the Galaxy. The usual tracers, in this sense, are giant and bright HII regions. We can use, then, Georgelin & Georgelin's (1976) catalog of the 100 brightest HII regions to perform a correlation analysis with the sample of low-latitude sources in the third EGRET catalog. When this is done, we find that 32 out of 75 sources are positionally overlapping HII regions. In order to quantify the statistical significance of this number we can make numerical simulations of random populations of galactic gamma-ray sources (subject to adequate boundary conditions) using the code developed by Romero et al. (1999) to produce synthetic gamma-ray source populations. Through thousands of simulations, we find that the expected number of chance coincidences is 13.2 ± 2.9. This implies a Poisson probability of chance correlation of 5×10^{-6}, i.e. the correlation is reflecting a physical relation with a confidence of $\sim 7\sigma$. The conclusion, consequently, seems to be that *there is a significant number of extreme*

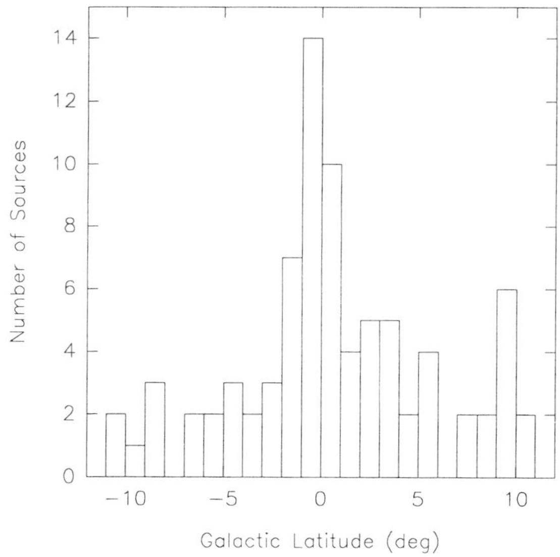

Figure 1 Distribution with galactic latitude of those unidentified gamma-ray sources in the 3EG catalog located at $|b| < 10°$.

Population I objects in the parent population of low-latitude gamma-ray sources. This result was already suggested by previous correlation studies using former gamma-ray catalogs presented by Montmerle (1979) and Yadigaroglu & Romani (1997), but the confidence only reaches overwhelming levels when data from the third EGRET catalog are used.

3. CORRELATIONS WITH GALACTIC OBJECTS

The main mechanisms for gamma-ray production in a galactic scenario are inverse Compton (IC) scattering of lower frequency photons, relativistic bremsstrahlung and π^0-decays from hadronic interactions. The common feature of all these mechanisms is that they require the presence of a population of relativistic particles (electrons or positrons in the first two cases, protons or ions in the latter). Consequently, if we look for γ-ray production sites in the Galaxy, we should look at sites where charged particles can be efficiently accelerated to high energies.

Basically, we have two types of scenarios where particles can be accelerated up to the required relativistic energies: 1) large sites where the acceleration is mediated by strong shock fronts in a first order, diffusive process, and 2) compact objects with very strong electromagnetic fields where the acceleration occurs in a single step. The first type of acceleration is expected to occur in supernova remnants (SNRs) and also at the strong shock that could be formed near very massive stars endowed with strong supersonic winds. The second type of acceleration should operate in pulsars and accreting black holes, where strong magnetic fields should be anchored in the surrounding accretion disks.

It is natural, then, to look for positional correlations between gamma-ray sources and galactic objects like SNRs, early-type stars, and OB associations (which are considered as pulsar tracers), and this has been done in the past as we have briefly mentioned in the Introduction. Regarding the sources in the third EGRET catalog, Romero et al. (1999) have performed a correlation analysis finding out that there is a suggestive number of spatial coincidences with Wolf-Rayet and Of stars, SNRs and OB associations. The probabilities of pure chance superposition are moderate ($< 10^{-2}$) for stars and quite negligible for remnants and star forming regions ($< 10^{-5}$). In the next sections we shall discuss these potential γ-ray emitters in more detail, with emphasis in stars.

4. STARS

Wolf-Rayet (WR) stars are very massive objects that have burnt their hydrogen and have entered in the final phases of their evolution. These stars present very strong supersonic winds with velocities of several thousands of km/s. The mass loss rate is as high as $10^{-4}\ M_{\odot}/\text{yr}$. Of stars are also early-type stars with strong winds; they are thought to be the progenitors of WR stars and, although their winds are not so strong, they have mass loss rates that can be 10^9 times higher than those presented by stars like the Sun.

The energy losses experienced by all these stars through their winds have a great impact in the surrounding interstellar medium. The winds swept up the ambient gas creating low-density cavities around the stars (e.g. Benaglia & Cappa 1999). A strong shock front is formed at the contact layer between the wind and the outer, colder ISM. These shocks are expected to accelerate particles up to high-energies through Fermi mechanism (e.g. Völk & Forman 1982). If a source of UV photons or a cloud are present near the acceleration site, significant amounts of gamma-rays could be produced through IC scattering, bremsstrahlung or hadronic interactions (Benaglia et al. 2000).

The winds of early-type stars are radiatively driven by absorption in spectral lines and are prone to undergo instabilities that can grow up to form strong shocks at the base of the outflow (Lucy & White 1980, Lucy 1982). These shocks can efficiently accelerate both electrons and protons up to energies of about a few GeV (White 1985), and these

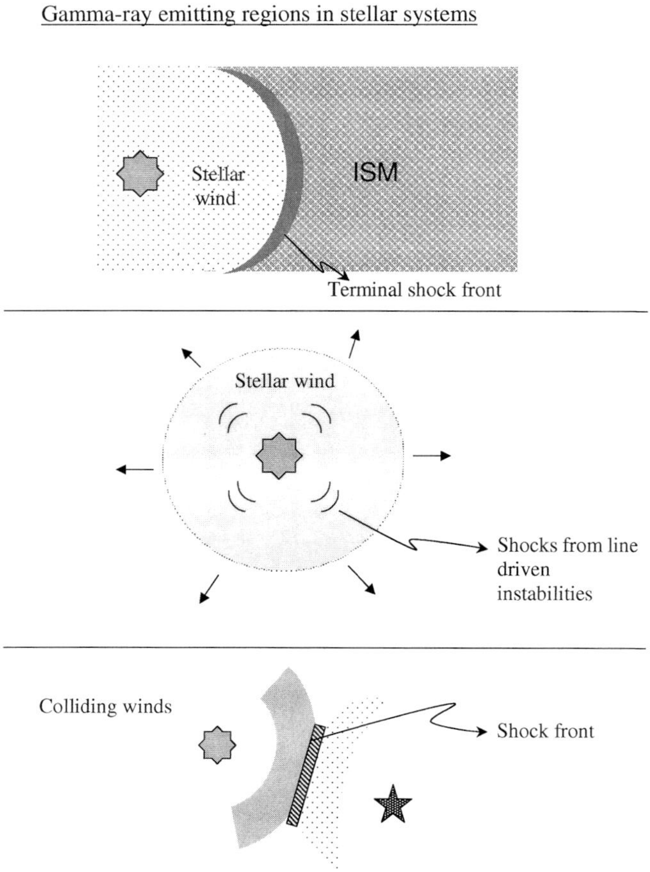

Figure 2 Sketch illustrating the different regions where gamma-rays can be produced in a stellar system. Top panel: Terminal shock region. Middle panel: The unstable base of the wind. Lower panel: Colliding winds region in a binary system.

particles, through interactions with stellar photons and ions, could yield gamma-ray emission in EGRET's energy range (Benaglia et al. 2000).

Another site where stellar high-energy emission can be generated is the colliding winds region in binary systems formed by two massive stars (Eichler & Usov 1993). Electrons are accelerated at the strong shock formed by the winds collision, as evidenced by the clear detection of non-thermal radio emission from the region between stars in several systems (e.g. Contreras et al. 1997). Gamma-rays should be produced by Comptonization of stellar photons in this region.

In Tables 1 and 2 we list those unidentified gamma-ray sources that are positionally coincident with WR and Of stars, respectively. A "B" letter marks those stellar systems that are confirmed binaries. From

Table 1 WR stars spatially correlated with 3EG sources.

γ-Source (3EG J)	Star	$\Delta\theta$ (deg)	r (kpc)	v_∞ (km s^{-1})	$\log\dot{M}$ ($M_\odot$ yr^{-1})	NT emission
0747 − 3412	WR 9 (B)	0.37	2.35	2200	−4.2	
1102 − 6103	WR 34	0.48	9.50	1200	−4.5	
	WR 35	0.30	9.58	1100	−4.3	
	WR 37	0.45	2.49	2150	< −4.1	
	WR 38	0.45	1.97	2400	< −4.2	
	WR 39	0.51	1.61	3600	< −4.0	Yes
1655 − 4554	WR 80	0.59	4.40	2000	−4.1	
2016 + 3657	WR 137 (B)	0.44	1.82	1900	−4.5	
	WR 138	0.50	1.82	1500	−4.7	
2021 + 3716	WR 142	0.15	0.95	5200	< −4.7	
2022 + 4317	WR 140 (B)	0.64	1.34	2900	−4.1	Yes

Table 2 Of stars spatially correlated with 3EG sources.

γ-Source (3EG J)	Star	$\Delta\theta$ (deg)	r (kpc)	v_∞ (km s^{-1})	$\log\dot{M}$ ($M_\odot$ yr^{-1})	NT emission
0229 + 6151	HD 15629	0.30	1.9	2900	−5.8	
0634 + 0521	HD 46150	0.64	1.3	2900	< −5.9	
	HD 46223	0.63	1.6	2800	-5.8	
1410 − 6147	HD 124314	0.25	1.0	2400	−4.7	Yes
2033 + 4118	Cyg OB2 5 (B)	0.27	1.8	1500	−4.43	Yes
	Cyg OB2 11	0.24	1.8	2500	−5.2	

left to right we provide the name of the gamma-ray source in the 3EG catalog, the star name, angular separation between the star and the best estimated position of the 3EG source, distance to the star, terminal wind velocity and mass loss rate. We also indicate (in the last column) whether non-thermal (NT) radio emission has been detected.

The presence of synchrotron radiation is important because it reveals the existence of a population of relativistic electrons that could be also responsible for the gamma-ray emission. In Figure 3 we show the non-thermal radio contours of the southern Of star HD 124314 observed with the Australia Telescope Compact Array (ATCA) by Benaglia et al. (2001). Similar observations for the remaining sources listed in Tables 1 and 2 are in progress. The source 3EG J2016+3657, positionally coincident with WR 138, has been recently studied by Mukherjee et al. (2000) who suggest that the counterpart is the blazar B2013+370.

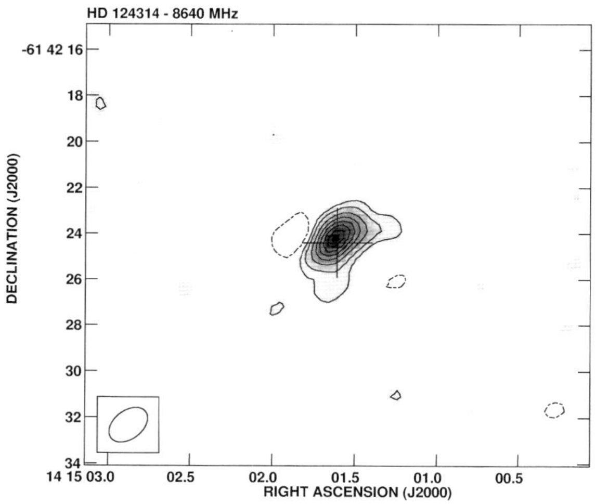

Figure 3 Nonthermal radio emission from HD124314 (from Benaglia et al. 2001). The same population of relativistic electrons that produces the observed synchrotron flux should produce gamma-rays through Comptonization of UV stellar photons.

The information on the relativistic electronic population obtained with interferometric radio observations can be used, along with the available information on the stellar parameters, to estimate the expected gamma-ray luminosity of individual, well-studied, stellar systems. Such work has been recently made by Benaglia et al. (2000) for the particularly interesting case of Cyg. OB 2 No. 5. This system is formed by three stars: an O7 Ia + Ofpe/WN9 contact binary and a B0 V star in a larger orbit. Benaglia et al. estimated the gamma-ray production in the colliding winds region, the terminal shock of the dominant wind, and the unstable zone at the base of the wind. Their results indicate that the high-energy emission is dominated by IC radiation from the colliding winds and π^0-decays from the inner wind region near the primary star. The expected luminosities are shown in Table 3 and can be considered as representative for similar systems.

Both *INTEGRAL* and *GLAST* missions will provide new and valuable elements to test our ideas about how gamma-rays can be generated in

Table 3 Expected gamma-ray production in Cygnus OB2 No. 5 in the EGRET energy range (from Benaglia et al. 2000)

Region	Mechanism	Luminosity (erg/s) Expected	Observed
Winds collision	IC scattering	$\sim 8 \times 10^{34}$	$\sim 2.4 \times 10^{35}$
	Bremsstrahlung	$\sim 3.4 \times 10^{30}$	
	π^0-decay	$\sim 5.2 \times 10^{24}$	
Terminal shock	π^0-decay	$\sim 2.3 \times 10^{32}$	
Base of the wind	IC scattering	–	
	π^0-decay[1]	$\sim 5 \times 10^{34}$	

1: See also White & Chen (1992)

stars. In particular, the Imager on Board *INTEGRAL* Satellite (IBIS), a coded masked detector designed to produce sky images in the 20 KeV – 10 MeV band with an angular resolution of 12 arcminutes, is expected to detect the IC flux density in a few, nearby stars that have been already found to be non-thermal radio sources. This instrument will also have a good spectral capability that could be used to differentiate stars where the high-energy emission comes from particles accelerated at a single shock (e.g. the shock formed by winds collision) from those where the emission is produced by particles accelerated at multiple shocks at the base of the wind (e.g. Chen & White 1991). The main limitations of IBIS are the sensitivity (long integrations of $\sim 10^6$ seconds will be necessary to detect even the nearest stars) and the fact that π^0-decays produce gamma-rays of energy well above instrument's energy range. *GLAST*, on the contrary, will detect this latter emission. Its improved sensitivity will allow to observe thousands of sources, including hundreds of stars. The determination of the spectral features and variability characteristics of these stars, along with new and more accurate measurements of their radio properties, will lead to a radical advancement in our understanding of high-energy stellar processes.

5. SUPERNOVA REMNANTS

Torres et al. (these proceedings) discuss in detail the case for the association of SNR and 3EG sources. If we exclude sources that are suspected to be artifacts, we have that 19 out 75 low-latitude EGRET detections are positionally coincident with SNR in Green's (2000) catalog (see also Romero et al. 1999). The probability of all these associations being the mere effect of chance is quite negligible: $\sim 10^{-5}$. However, the possibility that many sources result from the compact object left by the supernova explosion and not from the remnant itself cannot be ruled out.

In several cases, however, the gamma-ray emitter seems to be the SNR. This occurs when the remnant is interacting with nearby and dense molecular clouds. In these cases, as discussed by Aharonian et al. (1994) and Aharonian & Atoyan (1996), the probability of detection is dramatically increased. Some particular cases have been studied in detail by Esposito et al. (1996) and Combi et al. (1998, 2001). Cosmic rays are accelerated at the remnant shock front and transported into the cloud by convection. There, gamma-rays are produced by π^0-decays from pp interactions and relativistic bremsstrahlung. Additional contributions from IC scattering can be also important. The spectral shape at low and high energies can be used to infer the energy distributions of both hadrons and leptons in the remnant. Multifrequency studies of these SNRs can shed light on the details of the acceleration process of galactic cosmic rays at large, strong shock fronts (see Gaisser et al. 1998 for further details).

6. OB ASSOCIATIONS

OB associations are considered as tracers of young pulsar concentrations (Kaaret & Cottam 1996). Taking into account the distribution of pulsar transverse speeds derived by Lyne & Lorimer (1994), it has been estimated that a fraction of $0.25 - 0.4$ of all young pulsars (ages less than 10^5 yr) can be found within 1° from the center of OB associations (Kaaret & Cottam 1996).

A correlation analysis using the 3EG catalog and Mel'nik & Efremov's (1995) OB associations catalog indicates that 26 low-latitude gamma-ray sources are superposed to OB associations. This result is about 5σ away from what is expected by chance (probability of all coincidences being the effect of chance $\sim 10^{-5}$).

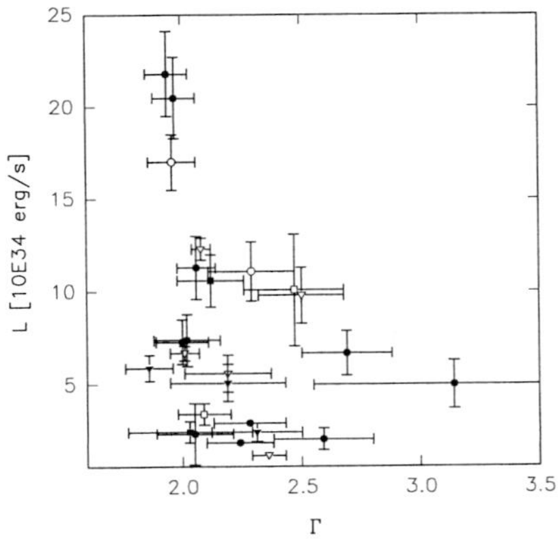

Figure 4 Expected luminosity versus high-energy spectral index for unidentified 3EG sources that are correlated with OB associations.

The idea that most of the gamma-ray sources in this group are pulsars is supported by the particularly hard spectra presented by them. All known gamma-ray pulsars are young objects with spectral indices smaller than 2.15 (Crab's) and a trend for spectral hardening with characteristic age (Fierro et al. 1993). In Figure 4 we show a plot of the expected luminosity of the gamma-ray sources (assuming for them the distance to the corresponding OB associations) versus the spectral index Γ of the high-energy emission ($F(E) \propto E^{-\Gamma}$). In this plot pulsars should be located towards the left side of the panel, precisely where most sources are concentrated. Three sources differentiate from the rest, with high luminosities, similar to the luminosity of Vela pulsar.

Recent modeling of gamma-ray pulsar populations by Zhang et al. (2000) also corroborates that most sources positionally associated with OB star forming regions are probably pulsars. The simulations made by these authors, based on the outer gap model proposed by Zhang & Cheng (1997), shows that a mixed population of radio-quiet and young radio pulsars can account for most (although not all) associations found by Romero et al. (1999) using the the 3EG catalog.

7. UNCORRELATED SOURCES

The correlation analysis made by Romero et al. (1999) showed that there is a significant number of low-latitude gamma-ray sources for which there is no positional superposition with any known galactic object thought to be capable of producing high-energy gamma rays. To clarify the nature of this group of sources is, perhaps, one of the most interesting problems aroused by EGRET.

A comparative analysis of the gamma-ray variability properties of these sources could provide some clues on their origin. Gamma-ray pulsars are expected not to vary over relatively short timescales, so the identification of clearly variable galactic sources could lead to the discovery of a new population of gamma-ray emitters in the Galaxy (Tavani et al. 1997).

A variability analysis for the group of sources without positional correlation has been performed by Romero et al. (2000) and Torres et al. (2001). They have used a statistical method widely applied to blazar studies (Romero et al. 1994). Although this method is applied to the catalog data, it is free of several problems that affect similar methods (see Reimer, this proceedings, and Tompkins 1999). Basically, a variability index $I = \mu_{\rm s}/ < \mu >_{\rm p}$ is introduced, where $\mu_{\rm s} = \sigma/ < F >$ is the fluctuation index of the gamma-ray source and $< \mu >_{\rm p}$ is the averaged fluctuation index of all known gamma-ray pulsars (which are usually considered as a non-variable population). Sources with $I > 1$ at 3σ or more are considered as possibly variable ones ($1\sigma = 0.5$). In the estimate of the fluctuation index of known gamma-ray pulsars only data from the 3EG catalog are considered. In the limits of very variable and non-variable sources this method yields similar results to the most comprehensive method applied by Tompkins (1999), who worked with the raw data and took into account background fluctuations, source contamination, and other sources of systematic error.

The basic results of the variability analysis of the low-latitude, uncorrelated gamma-ray sources in the 3EG catalog are shown in Figure 5 as a plot of variability index I versus the high-energy spectral index Γ. AGNs and pulsars are also shown in this plot for comparison. Known pulsars are located between the two solid lines at the bottom of the panel. Since they also have hard spectra, they are concentrated towards the left side of the frame. AGNs, on the contrary, mostly display high variability levels and in many occasions steep spectra. It can be clearly seen from the plot that there is a group of unidentified sources with a behaviour similar to that presented by AGNs. The surface density of these sources is, however, much higher than what could be expected through an extrapolation of the high-latitude AGN density towards the galactic plane. Actually, there seems to be a trend among these sources in the sense that those with the highest variability levels also present the steepest spectral indices. These sources seems to form a population of

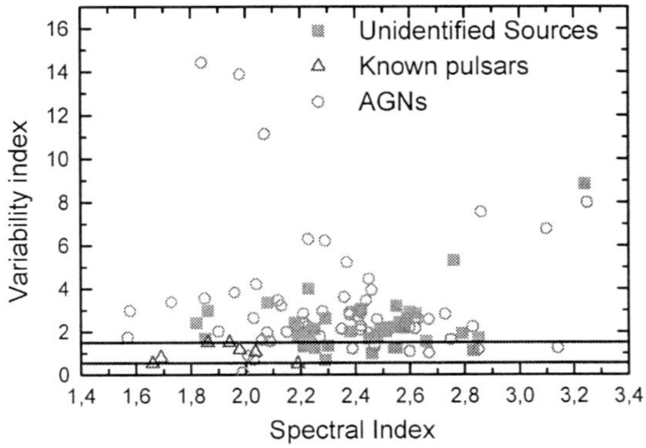

Figure 5 Plot of variability index I versus high-energy spectral index Γ for low-latitude 3EG sources without positional correlation with potential galactic gamma-ray emitters. Known gamma-ray pulsars are contained between the horizontal lines. Gamma-ray AGNs are shown for comparison.

galactic, variable sources, of which GRO J1838-04 (Tavani et al. 1997) and 3EG J1828+0142 (Punsly et al. 2000) are extreme examples.

8. VARIABLE GAMMA-RAY SOURCES IN THE GALAXY

The nature of the variable low-latitude sources remains a mystery. The large error boxes of the EGRET detections make very difficult the identification of potential lower frequency counterparts. A single radio field of 1° around the centre of the EGRET 95 % probability location contour can contain as many as 50 point-like weak radio sources, most of which are also of unknown nature (Torres et al. 2001). Although only the improved angular resolution of *GLAST* will allow to isolate the more promising candidates for counterparts and lead to a final identification, we can speculate about some possibilities on the nature of these sources. Among the main ones, we can mention:

- *Early-type stars with strong winds.* In our study of the positional correlation of sources in the 3EG catalog with massive stars we have only included extreme stars like WR and Of stars. However, other O and B stars can also be sources of gamma-ray emission strong enough as to be detected by EGRET, especially if they are forming binary systems (Eichler & Usov 1993, Benaglia et al. 2000) or are located in a very rich environment. The luminosity of these stars should not be very high, say around a few times 10^{34} erg s^{-1} (Benaglia et al. 2000), so only nearby sources would be detected. Variability is naturally expected in these stars due to geometric changes during the orbital evolution of the binaries and also due to the effect of wind instabilities. Stars are interesting candidates to explain mid-latitude sources probably linked to the nearby Gould belt (Gehrels et al. 2000).

- *Pulsars.* Pulsars, under certain circumstances, can be variable sources. For instance, a pulsar in a non-accreting binary system with a massive star, could be subjected to a changing UV photon bath from the star (e.g. Tavani & Arons 1997). Isolated pulsars also could be variable through some kind of quake-driven activity.

- *Faint microquasars.* This hypothesis is supported by the recent discovery of a persistent and faint microquasar (LS 5039) by Paredes et al. (2000, also these proceedings) positionally correlated with the source 3EG 1824–1514. Other, yet undetected microquasars could switch between high and low states due to periodic accretion instabilities. Recent evidence seems to show that LS 5039 displays short-term radio variability. If IC gamma-ray emission is associated with the radio emission, it should also be variable.

- *Isolated black holes accreting from the interstellar medium.* Bondi-Hoyle accretion of diffuse matter onto black holes of several solar masses can yield moderate gamma-ray luminosities (Dermer 1997). If the hole is moving through an inhomogeneous medium, variable emission could be observed.

- *Non-pulsating (NP) black holes.* Maximally rotating Kerr-Newman black holes (see Punsly, these proceedings, and references therein) can support a magnetosphere and, if they are located in a low-density medium, could, in principle, be gamma-ray sources (Punsly 1998a, b). Variability naturally results from jet instabilities. A model of this kind has been recently applied to the source 3EG J1828+0142 (Punsly et al. 2000).

Although this list does not exhaust all possibilities, it is enough to show that a variety of candidates can be postulated and that new observations of improved quality are required.

9. FINAL REMARKS

EGRET has confirmed the finding made by *COS-B* of a galactic population of gamma-ray sources and has shown that this population is not formed by a unique class of objects. Pulsars, supernova remnants, and early-type stars seem to be capable, under certain circumstances, to emit detectable gamma-rays. The possibility of other, more strange, gamma-ray emitting objects in the Galaxy remains open. The search for these objects will be one of the most interesting and challenging tasks of high-energy astrophysics in *GLAST* era.

Acknowledgments

The author is very grateful to Dr. Alberto Carramiñana for supporting his travel to Mexico as well as to all members of the Local Organizing Committee for a wonderful workshop. He also wishes to express his sincere thanks to the members of GARRA team, Paula Benaglia, Jorge Combi and Diego Torres, for many discussions and fruitful collaborations on the topics covered in this review. Comments by Isabelle Grenier are also gratefully acknowledged. High-energy astrophysics with G.E. Romero is supported by ANPCT through grant PICT 98 No. 03-04881 and Fundación Antorchas. Additional support is provided by CONICET.

References

Aharonian F.A., Drury L.O'C., Völk H.J. 1994, A&A 285, 645
Aharonian F.A., Atoyan A.M. 1994, A&A 309, 917
Benaglia P., Cappa C.E. 1999, A&A 346, 979
Benaglia P., Romero G.E., Stevens I.R., Torres D.F. 2000, A&A, in press [astro-ph/0010605]
Benaglia P., Cappa C.E., Koribalski B. 2001, in preparation
Chen W., White R.L. 1991, ApJ 381, L63
Bignami G.F., Hermsen W. 1983, ARA&A 21, 67
Combi J.A., Romero G.E., Benaglia P. 1998, A&A 333, L91
Combi J.A., Romero G.E., Benaglia P. 1999, AJ 118, 659
Combi J.A., Romero G.E., Benaglia P., Jonas J.L. 2001, A&A, in press
Contreras M.E., Rodríguez L.F., Tapia M., et al. 1997 ApJ 488, L153
Dermer C.D. 1997, Proc. of the Fourth Compton Symposium, C. D. Dermer, M. S. Strickman, and J. D. Kurfess Eds., AIP 410, 1275
Eichler D., Usov V. 1993, ApJ 402, 271
Esposito J.A., Hunter S.D., Kanbach G., Sreekumar P. 1996, ApJ 461, 820
Fierro J.M., et al. 1993, ApJ 413, L27
Gaisser T.K., Protheroe R.J., Stanev T. 1998, ApJ 498, 219
Gehrels N, Macomb D.J., Bertsch D.L., et al. 2000, Nature 404, 363
Georgelin Y.M., Georgelin Y.P. 1976, A&A 49, 57
Green D.A. 2000, A Catalog of Galactic Supernova Remnants, Mullard Radio Astronomy Observatory, Cambridge, UK (available on the World Wide Web at http://www.mrao.cam.ac.uk/surveys/snrs/)
Hartman R.C., Bertsch D.L., Bloom S.D., et al. 1999, ApJS 123, 79
Kaaret P., Cottam J. 1996, ApJ 462, L35

Lucy L.B., White R.L. 1980, ApJ 241, 300
Lucy L.B. 1982, ApJ 255, 286
Lyne A.G., Lorimer D.R. 1994, Nature 369, 127
McLaughlin M.A., Mattox J.R., Cordes J.M., Thompson D.J. 1996, ApJ 473, 763
McLaughlin M.A., Cordes J.M. 2000, ApJ 538, 818
Merck M., et al. 1996, A&AS 120, 465
Mel'nik A.M., Efremov Yu.N. 1995, Astron. Lett. 21, 10
Montmerle T. 1979, ApJ 231, 95
Mukherjee R., Gotthelf E.V., Halpern J., Tavani M. 2000, ApJ 542, 740
Paredes J.M., Martí J., Ribó M., Massi M. 2000, Science 288, 2340
Punsly B. 1998a, ApJ 498, 640
Punsly B. 1998b, ApJ 498, 660
Punsly B., Romero G.E., Torres D.F., Combi J.A. 2000, A&A, in press [astro-ph/0007465]
Romero G.E., Combi J.A., Colomb F.R. 1994, A&A 288, 731
Romero G.E., Banaglia P., Torres D.F. 1999, A&A 348, 868
Romero G.E., Torres D.F., Benaglia P., et al. 2000, in: Proceedings of the IV *INTEGRAL* Workshop, ESA-SP, in press
Sturner S.J., Dermer C.D. 1995, A&A 293, L17
Sturner S.J., Dermer C.D., Mattox J.R. 1996 A&AS 120, 445
Swanenburg B.N., et al. 1981, ApJ 243, L69
Tavani M., Mukherjee R., Mattox J.R., et al. 1997, ApJ 479, L109
Tavani M., Arons J. 1997, ApJ 477, 439
Tavani M., et al., 1998, ApJ 497, L89
Thompson D.J., et al. 1995, ApJS 101, 259
Thompson D.J., et al. 1996, ApJS 107, 227
Thompson D.J. 1996, in Pulsars: Problems and Progress, IAU Coll. 160, S. Johnston, M.A. Walker and M. Bailes Eds., ASP Conf Ser 105, 307
Tompkins W. 1999, Ph.D. Thesis, Stanford University
Torres D.F., et al. 2001, A&A, submitted [astro-ph/0007464]
Völk H.J., Forman M. 1982, ApJ 253, 188
Wallace P.M., Griffis N.J., Bertsch D.L., et al. 2000, ApJ 540, 184
White R.L. 1985, ApJ 289, 698
White R.L., Chen W. 1992, ApJ 387, L81
Yadigaroglu I.-A., Romani R.W. 1995, ApJ 449, 211
Yadigaroglu I.-A., Romani R.W. 1997, ApJ 476, 356
Zhang L., Cheng K.S. 1997, ApJ 487, 370
Zhang L., Zhang Y.J., Cheng K.S. 2000, A&A 357, 957

POPULATION STUDIES OF UNIDENTIFIED HIGH-ENERGY GAMMA-RAY SOURCES

Neil Gehrels, Daryl J. Macomb[1], David L. Bertsch, David J. Thompson, Robert C. Hartman, Barbara J. Mattson[2]
NASA/Goddard Space Flight Center
Mail Code 661
Greenbelt, MD 20771, USA

[1] *Also Universities Space Research Association*
Lanham, MD, USA
[2] *Also Emergent Information Technologies, Inc.*
Clinton, MD 20774, USA

Keywords: Gamma-ray: observations, Gamma-ray sources: unidentified

Abstract The third EGRET catalog contains $\sim$ 170 unidentified high energy gamma-ray point sources. Here we report that this group contains two populations with different spatial distributions, flux distributions and spectral properties. A cluster of $\sim$ 20 sources lies just north of the Galactic Center and is part of a larger population of faint objects at mid-latitudes. We show that these mid-latitude sources are distinct from the population of bright sources found at lower latitudes. The spatial distribution of the mid-latitude population suggests a Gould belt origin.

1. INTRODUCTION

One mystery which has intrigued gamma-ray astronomers for more than 20 years is the origin of a group of unidentified high energy gamma-ray point sources (Swanenburg et al. 1981). The third EGRET catalog (Hartman et al. 1999) has 172 unidentified sources which represent more than half of its total 271 sources. These sources are more numerous than all other high energy gamma-ray sources combined, yet counterparts in other wavebands remain elusive. Because the spacecraft carrying the EGRET instrument (the Compton Gamma Ray Observatory) has been

A. Carramiñana et al. (eds.),
The Nature of Unidentified Galactic High-energy Gamma-ray Sources, 81–88.

deorbited, more detailed observations of these sources will not be possible until the launch of GLAST in 2005. In the mean time, population studies of existing data may provide some clues as to the origin of these sources.

This paper summarizes results published recently in *Nature* by Gehrels, Macomb, Bertsch, Thompson & Hartman (2000).

2. TWO POPULATIONS OF STEADY UNIDENTIFIED SOURCES

The third EGRET catalog lists 172 unidentified sources from observations made between 22 April 1991 and 3 October 1995 (Hartman et al. 1999). The sources listed in the EGRET catalog have a statistical significance level of 5σ for sources within 10° of the Galactic plane or 4σ for all other sources. The angular resolution of EGRET depends on energy with full-width at half-maximum of $\sim 6^\circ$ at 100 MeV (decreasing at higher energies). This results in a typical positioning accuracy of $\sim 1^\circ$ for individual sources.

To limit our study to Galactic sources, we eliminate from the analysis those unidentified sources exhibiting high variability, which would be indicative of flaring active galactic nuclei. A source is defined as steady if the most significant EGRET catalog detection is for a timescale of years, rather than a single observation of a few weeks, and if the flux of that detection is within 3σ of the flux calculated for the full data set. Retaining only steady sources, we are left with 120 sources in our study; their distribution is shown in Figure 1. From this figure, two populations of sources become apparent: brighter objects (open circles) along the Galactic plane and fainter objects (filled circles) at mid-latitudes (extending up to $\sim 30^\circ$). In the next section, we will show that there is good evidence that these mid-latitude sources represent a population separate from those near the Galactic plane, where the separation between the two populations occurs at $|b| = 5^\circ$.

3. POPULATION PROPERTIES

Figure 2a shows a plot of the flux for each unidentified steady source as a function of absolute Galactic latitude. In addition, the distribution of weak and strong sources are plotted separately as functions of absolute Galactic latitude in Figures 2b and 2c. There is a distinct change at $|b| \approx 5^\circ$ between bright sources at low-latitudes and faint sources at mid-latitudes. This change does not occur at $|b| = 10^\circ$, where the source significance changes from 4σ to 5σ, which strongly suggests that this change at 5° is a real effect.

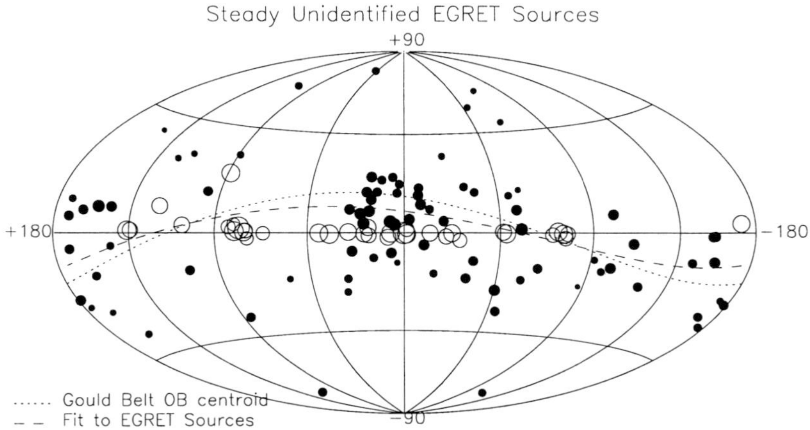

Figure 1 Sky plot of EGRET steady unidentified sources. The size of the plotted point for each source is proportional to the flux level of the source. The open circles are for sources brighter than 2.4×10^{-7} ph cm^{-2} s^{-1}, while the filled circles are for sources fainter than this value. The dotted line represents the approximate center (O and B centroid) of the Gould belt. The dashed line is the great circle that best fits the EGRET weak sources.

However, before concluding that the populations are distinct, we consider the possibility of a bias against detecting weak sources along the Galactic plane as a result of the diffuse background there being a factor of ten higher than in the mid-latitudes. Figure 1 plots the integral number of sources brighter than a given flux as a function of that flux (log N-log S plot) for each of the proposed populations. The $|b| < 5°$ data flatten sharply below 4×10^{-7} ph cm^{-2} s^{-1}, whereas the $|b| > 5°$ data flatten more gradually and at much lower flux levels ($< 1 \times 10^{-7}$ ph cm^{-2} s^{-1}). The solid line shown in both panels is the fit to the high-flux portion of the $|b| > 5°$ data (with the same slope, but different offset for the $|b| < 5°$ data). If these two populations were, indeed, the same, then the increased flattening in the $|b| < 5°$ data would be caused entirely by a detection bias and the true underlying shape would be represented by the solid line. The total flux from so many weak sources along the plane would be comparable to the total observed diffuse flux for E > 100 MeV from this region ($|b| < 5°$) of $\sim 2 \times 10^{-4}$ ph cm^{-2} s^{-1} (Hunter et al. 1997). However, the diffuse emission is understood, to within 10%, to be a result of cosmic ray interactions with the interstellar medium, so unresolved point sources could not contribute at the level

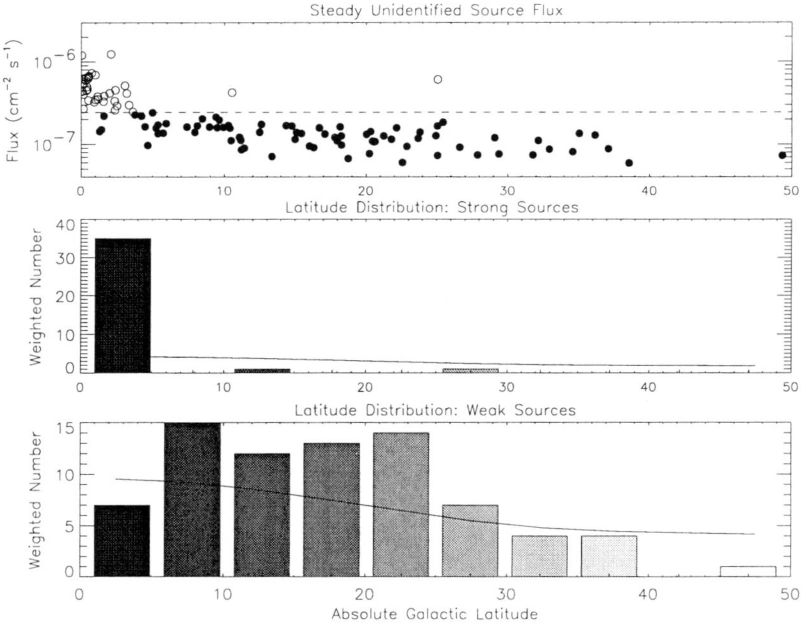

Figure 2 Latitude distribution of steady unidentified sources. **a**, The flux levels of the steady unidentified sources are shown as a function of absolute Galactic latitude. The open and filled circles have the same meaning as in Figure 1. The flux plotted is the average 1991-1995 flux greater than 100 MeV. The dashed line is the average over the sources shown. **b,c**, The distribution in absolute Galactic latitude of the unidentified steady sources with flux above and below the average (i.e. the dashed line in **a**). The curved line in each is the expected number based on the EGRET exposure.

necessary to keep the log N-log S shapes the same for both the low- and mid-latitude sources. Therefore, we conclude that the low-latitude sources and the mid-latitude sources have distinct intensity distributions: the low-latitude sources are predominantly bright sources and the mid-latitude sources are dominated by weak sources.

We also examine the spectral properties of the two populations of steady unidentified sources. The weighted average spectral index for the low-latitude group is 2.18 ± 0.04, and for the mid-latitude group it is 2.49 ± 0.04. These indices are different. The probability that these mean indices arise from the same sample is $\sim 10^{-6}$, according to a T-test. We investigate a possible bias in the spectral data which could arise from a correlation between gas density and spectral index. However, we find no difference of the magnitude necessary to suggest such a bias. For example, the difference between spectral index for low-latitude sources near the direction of the Galactic Center and those away from it is less

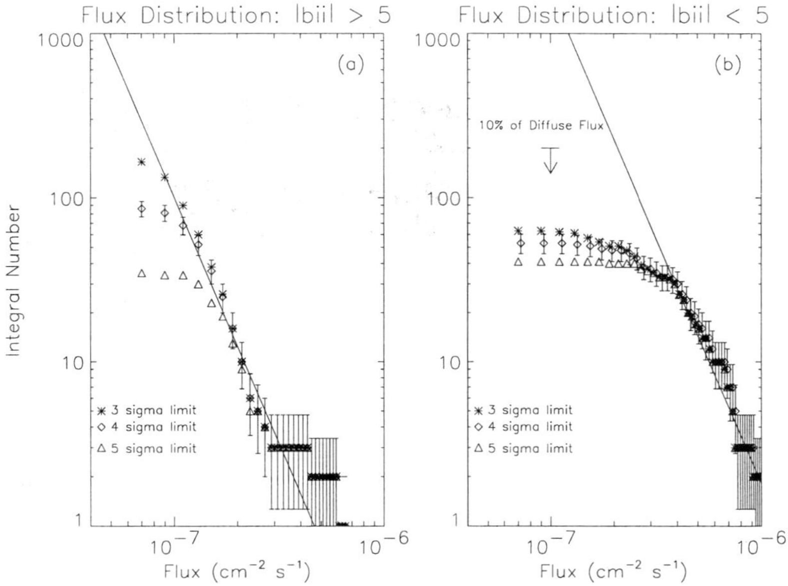

Figure 3 Flux distribution of low- and mid-latitude steady unidentified sources. The log N-log S (integral number, N, versus flux, S) distribution for the > 100 MeV EGRET unidentified steady sources for (**a**) $5^\circ < |b| < 30^\circ$ and (**b**) along the Galactic plane. Points are plotted for 3σ, 4σ and 5σ levels of significance, representing levels below those used for source detection in the catalog. Error bars are shown for only the 4σ sample (slightly offset in **b** for clarity); the error bars for 3σ and 5σ detections are of a similar size. The solid lines are a best-fit power law to the $|b| > 5^\circ$ data (**a**) and a same-slope power law drawn through the high flux points for $|b| < 5^\circ$ data (**b**). The upper limit shown in **b** is an approximate maximum number of unresolved point sources that could be hidden in the diffuse flux from the region.

than 0.1. This spectral evidence suggests that the low- and mid-latitude populations are quite different in their properties.

4. DIFFUSE GALACTIC GAMMA-RAY HALO

The cluster of weak sources north of the Galactic Center seen in Figure 1 is not an artifact of EGRET's observational exposure. In fact, EGRET's exposure of the Galactic Center is larger by about 50% than its exposure 20° north or south of the center. The spatial coincidence of this cluster of weak sources and the diffuse "Galactic Halo" in the same region reported by Dixon et al. (1998) bears further investigation. There are two possibilities for a relationship between these two components:

1 The weak point-sources are not real objects, but rathe tical fluctuations in the diffuse flux.

2 The diffuse flux consists of unresolved point sources that, individually, are below EGRET's detection threshold.

An analysis of the point spread distribution of the individual sources by comparing the χ^2 of fits to the EGRET point-spread distribution for identified and unidentified sources shows that the two distributions are consistent. Thus, the point sources are real, and not fluctuations in the diffuse flux. To investigate the possibility that the diffuse emission could be unresolved mid-latitude sources, we note that the strength of the diffuse flux in the $\sim 40°$-square excess region is $\sim 10^{-6}$ ph cm^{-2} s^{-1} (Dixon et al. 1998). There are 10 EGRET sources with fluxes greater than 10^{-7} ph cm^{-2} s^{-1} in this region, so even a modest extrapolation to weaker sources could produce enough flux to account for the diffuse emission. We therefore conclude that the diffuse flux is likely to consist of unresolved point sources from the cluster of faint sources above the Galactic Center that we report here.

5. GOULD BELT ORIGIN

The presence of a population of weak sources at mid-latitudes suggests that these sources could originate in the Gould belt(Grenier & Perrot 1999). The Gould belt is an asymmetric structure in a great circle in the sky tilted at $\sim 20°$ from the Galactic plane. The types of sources found in the Gould belt are massive (Stothers & Frogel 1974) and late-type stars (Guillout et al. 1998), clusters of molecular clouds (Taylor, Dickman & Scoville 1987), and expanding interstellar gas (Lindblad, Grupe, Sandqvist & Schuber 1973). The distances of sources range from 100 to 400 pc. The dotted line in Figure 1 shows the center of O and B stars associated with the Gould belt (Stothers & Frogel 1974). The dashed line in the same figure represents the best-fit to the positions of EGRET weak sources (i.e. the filled circles in Figure 1) to a line of the form $b = A + B\sin(l) + C\cos(l)$, where the fit gives $A = -0.37 \pm 0.11$, $B = 6.62 \pm 0.19$ and $C = 10.02 \pm 0.14$. The two lines agree well in both phase and amplitude; however, the geometry of the Gould belt in terms of ascending nodes depends on the class of sources under consideration (Guillout et al. 1998; Olano 1982). We compare this fit to a constant latitude fit and find a 2% probability that the improvement to the fit due to the added two degrees of freedom of the great circle model is by chance, according to an F-test.

The fit of unidentified sources (dashed line) in Figure 1 has a smaller amplitude than the Gould belt curve (solid line). This smaller amplitude may be due to a combination of factors. EGRET has had higher exposure at the Galactic Center than at 20° north of the center, which would lead to an increased density of detected sources at lower Galactic latitude and would decrease the amplitude of the data curve. In addition, the few low-latitude Galactic plane sources in the weak-flux sample will tend to pull the data curve down. If the weak unidentified sources are indeed originating in the Gould belt, then the concentration of weak sources above the Galactic center could be explained by the fact that this is the direction of the closest portion of the belt (Guillout et al. 1998). We examine this further by separting the weak sources into two groups: the 29 sources with $60° < l < 240°$ and the 54 sources with $240° < l < 60°$. The group around the Galactic center has an average flux of 13.5 ± 0.6 while the group around the anticenter has an average of 11.9 ± 0.7. The T-test likelihood that these means arise from the same parent sample is 5%. The Galactic center sources are indeed brighter than the anticenter sources, as would be expected from their closer Gould belt distance. Based on these findings, we conclude that there is evidence, though not proof, that the mid-latitude sources are associated with the Gould belt.

6. IMPLICATIONS AND FUTURE

We find that there are two distinct populations of steady unidentified gamma-ray sources exhibiting different spatial distributions, flux distributions and spectral properties. The bright, low-latitude sources lie in the Galactic plane and therefore are at large (kpc) distances; however, the weaker, mid-latitude sources lie off the Galactic plane and are presumably at the 100-400 pc distances of the Gould belt. The fact that the mid-latitude population is > 5 times nearer and ~ 3 times fainter than the low-latitude population implies that the luminosity of the mid-latitude population is a factor of > 75 less than that of the low-latitude sources.

The EGRET error circles of 1° - 2° are too large to search for individual source identifications. Likely candidates for these unidentified gamma-ray sources include:

- *Radio-quiet pulsars* (e.g. Gemingas). Gamma rays are produced in the particle acceleration regions of the pulsar magnetosphere (Romani & Yadigaroglu 1995, Thompson et al. 1994, Kaaret & Cottam 1996).

- *Massive stars.* Gamma rays are produced in the outflowing winds from the stars (Romero, Benaglia & Torres 1999; Montmerle 1979).

- *Molecular clouds.* Cosmic rays (CRs) interact with the enhanced gas concentrations in the cloud to produce gamma-ray enhancements in the sky (Houston & Wolfendale 1983).
- *Supernova remenants.* CRs accelerated in the supernova explosion interact with surrounding gas to produce gamma rays (Sturner & Dermer 1995; Esposito et al. 1996; Romero, Benaglia & Torres 1999; Montmerle 1979).

The Gould belt is known to have an enhanced concentration of massive stars and molecular clouds, so we speculate that the mid-latitide sources are produced by these two types of objects. The GLAST mission (Gehrels & Michelson 1999), scheduled for launch in 2005, will be more sensitive and have higher angular resolution (arcmin error circles) which will allow individual identifications of sources.

References

Dixon, D. D. et al. (1998) *New Astronomy*, **3**, 539
Esposito, J. A. et al. (1996) *ApJ*, **461**, 820
Gehrels, N., Michelson, P. (1999) *Astroparticle Physics*, **11**, 277
Gehrels, N., Macomb, D. J., Bertsch, D. L., Thompson, D. J., Hartman, R. C. (2000) *Nature*, **404**, 363
Grenier, I. A., Perrot, C. (1999) *Proc. XXVI Int. Cosmic Ray Conf. Salt Lake City*, **3**, 476
Guillout et al. (1998) *A&A*, **337**, 113
Hartman, R.C., et al. (1999) *ApJS*, **123**, 79
Houston, B. P.,Wolfendale, A. W. (1983) *A&A*, **126**, 22
Hunter, S. D. et al. (1997) *ApJ*, **481**, 205
Kaaret, P., Cottam, J. (1996) *ApJ*, **462**, L35
Lindblad, P. O., Grupe, K., Sandqvist, Aa.,Schuber, J. (1973) *A&A*, **24**, L309
Montmerle, T. (1979) *ApJ*, **231**, 95
Olano, C. A. (1982) *A&A*, **112**, 195
Romani, R. W., Yadigaroglu, I.-A. (1995) *ApJ*, **438**, 314
Romero, G. E., Benaglia, P. & Torres, D. F. (1999) *A&A*, **348**, 868
Stothers, R. & Frogel, J. A. (1974) *AJ*, **79**, 456
Sturner, S. J. & Dermer C. D. (1995) *A&A*, **293**, L17
Swanenburg, B. N. et al. (1981) *ApJ*, **243**, L69
Taylor, D. K., Dickman, R. L., Scoville, N. Z. (1987) *ApJ*, **315**, 104
Thompson, D. J. et al. (1994) *ApJ*, **436**, 229

POPULATION STUDIES OF THE EGRET SOURCES

Sandro Mereghetti, Andrew Chen, Alberto Pellizzoni, Marco Tavani, Stefano Vercellone

Istituto di Fisica Cosmica G.Occhialini, CNR

via Bassini 15, Milano, I-20133, Italy

sandro@ifctr.mi.cnr.it

Keywords: EGRET, Unidentified Gamma-ray sources

Abstract We present preliminary modelling of the populations of gamma-ray point sources observable by AGILE as constrained by the 3rd EGRET source catalog. A simple analysis of latitude distributions shows that no more than half of the EGRET sources are derived from an isotropic distribution. To analyze the two-dimenstional distribution, we wrote Monte Carlo simulations to produce model populations which we compared to the EGRET catalog. We examined models containing an isotropic component and one or two galactic components; preliminary results indicate that while one galactic component produces a spatial distribution consistent with the EGRET sources, the resulting flux distributions are inconsistent. Simultaneous comparison of spatial and flux distributions will allow physical models to be constrained.

1. INTRODUCTION

Two different approaches have been successfully adopted to understand the nature of gamma-ray sources, in the lack of obvious identifications as provided, e.g., by the detection of periodic variability.

The first one is based on observations at lower energies, to look for candidate counterparts in the gamma-ray error boxes. Typically such searches start from the X-ray and/or radio band, where it is easier to spot potentially unusual objects. Optical/IR follow-ups, that would be very inefficient in the large gamma-ray error boxes, can then be targeted to better study a few selected objects. A very instructive example of this approach is given by the more than 20 years long chase to Geminga (see Bignami & Caraveo 1996 for an excellent historical review). Several

A. Carramiñana et al. (eds.),
The Nature of Unidentified Galactic High-energy Gamma-ray Sources, 89–96.

other unidentified EGRET sources studied with this multi-wavelength approach have been presented at this Workshop.

The second method to unveil the nature of gamma-ray sources consists in the study of their collective properties. These "population studies" yield information like average flux, spectral and variability properties, spatial distribution, etc.., that can be compared to those of other known populations of astrophysical objects. A classical, early example of this is given by Swanenberg et al.(1981), who deduced from the second COS-B Catalog the existence of a class of Galactic sources with luminosities between 0.4 and 5×10^{36} erg s^{-1}. Recent, more sophisticated studies of this kind have been done, e.g., by Gehrels et al. (2000), Zhang, Zhang & Cheng (2000), Romero et al. (1999), Grenier (2000).

We are undertaking a population study of the EGRET sources that, in several respects, differs from other analysis of this kind performed in the past. In Section 2 we discuss the motivations of our work, and we illustrate our method with a simple example in Section 3. Some preliminary results are then presented in Sections 4 and 5. Although for the moment the derived conclusions are not different from those reported by other authors, we believe that our approach is less affected by biases and uncertainties and will probably lead to more robust results.

2. OUR APPROACH TO POPULATION STUDIES

Our method is based on the following guidelines:

- Avoid data binning.

  We are dealing with relatively small samples of sources. The use of binned distributions leads to histograms with only a few objects per bin. The use of traditional techniques to compare the data with the models, such as those based on the χ^2 test, is not correct if the statistical errors are not normally distributed. This problem can be solved by the use of more appropriate statistical tests, but the data binning leads also to other disadvantages. In fact, it inevitably causes a loss of information and involves a subjective, arbitrary choice of the bin boundaries.

  An alternative representation of the data is possible by means of unbinned integral distributions, that can be compared to the expectations by means, e.g., of the Kolmogorov-Smirnov (KS) test.

- Avoid cuts in the sources population (i.e. model the whole 3EG Catalogue)

Most authors concentrated on population studies of the *unidentified* gamma-ray sources. Especially when dealing with the spatial distribution of the sources, one must take into account that the process leading to identifications is affected by the source location. It is obvious, for instance, that high galactic latitude sources are more easily identified with AGN's. This fact has two consequences. First, there is a relatively higher probability of wrong AGN's identifications among high latitude sources; second, a larger fraction of the unidentified sources in the Galactic plane are likely AGN's. Though these biases are quite obvious, it is very difficult to properly quantify them.

We therefore decided to model the spatial (and flux) distribution of the *whole sample of sources* of the 3EG Catalogue, independent of their identification.

- Take properly into account the sensitivity variation across the sky.

  The non uniform sky exposure and the presence of the diffuse galactic gamma-ray emission result in large spatial variations of the limiting flux achieved by gamma-ray satellites. Due to the large point spread function, the sensitivity at any sky positions is also affected by the presence of nearby (strong) sources. These effects have to be properly corrected.

  In our analysis we model the 3-D distribution of potential sources in the galaxy and select those that would be visible given the sensitivity of the instrument, which varies as a function of direction due to instrument exposure, the intensity of the diffuse background, and source confusion.

- Use extensive Monte Carlo simulations to obtain reliable estimates of significance.

  This is the most time consuming task in this kind of studies. On the other hand, it is essential in order to assess correctly the confidence levels at which a given model can be accepted or rejected.

3. A SIMPLE EXAMPLE

To illustrate the above guidelines with a simple example we consider the galactic latitude distribution of the sources in the 3EG catalogue (see also Mereghetti & Pellizzoni 1997). In Fig. 1 the integral distribution of |b| is plotted as a function of sin(|b|). With this choice for the X axis in the plot, that compensates for the effect of decreasing solid angle toward the galactic poles, an isotropic distribution observed with a uniform

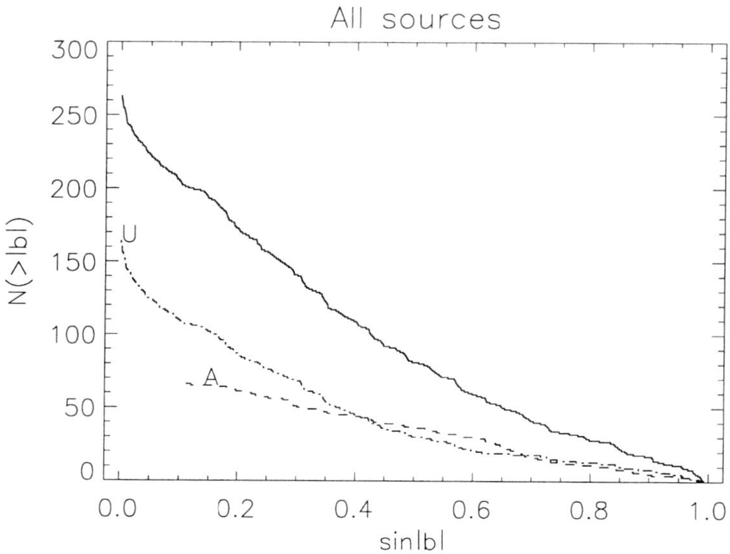

Figure 1 Integral distribution of |b| for the 264 sources of the 3EG catalog. The solid line refers to the whole sample; the dot-dashed line (U) to the 165 unidentified sources and the dashed one (A) to the 67 AGN's.

sensitivity all over the sky would appear as a straight line. This is obviously not the case for real data, as shown, for example, by the distribution of AGN identified sources (dashed lines in Fig. 1 and 2) that presents a clear flattening at low |b| values.

The expected curve for an isotropic sky distribution can be computed taking into account the effective sensitivity reached by EGRET at different coordinates. We have done this by considering the map of flux upper limits from the 3rd EGRET catalog as a measure of the sensitivity of EGRET. At each latitude the hypothetical isotropic source distribution was weighted by the flux upper limit raised to the (−1.5) power. The resulting distribution, as a function of $\sin|b|$, curves downward at low latitudes. These curves are shown in Fig. 3.

Comparing the observed distribution of the *total sample* with these expected distributions corresponding to different numbers of isotropically distributed sources (see Fig. 3) we can obtain robust constraints on the maximum allowed fraction of the isotropic component. We obtain a 2 σ upper limit to the number of isotropically distributed sources of 135, corresponding to ∼50% of the total sample. This result does not change considering only the 124 sources detected at more than 4 σ in the Viewing Periods 1 to 5, for which the corresponding upper limit is 60.

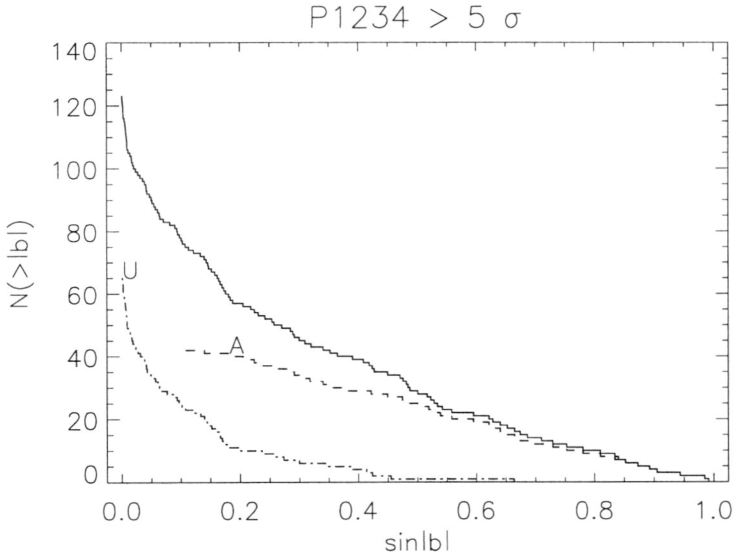

Figure 2 Same as Fig.1 for the 124 sources detected above 5σ in VP1234. (U: 67 unidentified, A: 43 AGN's).

4. MODELING THE SKY DISTRIBUTION

A method similar to that presented in the previous section has been applied to the two dimensional (l,b) distribution of the 3EG sources. To compare the observed distribution with the ones expected in different models, we used a generalization of the KS test introduced by Peacock (1983). While no formal proof of the robustness of the Peacock method exists as it does for the KS test, extensive Monte Carlo simulations as well as use in other contexts have demonstrated its robustness in a wide variety of physical situations, including (l,b) distributions.

Our first objective was to test the sky distributions against a very simple two components model, consisting of an isotropic and a disk component. The expected (l,b) distributions have been obtained with MC simulations in which we have taken into account the effect of sky exposure and galactic background as described above in Section 2.

As expected, the results of this analysis show that single component models (i.e. disk or isotropic only) are clearly rejected. We find that models based only on 2 components cannot be ruled out based only on the spatial distribution.

For example, a two-component model in which 65% of the sources are from an exponential disk with scale height 2.5 kpc and radius 29 kpc,

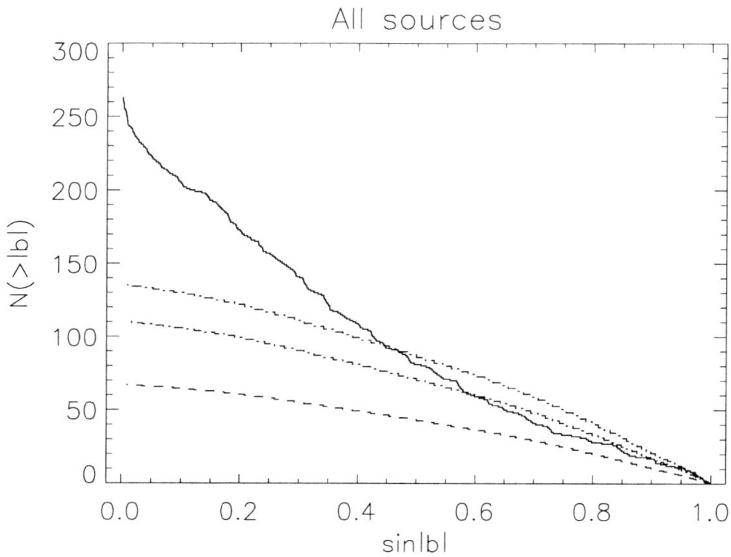

Figure 3 Comparison of the observed latitude distribution of the total sample with that expected for an isotropic distribution of 67 AGNs (dashed line) corrected for the effects of sky exposure and galactic background. The 2 σ upper limit to the number of isotropically distributed sources is 135, corresponding to ~50% of the total sample. The 1 σ upper limit is also shown.

the rest from an isotropic component, was consistent with the spatial distribution of sources (Fig. 4).

One might be tempted to speculate about the physical implications of such a model, but, as we see in the next section, the resulting flux distributions is inconsistent with that of the observed sources. In fact, in this analysis we have not considered another important information available in the data, i.e. the flux distribution of the sources.

5. 3-D MODELING: L, B, AND FLUX

If we compare the models expectations with the data also in terms of flux distributions, we must adopt a further generalization of the KS statistical test to the case of multi-dimensional distributions. We used the method proposed by Gosset (1987). While the Peacock statistic uses the number of sources in the four quadrants defined by the coordinates of each point, taking the maximum difference in each case, the Gosset statistic generalizes this approach to the eight octants defined by the three coordinates.

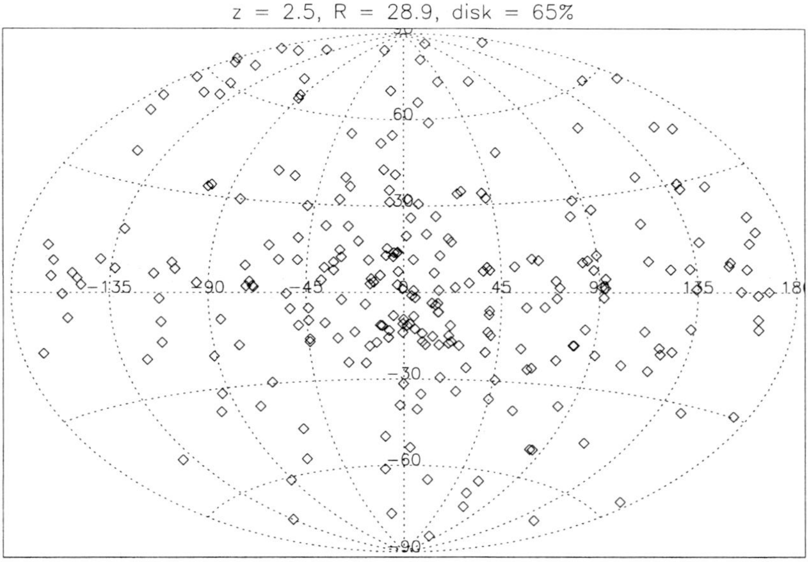

Figure 4 A model distribution consistent with the EGRET sources, taken from a parent distribution with an isotropic component and a single axisymmetric Galactic disk component with the parameters listed in the title. The distances are in kpc.

Given a theoretical, 3-dimensional parent distribution of source positions, and luminosities, we generate sample source distributions, taking into account instrument exposure and source detection sensitivity. These are compared with the EGRET source distribution using the Gosset statistic. We start by ruling out the simplest models, progressing to more complex models. For example, in the 2-D case we ruled out the one-component model before validating the two-component model.

Note that a high statistic, failing to rule out a theoretical model, does not indicate that the sources are actually derived from that model. It simply indicates that one cannot prefer other models on purely statistical grounds. For example, if it is found that a model with two disk components and an isotropic component is consistent with the data, and the flux distributions for each are physically plausible, then in order to require, for example, an association with star-forming regions or an asymmetrical warp, one would have to use other considerations, such as variability or spectra.

Extensive Monte Carlo simulations are required to properly characterize the distribution of the Gosset statistic. This is because preliminary calibrations indicate that hypothetical source distributions drawn from identical parent distributions frequently produce statistics far from the mean. Therefore, ruling out a given model requires a high degree of confidence, and, because exact analytic expressions do not exist for the distributions of the multi-dimensional KS-type statistics, quantifying that confidence level necessitates the Monte Carlo simulations.

This work is still in progress. The preliminary results seem to indicate that when fluxes are taken into consideration, a simple two-component model is not consistent with the data.

6. CONCLUSIONS AND FUTURE WORK

We have shown in Section 3 how, in a very simple way, a robust upper limit on the fraction of isotropically distributed sources can be derived. They account at most to 50% of the total 3EG catalog. If the only isotropic component were due to AGN's this would be a significant result. Unfortunately we cannot exclude the presence of different classes of galactic sources, possibly extending also at medium and high galactic latitudes.

This will be the object of our future work. We would like to verify that it is impossible to reproduce the observed flux and spatial distribution with two components only. The next step will be a three-component model with two disks and an isotropic component. Many physical models would be consistent with this framework, for example a thick disk of isolated pulsars and a thin disk containing both young pulsars and SNRs emitting cosmic rays interacting with high-density nearby material. If a simple three-component model is consistent, we will explore the physical implications of the parameter space of each component, while looking at other features such as spectra and variability. If not, we will test more sophisticated models, where one component is correlated with spiral arms, OB associations, or the Gould belt.

References

Bignami G.F. & Caraveo P.A. (1996), ARA&A, 34, 331

Gehrels, N. et al. (2000), Nature, 404, 363

Gosset, E. (1987), A&A, 188, 258

Grenier, I. (2000), A&A, 364, 552

Mereghetti S. & Pellizzoni A. (1997), in The Transparent Universe, Proceedings of the 2nd INTEGRAL Workshop, ESA SP-382, 283

Peacock, J.A. (1983), MNRAS, 202, 615

Romero, G.E., Benaglia, P., & Torres, P.F. (1999), A&A 348, 868

Swanenburg, B.N. et al. (1981), ApJ, 243, L69

Zhang. L., Zhang, Y.J., & Cheng, K.S. (2000), A&A 357, 957

POSITIONAL CORRELATION BETWEEN LOW-LATITUDE γ-RAY SOURCES AND SUPERNOVA REMNANTS

Diego F. Torres, Jorge A. Combi, Gustavo E. Romero and P. Benaglia
Instituto Argentino de Radioastronomía
C.C. 5, (1894) Villa Elisa, Buenos Aires
Argentina
dtorres@venus.fisica.unlp.edu.ar

Keywords: gamma-rays: observations, ISM: supernova remnants, stars: pulsars

Abstract We present the results of a spatial correlation analysis between unidentified gamma-ray sources in the 3EG catalog and candidates to supernova remnants. This work extends a previous study made by Romero, Benaglia & Torres (1999). We also consider the gamma-ray emission variability and the spectral index for the sources with positional coincidence.

1. SUPERNOVA REMNANTS AS GAMMA-RAY EMITTERS

Supernova remnants (SNRs) are usually considered as the main sources of cosmic rays with energies below $\sim 10^{15}$ eV. Both electrons and protons are believed to be accelerated by Fermi mechanism in the expanding shock front of these objects. The electrons produce synchrotron emission detected at radio wavelengths whereas the interactions between relativistic protons and ambient nuclei can produce neutral pions, which quickly decay yielding γ-ray emission at energies $E \geq 100$ MeV, in the EGRET range. The expected γ-ray flux at Earth was given by Drury et al. (1994),

$$F(\geq 100\text{MeV}) \sim 4.4 \times 10^{-7} \theta \frac{E_{SN}}{10^{51}\text{erg}} \left(\frac{d}{\text{kpc}}\right)^{-2} \left(\frac{n}{\text{cm}^{-3}}\right) \text{cm}^{-2}\text{s}^{-1}, \quad (8.1)$$

where E_{SN} is the energy of the SN in ergs, θ is the fraction of the total energy of the explosion converted into cosmic ray energy, and n and d have their usual meaning, number density and distance, respectively.

A. Carramiñana et al. (eds.),
The Nature of Unidentified Galactic High-energy Gamma-ray Sources, 97–104.

In most cases, the expected flux at GeV energies is far too low to be detected in the range of EGRET, but the presence of nearby clouds can produce a significant enhancement of the γ-ray emission. Such scenario has been recently studied by Combi et al (1998) in relation with the source 3EG J1659-6251.

As showed by Aharonian et al. (1994), when a SNR hits a cloud a part of the proton population can be transported into it by convection, and illuminate the cloud by subsequent p-p interaction, via pion decay. The resulting flux is

$$F(\geq 100\text{MeV}) \sim \frac{q_\gamma M_{cloud}}{4\pi d^2 m_H}, \tag{8.2}$$

where q_γ is the γ-ray emissivity per H-atom in the cloud. This parameter can be related with its value in the vicinity of Earth by $q_\gamma = kq_\odot$, with $k > 1$.

Other mechanisms, like relativistic bremsstrahlung and inverse Compton losses, associated with the leptonic component, can also play a role if the electron density and/or the photon fields are high enough, see for instance Pohl (1996). Then, there are many theoretical reasons to expect a positional correlation between SNRs and unidentified γ-ray sources. We devote the rest of this work to analyze this correlation.

2. POSITIONAL CORRELATION

Possible physical correlation between SNRs and unidentified EGRET sources, on the basis of two dimensional positional coincidence, has been proposed since the release of COS-B data (Montmerle 1979) and the first EGRET catalog. Sturner & Dermer (1995) suggested that some of the unidentified sources lying at low galactic latitudes $|b| < 10^o$ might be associated with SNRs: of 37 detections, 13 overlapped SNR positions in the first EGRET catalog. However, the statistical significance was not too high as to provide a strong confidence. Using the 2EG catalog, Sturner et al. (1996) repeated the analysis, and showed that 95% confidence contours of 7 unidentified EGRET sources overlapped SNRs, some of them appearing to be in interaction with molecular clouds. Similar results were independently reported by Esposito et al. (1996), considering only radio-bright SNRs, and Yadigaroglu and Romani (1997), although they did not assessed the overall chance probability of these 2EG-catalog findings. The evolution in the number of coincidences and SNRs considered in the different studies is shown in Table 1. Note that from the First to the Second EGRET catalogs, fourteen unidentified sources were discarded. In the Third catalog, 6 unidentified sources are possibly artifacts produced by the strong emission of the Vela pulsar: these sources disappear in a map where the pulsed Vela emission is suppressed.[1]

In Table 2 we show the 3EG sources that are positionally coincident with SNRs listed in the latest version of Green's catalog. From left to right we provide the γ-ray source name, the measured flux, the photon spectral index Γ, the variability index I (see below), the SNR identification, the angular distance between the best γ-ray source position and the center of the remnant, the size of the remnant in arcminutes, the SNR type (S for shell, F for filled-centre, and C for composite), and other positional coincidences found in our previous study (Romero, Benaglia and Torres 1999). It is interesting to note that in the 3EG catalog, not all the positional coincidences with SNRs are SNOBs, as was the case reported by Montmerle (1979) and Yadigaroglu and Romani (1997) using previous samples.

We also consider whether some of the sources in our sample may be associated with recently proposed candidates to supernova remnants, presently not catalogued by Green (1998). Our interest in this search resides in the fact that young stellar objects, like recently formed black holes and pulsars, can still be associated with the gaseous remnant of the original supernova that created them. The diffuse non-thermal emission of the galactic disk, originated in the interaction of the leptonic component of the cosmic rays with the galactic magnetic field, is surely veiling many remnants of low surface brightness. Recent observational studies using filtering techniques in the analysis of large-scale radio data have revealed several new SNR candidates that are not yet included in the latest issue of Green's catalog. In general, these new candidates are much more extended than those previously known. There are 101 of these weak non-thermal structures detected so far in the Galaxy. This number significantly extends Green's (1998) catalogue. The list of these new candidates, and the references from where they were compiled, can be obtained from the paper by Torres et al. (2000).

We have found that only 7 gamma-ray sources in our sample are positionally coincident with non-thermal radio structures. The positional coincidences thus obtained are shown in Table 3, where we provide similar information as that given in Table 2.

In order to estimate the statistical significance of these coincidences, we have numerically simulated a large number of synthetic sets of EGRET sources using the code described in the paper by Romero, Benaglia and Torres (1999). The results of this study are shown in Table 4 where we provide results for different samples of unidentified sources and SNRs catalogs: the original 81 unidentified EGRET sources, the 75 high confidence ones, and the 40 sources without any positional corre-

Table 1. Evolution in the number of positional coincidences between SNRs and unidentified EGRET sources

Catalog EGRET	Unidentified detections	Real coincidences	Number of SNRs in Green's catalog	Significance (statistical)
1EG[a]	37	13(35%)	182	1.8σ
2EG[b]	32	7(22%)	194	?
2EG[c]	32	5(16%)	14[g]	?
2EG[d]	33	10(30%)	194	h
3EG[e]	81	22(27%)	220	5.7σ
3EG[f]	75	19(25%)	220	4.8σ

Table 2. Positional coincidences with supernova remnants.

γ-source (3EG J)	F_γ (10^{-8} ph cm^{-2} s^{-1})	Γ	I	SNR	$\Delta\theta$ (deg)	Size (arcmin)	Type	Other coincidences
0542+2610	14.7±3.2	2.67±0.22	3.16	G180.0−1.7	2.04	180	S	
0617+2238	51.4±3.5	2.01±0.06	1.68	G189.1+3.0	0.11	45	S	OB
0631+0642	14.3±3.4	2.06±0.15	1.52	G205.5+0.5	1.97	220	S	
0634+0521	15.0±3.5	2.03±0.26	1.02	G205.5+0.5	2.03	220	S	Of/OB
1013−5915	33.4±6.0	2.32±0.13	1.63	G284.3−1.8	0.65	24	S	
1102−6103	32.5±6.2	2.47±0.21	1.86	G290.1−0.8	0.12	19 ×14	S	WR/OB
1410−6147	64.2±8.8	2.12±0.14	1.22	G312.4−0.4	0.23		S	OB/Of
1639−4702	53.2±8.7	2.5±0.18	1.95	G337.8−0.1	0.07	9 ×6	S	OB
				G338.1+0.4	0.65	15	S	
				G338.3+0.0	0.57	8	S	
1714−3857	43.6±6.5	2.30±0.20	2.17	G348.5+0.0	0.47	10	S	
				G348.5+0.1	0.50	15	S	
1734−3232	40.3±6.7	-	2.90	G355.6+0.0	0.16	6 ×8	S	OB
1744−3011	63.9±7.1	2.17±0.08	1.80	G359.0−0.9	0.41	23	S	
				G359.1−0.5	0.25	24	S	
1746−2851	119.9±7.4	1.70±0.07	2.00	G0.0+0.0	0.12	3.5 ×2.5	S	
				G0.3+0.0	0.19	15.8	S	
1800−2338	61.3±6.7	2.10±0.10	1.60	G6.4−0.1	0.17	42	C	
1824−1514	35.2±6.5	2.19±0.18	3.00	G16.8−1.10	0.43	30 ×24		OB
1837−0423	19.1	2.71±0.44	5.41	G27.8+0.6	0.58	50 ×30	F	
1856+0114	67.5±8.6	1.93±0.10	2.92	G34.7−0.4	0.17	35 ×27	S	
1903+0550	62.1±8.9	2.38±0.17	2.28	G39.2−0.3	0.41	8 ×6	S	
2016+3657	34.7±5.7	2.09±0.11	2.06	G74.9+1.2	0.26	8 ×6	F	WR/OB
2020+4017	123.7±6.7	2.08±0.04	1.12	G78.2+2.1	0.15	60	S	OB

Table 3. Positional coincidences with candidates to SNRs.

γ-source (3EG J)	F_γ (10^{-8} ph cm^{-2} s^{-1})	Γ	I	SNR	$\Delta\theta$ (deg)	Size (arcmin)	Type
0903−3531	22.8±4.6	2.66±0.24	1.55	260.20+1.40	6.05	1560	S
1631−4033	12.7±3.0	2.25±0.27	1.26	342.60+8.20	3.11	360	S
1638−5155	67.3±14.2	2.56±0.21	2.44	333.00+0.00	3.50	360	S
1704−4732	117.8±27.1	1.86±0.33	2.96	340.80−4.80	1.22	174	S
1717−2737	17.8±3.6	2.33±0.15	2.00	356.90+8.50	2.66	570	S
1834−2803	16.2±2.9	2.62±0.20	2.83	7.30−5.30	3.91	432	S
1837−0606	49.6±7.9	1.82±0.14	2.41	27.00+0.50	1.14	420	S

lation with known galactic gamma-ray emitters (as reported by Romero et al. (2000)). For the latter sample we are interested in checking

Table 4. Statistical results. G stands for Green's SNRs and C for candidates.

Number of SNRs	Unidentified Sources	Real coincidences	Simulated coincidences	Poisson probability
220 (G)	81	22	7.8 ± 2.5	1.5×10^{-5}
220 (G)	75	19	7.0 ± 2.4	9.9×10^{-5}
321 (G + C)	75	30	22.4 ± 3.8	0.02
101 (C)	40	7	10.4 ± 2.7	0.07

the possible association with candidates to SNRs. We conclude that there is strong statistical evidence suggesting that some 3EG unidentified EGRET sources must be physically associated with SNRs in Green's Catalog. At the same time, there is no statistical evidence suggesting that the 3EG sources analyzed in our sample are physically associated with candidates to SNRs: the number of real positional coincidences with them is totally compatible with, and even lower than, the result of a random association. This latter result appear to be the consequence of the candidates to SNRs being much more extended objects, thus improving the random coincidences with any given population, and does not discard in itself that some of the coincidences could be physical ones.

3. SPECTRAL AND VARIABILITY INDICES

In Figure 1 we present the distribution of the spectral index for all unidentified sources which resulted to have positional correlation with SNRs and candidates. All values are compatible with Fermi-like acceleration processes that could happen in the strong shocks generated by the explosion of the supernova, or be the outcome of the interaction between relativistic material and a nearby cloud, or –for the lowest ones– be associated with gamma-ray emission by pulsars. The distribution peaks around $\Gamma = 2.1$, but some sources have a large spectral index of about 2.8. Recall that the steepest measured spectral index for pulsars is around 2.2, so this is pointing against a pulsar origin for several of these gamma-ray sources.

We now assess the possible long term variability of the sources. We define a mean weighted value for the EGRET flux as:

$$\langle F \rangle = \left[\sum_{i=1}^{N_{\rm vp}} \frac{F(i)}{\epsilon(i)^2} \right] \times \left[\sum_{i=1}^{N_{\rm vp}} \frac{1}{\epsilon(i)^2} \right]^{-1} . \tag{8.3}$$

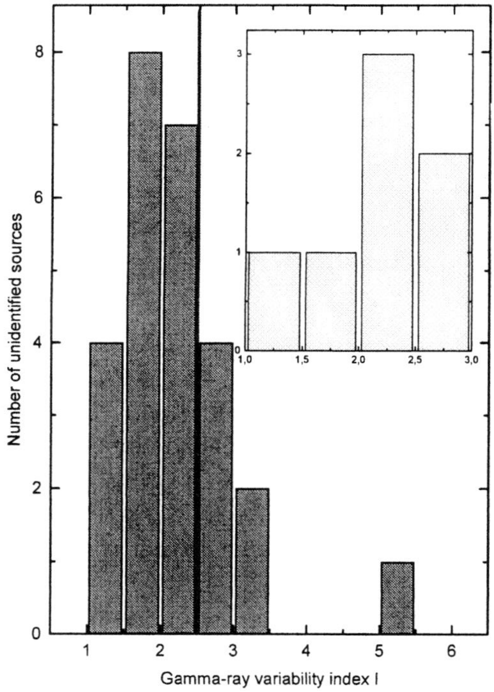

Figure 1 Distribution of spectral index for the sources with positional coincidences with SNRs. The clearer boxes stand for the sources coincident with candidates to SNRs.

$N_{\rm vp}$ is the number of viewing periods for each gamma-ray source. $F(i)$ is the observed flux in the $i^{\rm th}$-period, whereas $\epsilon(i)$ is the corresponding error in the observed flux.[2] We then define the fluctuation index μ as: $\mu = 100 \times \sigma_{\rm sd} \times \langle F \rangle^{-1}$. In this expression, $\sigma_{\rm sd}$ is the standard deviation of the flux measurements. In order to remove as far as possible any spurious variability introduced by the observing system, we computed the fluctuation index μ for the confirmed gamma-ray pulsars in the 3EG catalog. The identification by Kuiper et al. (2000) is not included because of the blazar contamination of the EGRET flux, observed by these authors. We adopt the physical criterion that pulsars are non-variable gamma-ray sources. Then, any non-null μ-value for pulsars is attributed to experimental uncertainty. We then define an averaged statistical index of variability, I, as

$$I = \frac{\mu_{\rm source}}{< \mu >_{\rm pulsars}} = \frac{\mu_{\rm source}}{26.9}. \qquad (8.4)$$

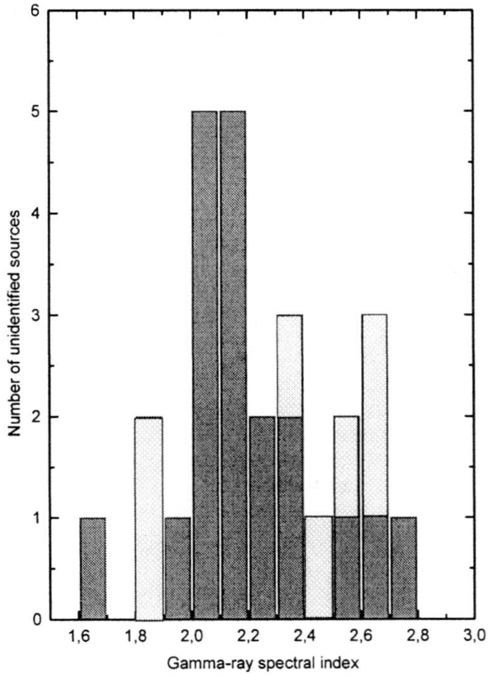

Figure 2 Distribution of the variability index for the sources with positional coincidences with SNRs. The upper box is the distribution only for the sources with positional coincidences with candidates to SNRs. The vertical line separates the sources which are variable (to the right) from the rest.

In terms of the averaged index I, the adopted variability criterion is then that variable sources will be those with $I > 2.5$, which is 3σ away from the statistical variability of pulsars.[3]

In Figure 2 we show the distribution for the variability index I of those sources that are positionally coincident with SNRs. We see that most of the 19 sources positionally related with Green's SNRs have $I < 2.5$, being their mean value 2.17. Indeed, 12 sources out of 19 have $I < 2$. There are three sources with very high I-index: 3EG J1824-1514, J1837-0423 and 3EG J0542+2610, with $I = 3.00$, $I = 5.41$ and $I = 3.16$, respectively. These sources show spectral indices of 2.19, 2.71 and 2.67. The source 3EG J1824-1514 has been recently proposed by Paredes et al. (2000) as a faint microquasar detected through VLBI observations. This and other sources with $I > 2.5$, are very unlikely physically associated with the SNRs in the usual sense, i.e. being pulsars or nearby clouds in interaction with the swept up material. Nevertheless, compact objects like accreting black holes, or even isolated Kerr-Newmann black holes (Punsly et al. 2000) are interesting possibilities, they surely requires additional analysis.

Acknowledgments

This work was partially supported by CONICET, ANPCT (PICT 98 No. 03-04881), and Fundación Antorchas (through separate grants to D.F.T., J.A.C. and G.E.R.). Additional on-line data can be obtained at URL: www.iar.unlp.edu.ar/garra. Authors would like to thank the Organizers for their kind invitation to this meeting.

Notes

1. Notes corresponding to Table 1 are as follow: a: Sturner and Dermer (1995). b: Sturner, Dermer and Mattox (1996). c: Esposito, Hunter, Kanbach and Sreekumar (1996). d: Yadigaroglu and Romani (1997). e: Romero, Benaglia and Torres (1999). f: This work. g: Only radio-bright SNRs, flux at 1 GHz greater than 100 Jy, were used. h: Computed for pairs.

2. For those observations in which the significance ($\sqrt{TS}$ in the EGRET catalog) is greater than 3σ, we took the error as $\epsilon(i) = F(i)/\sqrt{TS}$. However, many of the observations are in fact upper bounds on the flux, with significance below 3σ. For these ones, we assume both $F(i)$ and $\epsilon(i)$ as half the value of the upper bound.

3. Detailed comments on how this variability index compares with others, especially with the analysis made in W. Tompkins' Ph.D. thesis, are given elsewhere (Torres et al. 2000).

References

Aharonian F.A., Drury L.O'C., Völk H.J. 1994, A&A 285, 645
Combi J.A., Romero G.E., Benaglia P. 1998, A&A 333, L91
Drury L.O'C., Aharonian F.A., Völk H.J. 1994, A&A 297, 959
Esposito J.A., Hunter S.D., Kanbach G., Sreekumar P. 1996, ApJ 461, 820
Green D.A. 1998, A Catalog of Galactic Supernova Remnants, Mullard Radio Astronomy Observatory, Cambridge, UK (available on the World Wide Web at http://www.mrao.cam.ac.uk/surveys/snrs/)
Kuiper L., Hermsen W., Verbunt F., et al. 2000, A&A 359, 615
Montmerle T. 1979, ApJ 231, 95
Paredes J. M., Marti J., Ribo M., Massi M. 2000, Science, 288, 2340
Pohl M., 1996, A&A, 307, 57
Punsly B., Romero G.E., Torres D.F., Combi J.A. 2000, A&A 364, 552
Romero G.E., Benaglia P., Torres D.F. 1999, A&A 348, 868
Romero G.E., Torres D.F., Combi J.A. 2000, Proceedings of the 4th Integral Workshop, ESA-SP Publications, in press.
Sturner S.J., Dermer C.D. 1995, A&A 293, L17
Sturner S.J., Dermer C.D., Mattox J.R. 1996 A&AS 120, 445
Torres D.F., Romero G.E., Combi J.A., et al., 2001, A&A 370, 468
Yadigaroglu I.-A., Romani R.W. 1997, ApJ 476, 356

III

HIGH-ENERGY PULSARS

GEMINGA PULSARS

Alberto Carramiñana

Instituto Nacional de Astrofísica, Óptica y Electrónica

Luis Enrique Erro 1, Tonantzintla, Puebla 72840, MEXICO *

alberto@inaoep.mx

Keywords: Neutron stars, Pulsars: radio-quiet

Abstract Multiwavelength astronomy has shown that Geminga is a particular kind of object: a radio-quiet γ-ray loud pulsar. While the enigma of why the observed radio flux of Geminga is so low persists, the discovery of over a hundred unidentified γ-ray sources by the EGRET telescope has pointed to Geminga-like pulsars as a natural explanation for these sources. Some unidentified sources display properties similar to those of Geminga and there are at least four well documented cases of objects likely to contain a radio-quiet or radio-weak pulsar. While *GLAST*, the prototype of the next generation of γ-ray telescopes, will be able to perform direct timing studies of these pulsars, the possibility of finding more radio-quiet pulsars in the EGRET data remains open.

1. INTRODUCTION

Twentieth century astronomy was marked by the opening of the different windows of the electromagnetic spectrum, from radio waves to gamma-rays. Astronomers learned, and are still learning, to complement the distinct information acquired at each spectral regime in order to disentangle the nature of celestial objects. Observational multi-wavelength astronomy has brought to light objects and phenomena predicted by theoretical physicists decades before, many of them considered until then as intellectual exercises.

The last of the major electromagnetic bands to be opened was the gamma-ray regime, first explored in the seventies and early eighties by the *SAS-II* and *COS-B* telescopes. The EGRET experiment on board of the *Compton Gamma-Ray Observatory (CGRO)* provided for the first

*Project supported by CONACyT grant 25539-E.

A. Carramiñana et al. (eds.),
The Nature of Unidentified Galactic High-energy Gamma-ray Sources, 107–134.

time all-sky coverage and culminated several years of observations with its Third catalog of high-energy gamma-ray sources (Hartman et al. 1999). Most of the 271 objects included in this catalog, presently the most complete database for gamma-ray astronomy, remain unidentified and their nature is an open question.

Every time a new spectral window has been opened unexpected discoveries have arrived and the γ-ray sky has not been the exception. Multiwavelength astronomy brought to light that Geminga, the second brightest object in the sky as observed above 100 MeV, is an unexpected type of celestial object: a γ-ray loud and radio-quiet pulsar. To deduct that this type of object must be present among unidentified γ-ray sources is just the mandatory next step forward. Unidentified EGRET sources are distributed as a Galactic objects, some very close to the Galactic plane and are frequently associated with supernova remnants. Questions one can presently address are: which individual EGRET sources are most likely to be Geminga-like pulsars? What can we learn from them before the next generation of γ-ray satellites goes into orbit? What can we expect from future γ-ray telescopes like *GLAST*?

2. GEMINGA

Geminga promptly appeared in some of the first data taken with γ-ray telescopes in orbit: 121 photons lead to the serendipitous discovery of the bright gamma-ray source γ195+5 by the *SAS-II* telescope in the mid-seventies (Thompson et al. 1977). Its identification is the example of multiwavelength strategy still followed today. It was first positionally identified with the *Einstein* X-ray source 1E 0630+178 by Bignami, Caraveo & Lamb (1983). Optical observations of the X-ray source lead to discard some possible counterparts (the G and G' stars) and to the identification of the G" star as the preferred optical counterpart on grounds of its colour (Halpern & Tytler 1988), proper motion (Bignami, Caraveo & Mereghetti 1993) and parallax (Caraveo et al. 1996), all consistent with a nearby neutron star (Table 1). The final identification of the X-ray and γ-ray sources as a pulsar was made on basis of coherent pulsations with common period (Halpern & Holt 1992; Bertsch et al. 1992). The 237-ms pulse was then confirmed in the *COS-B* and *SAS-II* data (Hermsen et al. 1992; Mattox et al. 1992), providing a timing baseline of more than 20 years. A confirmation of the report of optical pulsations by Shearer et al. (1998) would make the association of G" with Geminga definitive. More than half a dozen satellite and ground based telescopes were involved in the identification of Geminga through observations in the radio, optical, X-ray and γ-ray regimes.

Table 1 Properties of the G" star

Magnitude	B = 26.35±0.35
	V = 25.37±0.20
	R = 25.54±0.30
Colour	$g - r = -0.32$
Proper motion	0.17"/year
Parallax	$(6.4\pm1.7) \times 10^{-3}$ arcsec
Parallactic distance	157^{+59}_{-34} pc

2.1. RADIO-SILENCE, RADIO-QUIETNESS

- **silent: 1.** characterized by an absence or near absence of sound: *a silent house.* **2.** tending to speak very little. **3.** failing to speak or communicate when expected: *the witness chose to remain silent.* **4.** not spoken: *silent disaproval.* **5** (of a letter) used in the spelling of a word but not pronounced, such as the *k* in *know.* **6.** denoting a film that has no soundtrack.

- **quiet: 1.** characterized by an absence of noise. **2.** calm or tranquil: *the sea is quiet tonight.* **3.** untroubled: *a quiet life.* **4.** not busy: *business is quiet today.* **5.** private; secret: *a quiet word with someone.* **6.** free from anger, impatience or other extreme emotion. **7.** not showy: *quiet colours; a quiet wedding* **8.** modest or reserved: *quiet humour.* **9.** the state of being silent, peaceful or untroubled.

"Collins New Compact English Dictionary", 2nd edition.

Geminga is the first known example of a radio-quiet pulsar. Following Brazier & Johnston (1999), a radio-*quiet* pulsar is understood here as a rotation-powered pulsar which has not been detected at 400 MHz or 1400 MHz and which has a low *inferred* luminosity at these frequencies, while a radio-*silent* pulsar has no radio emission. Because radio emission in pulsars is known to be beamed, objects like Geminga are radio-quiet but might not be radio-silent. Caraveo, Bignami & Trümper (1996) reviewed the observational evidence for radio-quiet isolated neutron stars. Except for Geminga, all the seven candidates been originally found in X-rays, where thermal or magnetospheric emission occurs. Brazier & Johnston (1999) added to the list the X-ray source associated with the Supernova remnant 3C58 and two Geminga-like candidates,

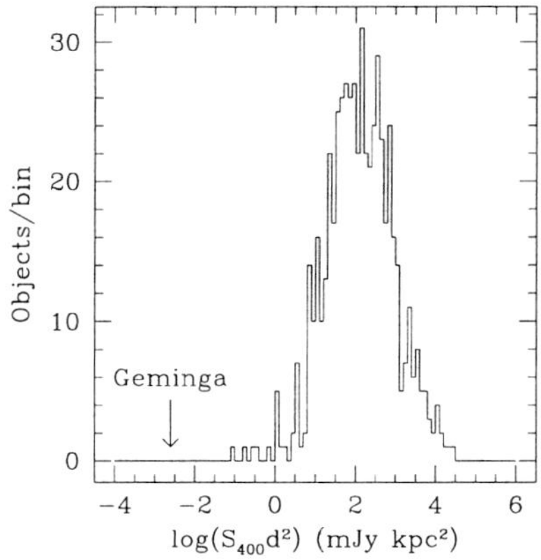

Figure 1 Distribution of $S_{400}\Delta\nu\ d^2$ for 561 radio pulsars. The arrow marks the upper limit of $S_{430}\Delta\nu\ d^2$ for Geminga, at least two orders of magnitude below the lowest radio luminosity observed for radio pulsars, and six orders of magnitude below common values.

RX J0007.0+7302 in CTA-1 and RX J2020.2+4026 in G078.2+2.1, revealed through the γ-ray data. It is becoming apparent that both X-rays and γ-rays are good alternative tools to search for isolated neutron stars.

Several radio studies at the position of G" have been performed (Seiradakis 1992; Kassim & Lazio 1999 and references therein). Most have provided upper limits and the reported (weak) detections have been made at low frequencies, near and below 100 MHz (Malofeev & Malov 1997; Shitov & Pugachev 1997). While the Puschino detections indicate $S_{102}\Delta\nu\ d^2 \simeq 10^{22}$erg/s, where S_{102} is the flux density at 102 MHz. The most stringent upper limits (Burderi, Fauci & Boriakoff 1999; McLaughlin et al. 1999) imply $S_{430}\Delta\nu\ d^2 \leq 10^{21}$erg/s at 430 MHz (Figure 1). The different radio measurements can only be reconciliated through a very steep spectral index ($\lesssim -5$), atypical for a pulsar, or if the measurements affected by variability, either intrinsic to the pulsar or due to scintillation in the propagation of low frequency radio waves through the interstellar medium.

There are no clear clues on why is Geminga radio-quiet, if not radio-silent. One interpretation is that Geminga has a radio beam which misses the Earth, indicating that the radio beam is either smaller than the gamma-ray beam, or misaligned. The other known gamma-ray pulsars have radio emission and therefore the argument can only be partial in that the radio beam might be smaller but comparable or only slightly disaligned. The quantification of any of these possibilities requires a better knowledge of the population of radio-quiet pulsars. A second,

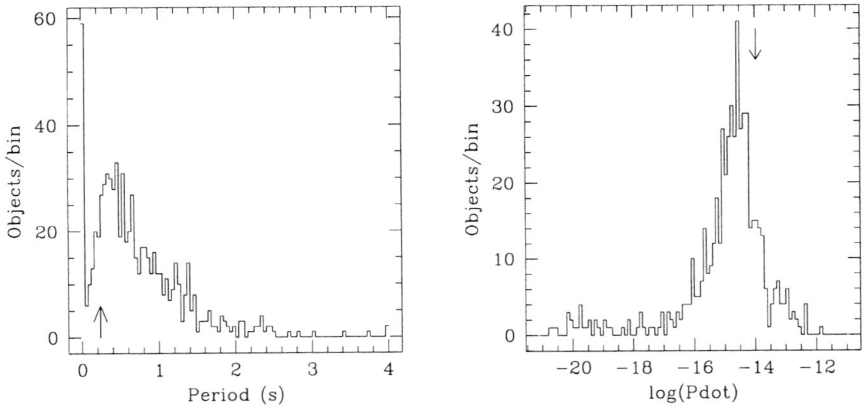

Figure 2 Geminga period (*left*) and period derivative (*right*), indicated by the arrows, compared to the distribution of values of pulsars in the Princeton catalog. Geminga is a fairly rapid pulsar with large $\dot{P}$, but not abnormal. Millisecond pulsars are evident in the bin of lowest periods.

less preferred, interpretation is that Geminga is radio-silent, i.e. its spectral energy distribution is intrinsically different from that of other known pulsars, lacking (at least practically) of radio emission. Lack of radio emission in the magnetosphere could occur for pulsars with certain characteristics, like the magnetic field strength or the relative geometry of the magnetic field, rotation axis and line of sight. One could then hope to find some observational sympton, like peculiar $P - \dot{P}$ location or unusual light curves. In that respect it is relevant that Geminga is mostly similar to other radio or γ-ray pulsars, with very few distinctive characteristics.

2.2. PSR J0630+1746

The timing parameters of PSR J0630+1746, the Geminga pulsar, are not unusual compared to those of common radio pulsars (Fig 2). The dynamical age, $P/2\dot{P}$, is about 3.42×10^5 years and the derived (vacuum dipole) magnetic field is $\approx 1.6 \times 10^{12}$ Gauss, fairly typical values for pulsars. Compared to the 706 radio pulsars of the Princeton catalog (Taylor, Manchester & Lyne 1993), the only distinctive feature of Geminga, other than the lack of radio signal, is its close distance (Fig 3).

The comparison with γ-ray pulsars is more difficult as there are just a few of them, and their properties seem to differ (see below). The timing characteristics, spectrum and (lack of) variability of the γ-ray emission from Geminga are similar to those of the other γ-ray detected radio pulsars. The main distinction is the observed peak-to-peak separation

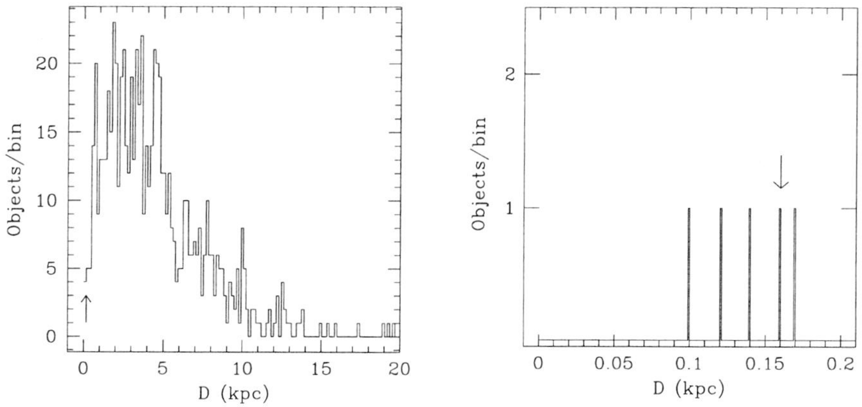

Figure 3 Geminga distance, indicated by the arrows, compared to the distribution of values of pulsars in the Princeton catalog. Geminga is one of the closest pulsars known (*right*).

of the light curve, equal to 0.50 in phase, while the (radio-loud) Crab and Vela pulsars have a smaller separation between peaks, 0.40 ± 0.02 and 0.424 ± 0.002 respectively (Bertsch et al. 1992; Nolan et al. 1993; Kanbach et al. 1994). In this respect it would be interesting to see whether the outer-gap and polar-cap models can account simultaneously for beaming differences (or radio-silence) and the distinctive peak-to-peak separation.

3. GALACTIC EGRET SOURCES

The Third EGRET Catalog (3EG Catalog; Hartman et al. 1999) represents the most important database of high-energy gamma-ray sources available for the next few years. It lists 271 sources, 163 of them classified as unidentified (after excluding 7 possible or likely Geminga and Vela artifacts). The spatial distribution of unidentified sources indicates that most of them are Galactic, with two distinct classes of γ-ray sources (Gehrels et al. 2000). The only type of Galactic source identified so far in the EGRET data are pulsars (and a solar flare).

3.1. γ-RAY PULSARS

Five known γ-ray pulsars are included in the 3EG Catalog: Vela, Geminga, Crab, PSR B1706–44 and PSR B1055–52. PSR B1046–58 is coincident with the unidentified source 3EG J1048–5840 and a marginal detection of pulsations has been made (Kaspi et al. 2000). This contrasts with the case of PSR B1951+32, where there is solid evidence for the pulsations in the EGRET data (Ramanamurthy et al. 1995) but no corresponding 3EG source is included in the Catalog. The high Galactic background in that region ($l = 69^\circ, b = 3^\circ$) makes the detected pulsed flux and Catalog upper limits compatible. Marginal evidence for EGRET detections have been provided for PSR B0656+14 (Ramanamurthy et al. 1996) and for the millisecond pulsar PSR J0218+4232 (Kuiper et al. 2000). At lower γ-ray energies, COMPTEL -apart from solid detections of the Crab and Vela pulsars- has given reasonable evidence for detections of PSR B1509–58 and PSR B1951+32 (Kuiper et al. 1999a; 1999b) in the 1-30 MeV range.

Known γ-ray pulsars have distinct properties like hard spectral indices and fluxes consistent with no variability (Table 2). Although differences in light curves and other properties make the classification of these few objects difficult, some patterns have been identified. It appears that the efficiency to convert rotational energy into high-energy quanta increases with age, measured as $P/2\dot{P}$, while the spectra tend to get harder (Figure 4). These patterns are strong guidance to theoretical models of high energy photon production.

Other radio pulsars are positionally coincident with EGRET sources, although no pulsed γ-ray emission has been proven (Nel et al. 1996). Some of these are related to supernova remnants (SNR) and both, pulsar or SNR, are potential γ-ray sources (Zhang & Cheng 1998). If the pulsar is the emitter, pulsations might be below detection level, although unpulsed emission from pulsars has also been studied theoretically (Kwok, Cheng & Lau 1991). Extended γ-ray emission from the SNR might be suggested by the softer spectra of the coincident pulsar. Except for 3EG J1856+0114, these sources have spectra harder than those of the 3EG sources related to detected pulsars and only two of them are consistent with a non variable source (Table 2). Note that spectral indices of γ-ray pulsars and their corresponding 3EG Catalog sources normally differ, the pulsed component been harder. Also, all six catalogued pulsars are included in the GeV catalog of Lamb & Macomb (1997).

Two main families of models for pulsar γ-ray emission have been developed: the polar cap (Sturrock 1971; Ruderman & Sutherland 1975) and the outer gap (Cheng, Ho & Ruderman 1986) models. Both rely on charge replenishment of the magnetosphere through $e^{\pm}$ pairs. The main difference between the two scenarios is the region inside the magnetosphere where γ-rays are actually produced: in the polar cap model the neutron star supplies the pairs and these are accelerated at the vicinity of the star, while in the outer gap model particle pairs are created and accelerated near to the intersection of the charge free surface, $\rho \propto \vec{\Omega} \cdot \vec{B} = 0$, with the last closed magnetic field line, in the outer magnetosphere. Both

Table 2 Pulsars identified or positionally coincident with 3EG sources (Zhang & Cheng 1998). All the positionally coincident pulsars with no pulsed emission observed have a SNR associated. The variability parameter τ is defined in Thompkins (1999) and briefly described in §3.2. The intervals quoted indicate 68% confidence lower and upper limits.

Pulsar PSR	γ-ray pulsar spectral index	γ-ray pulsar spectral break	3EG source J Name	3EG source spectral index	τ range
B0531+21	2.15±0.04	none	0534+2200	2.19±0.02	[0.00,0.10]
J0633+1746	1.50±0.01	$\gtrsim$ 2 GeV	0633+1751	1.66±0.01	[0.10,0.16]
B0656+14	2.8±0.3	....	....	....	....
B0833–45	1.70±0.02	2-4 GeV	0834–4511	1.69±0.01	[0.12,0.24]
B1046–58	....	....	1048–5840	1.97±0.09	[0.00,0.18]
B1055–52	1.73±0.16	....	1058–5234	1.94±0.10	[0.00,0.25]
B1706–44	1.27±0.09	$\gtrsim$ 1 GeV	1710–4439	1.86±0.04	[0.00,0.27]
B1951+32	1.74±0.11	> 30 GeV	....	....	....
J1105–6107	....	....	1102–6103	2.47±0.21	[0.00,0.90]
B1758–23	....	....	1800–2338	2.10±0.10	[0.00,0.32]
B1823–13	....	....	1823–1314	2.69±0.19	[0.40,1.37]
B1853+01	....	....	1856+0114	1.93±0.10	[0.50,1.51]
B1900+05	....	....	1903+0550	2.38±0.17	[0.18,0.60]

models accelerate the primary charges through an electric field component parallel to the magnetic field ($\vec{E} \cdot \vec{B} \neq 0$), generally considered to be a small fraction of the electrostatic potential induced in the vacuum solution. In both models acceleration of the primary pairs is followed by electromagnetic cascades with copious γ-ray production.

In the polar cap model magnetic photon processes are important and electromagnetic cascades usually proceed either through curvature radiation and magnetic pair production (Daugherty & Harding 1982) or through inverse-Compton scattering of thermal photons emitted by the neutron star and magnetic pair production(Dermer & Sturner 1994). Absorption processes are important near the neutron star, where B can be close to $B_{cr} = m^2c^3/e\hbar \approx 4.414 \times 10^{13}$ G, specially in high magnetic field pulsars, like PSR B1509–58. In outer gap models $e^{\pm}$ pairs are created and accelerated in gaps located near the free charge regions of the outer magnetosphere. Electromagnetic cascades, through direct pair production with X-ray (keV) photons from the star and further curvature

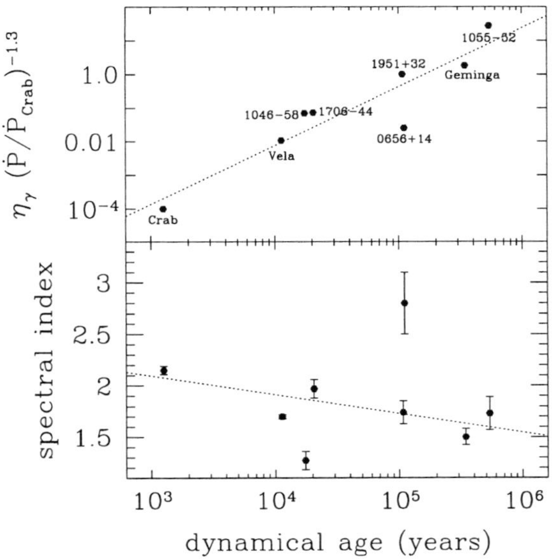

Figure 4 Efficiency multiplied by $\dot{P}^{-1.3}$ (*top*) and spectral index (*bottom*) versus dynamical age, $\tau = P/2\dot{P}$, for the known gamma-ray pulsars. The efficiency is calculated as $\eta_\gamma = f(E_\gamma \geq 100 \text{ MeV})d^2/\dot{E}_{rot}$ using a moment of inertia of 10^{45} g cm^2. The spectral index is from the pulsed emission and differs from that of the Third EGRET Catalog source. The fitting lines correspond to $\eta_\gamma \propto \tau^{0.8}$ and $\eta_\gamma \propto \tau^{-0.18}$ respectively.

emission, follow. Magnetic attenuation processes do not play a significant role in these models. The geometrical aspect is more favorable: as Romani (1996) shows, the efficiency is optimal for inclination angles $\gtrsim 65°$. The typical phase averaged spectrum for a young pulsar shows a strong curvature radiation component dominating from ~ 30 MeV to ~ 10 GeV together with a smaller synchrotron component peaking at $\lesssim 10$ MeV and decaying afterwards. An inverse Compton component in the TeV range is predicted by the model but still remains to be observed.

3.2. UNIDENTIFIED GALACTIC SOURCES

The main evidence that the majority of the unidentified EGRET sources are Galactic is their spatial distribution. Using the data provided in the First EGRET catalog, Mukherjee, Bertsch & Dingus (1995) showed that the unidentified sources are distributed according to a Galactic population, included those with $|b| > 10°$. Gehrels et al. (2000) used the 3EG Catalog to establish that the spatial distribution of steady sources in the 3EG Catalog has clear evidence of two distinct types of objects (Figure 5): high γ-ray flux sources at low-latitudes ($|b| < 5°$) and faint sources at high Galactic latitudes ($|b| > 5°$). The result found by Gehrels et al. (2000) for steady sources is robust and only two bright steady EGRET sources have $|b| > 10°$: 3EG J0010+7309 and 3EG J1835+5918 (see §4.1 and §4.2). The superposition of flux and distance makes the bright low-latitude sources a factor of 75 more

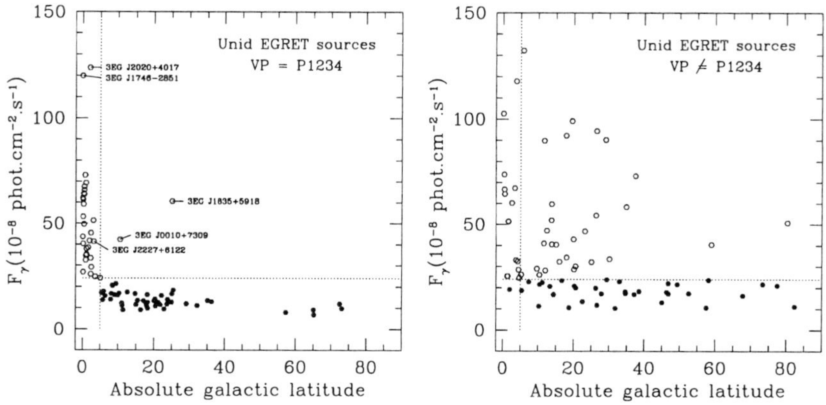

Figure 5 Photon flux *versus* latitude for: *left* 80 unidentified EGRET sources detected in *CGRO* periods 1 to 4; *right:* 82 unidentified EGRET sources for which the most significant detection is not from *CGRO* periods 1 to 4 added. Open circles denote $F > 24.0 \times 10^{-8}$ phot cm^{-2} s^{-1} while filled circles denote sources with fluxes below the same value, indicated by the dotted horizontal line. The dotted vertical line corresponds to $|b| = 5°$. Sources discussed in the text are indicated. The correlation between photon flux and Galactic latitude is observed only for steady (P1234)sources.

luminous than the dimm high-latitude ones. γ-ray pulsars come as an extension of the bright sources to higher fluxes, suggesting the presence of Geminga-like pulsars in the low-latitude sources. Based on the outer gap model, Romani & Yadigaroglu (1995) and Yadigaroglu & Romani (1995; 1997) predicted about 22 pulsars detectable by EGRET within its 1st Catalog. It is clear that not all Geminga-like pulsars are going to be nearby. In fact Geminga might just be the closest of its kind. Radio-quiet pulsars within the low-latitude population are likely to be more energetic, and probably younger, than Geminga.

High-latitude sources are subject of on-going studies (Grenier 2000; also in these proceedings). Gehrels et al. (2000) pointed out the spatial correlation with the Gould belt and proposed molecular clouds, SNR, massive stars and pulsars as possible counterparts. Inspired by the reported EGRET detection of the millisecond pulsar PSR J0218+4232 (Kuiper et al. 2000), Grenier (2000) proposed millisecond pulsars, together with accreting black holes, as likely sources, leading to the idea that the different spatial distributions reflect differences in ages between the types of sources. The presence of Geminga-like pulsars among the γ-ray sources off the Galactic plane is an open question.

Gehrels et al. (2000) defined as *steady* sources those for which the most significant EGRET detection is given by the entire phase 1 to 4

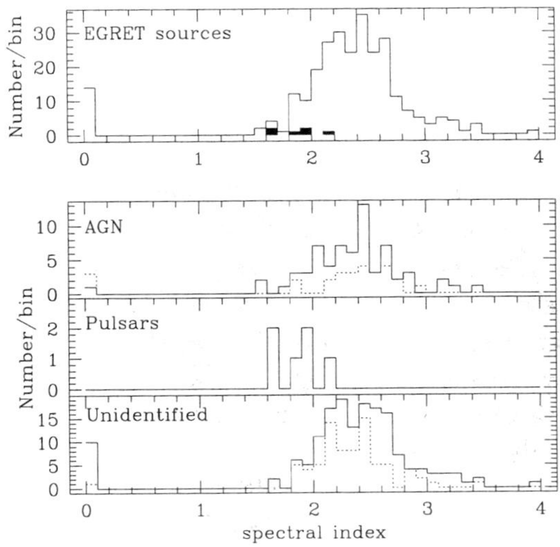

Figure 6 Distribution of spectral indices for EGRET sources, AGN (weak identifications in dotted line), pulsars and unidentified sources (steady sources in dotted line). The distribution of indices for pulsars is also shown in black in the *top* figure. They have distinctively lower indices. The sources in the bin of null spectral index are those too weak for a spectral fit.

dataset (P1243 in the VP catalog entry). This is a simple criterion which overcomes most of the difficulties inherent to estimate the time variability of EGRET sources and I will use the term *steady* in the same sense. An analytical criterion to assess whether a source is variable or not was proposed by McLaughlin et al. (1996) who used a χ^2 statistics on variance-weighted photon fluxes and special handling of upper limits to calculate a variability parameter V for the sources within the 2EG Catalog. This method is relatively easy to implemented and can be adapted to the entries in the 3EG Catalog. This method come is mainly qualitative: a low (large) V indicates that the flux values are (are not) consistent with a constant flux but V does not quantify the degree of variability. Thompkins (1999) introduced a τ parameter which does quantify source variability from the EGRET data, i.e. $\tau = 0.5$ indicates a source with flux variations of half of its mean value. An important feature of τ is that it is derived from the EGRET data and therefore incorporates most of the peculiarities of those data. Thompkins (1999) estimated τ values and, probably more important, their 68% confidence bands for 3EG Catalog sources. From these values Thompkins (1999) identified low-variability sources, those for which the 68% upper limit of τ is below 0.30, and high-variability sources, for which the 68% lower limit of τ is above 1.0. Given a 10% systematic error in the EGRET data, all 68% confidence lower limits on τ for pulsars in the 3EG Catalog are consistent with a constant flux. For Vela the 68% lower limit of τ is 0.12, barely indicative of any variability, specially considering the artifacts in the data caused by such a strong source.

Spectral indices, together with flux stability and spectral breaks, were previously used by Merck et al. (1996) to select pulsar candidates among $|b| < 10°$ sources of the 2nd EGRET Catalog. Following these authors, one can identify pulsar candidates as those having spectral indices below 2.2 in the 3EG Catalog (Table 3). Figure 6 shows the comparison of spectral indices of γ-ray pulsars with all EGRET sources, those associated to AGNs and unidentified sources, using for the pulsars the index corresponding to the 3EG Catalog entry. Pulsars have a clear tendency towards lower indices. The 68% confidence limits for τ are included in Table 3, indicating variability through a more conservative threshold (that the 68% confidence lower limit of τ be ≤ 0.20, considering systematics make for at least 10%). I considered sources with VP=P1234 or τ compatible with ≤ 0.20. Note that theoretical models for γ-ray pulsar emission *can* accommodate variability.

4. NOTORIOUS GEMINGA CANDIDATES

Several unidentified sources from the 3EG Catalog have deserved individual attention, with examples of sources tentatively related to extragalactic objects (Bloom et al. 1997) or Galactic microquasars (Paredes et al. 2000). Relevant to the search of Geminga-like pulsars are those studies indicating X-ray point sources with no obvious optical counterpart, as expected for a pulsar. At least four EGRET sources with γ-ray Geminga-like properties have been found to have such counterparts, pointing to the presence of a radio-quiet pulsar: 3EG J0010+7309, 3EG J1835+5918, 3EG J2020+4017 and 3EG J2227+6122. These sources are briefly reviewed in the next sections.

4.1. 3EG J0010+7307

The first entry in the Third EGRET catalog, this is one of the rare steady unidentified sources with strong photon flux and $|b| \geq 10°$. It has a pulsar like spectral index of 1.85 ± 0.10 from the 3EG Catalog or 1.58 ± 0.18 according to Brazier et al. (1998). Although McLaughlin et al. (1996) put it just above the variability threshold ($V = 1.28$ where $V = 1.0$ defines the threshold between variable and non variable sources), Brazier et al. (1998) estimated $V = 0.55$ based on phase 1-4 data. Thompkins (1999) give $\tau = 0.31$ with a 68% confidence lower limit of 0.16. Given the systematics inherent to the EGRET data, this values are only marginally indicative of low level variability. At present there is no compelling evidence that this source is variable.

The original association with QSO B0016+731 was found to be inconsistent with the refinements of the EGRET location made by Brazier et

Table 3 Steady (P1234) or possibly non-variable (LowLim(τ) $\leq$ 0.20) unidentified EGRET sources with spectral indices below 2.0 (*upper panel*) and 2.2. Note spectral index selections do not separate strong/low-latitude and weak/mid-latitude sources.

Jname (1)	l , b (°)	F_γ (2)	spectral index	τ (3)	Tent. assoc.
0010+7309	119.9, 10.5	42.3	1.85±0.10	[0.16,0.55]	CTA-1
0613+4201	171.3, 11.4	9.0	1.92±0.26	$[0.68,1.24]^{v}$	
1027–5817	284.9, -0.5	65.9	1.94±0.09	[0.09,0.45]	
1337+5029	105.4, 65.0	9.2	1.83±0.29	[0.00,1.35]	
1746–2851	0.1, -0.0	119.9	1.70±0.07	$[0.36,0.69]^{v}$	Gal Cen
1835+5918	88.7, 25.1	60.6	1.69±0.07	[0.00,0.32]	ns-id
1837–0606	25.9, 0.4	49.6	1.82±0.14	[0.00,0.49]	
1856+0114	34.6, -0.5	67.5	1.93±0.10	$[0.50,1.51]^{v}$	Psr/W44
1958+2909	66.2, -0.2	26.9	1.85±0.20	[0.15,0.98]	
2021+3716	75.6, 0.3	59.1	1.86±0.10	[0.11,0.53]	
2033+4118	80.3, 0.7	73.0	1.96±0.10	[0.00,0.37]	
0348+3510	159.1,-15.0	11.5	2.16±0.27	[0.00-1.41]	
0426+1333	182.0,-23.8	14.0	2.17±0.25	[0.00-0.71]	
0510+5545	154.0, 9.4	21.3	2.19±0.20	[0.00-0.26]	
0616–3310	240.4,-21.2	12.6	2.11±0.24	[0.00-0.75]	
0617+2238	189.0, 3.0	51.4	2.01±0.06	[0.15-0.38]	IC 443
0910+6556*	148.3, 38.6	18.3	2.20±0.26	[0.00,1.14]	
1234–1318*	296.4, 49.3	21.6	2.09±0.24	[0.12,0.81]	
1249–8330*	302.9,-20.6	19.9	2.14±0.27	$[0.18,2.74]^{lkv}$	
1410–6147	312.2, -0.4	64.2	2.12±0.14	[0.16,0.55]	G312-0.4
1631–1018	5.6, 24.9	12.7	2.20±0.27	[0.00,0.89]	
1634–1434	2.3, 21.8	11.5	2.15±0.23	[0.00,0.55]	
1655–4554	340.5, -1.6	38.5	2.19±0.24	$[0.46,2.27]^{v}$	
1719–0430	17.8, 18.2	16.2	2.20±0.24	[0.00,0.39]	
1744–3011	358.8, -0.5	63.9	2.17±0.08	$[0.20,0.62]^{lkv}$	
1800–2338	6.2, -0.2	61.3	2.10±0.10	[0.00,0.32]	Psr/W28
1809–2328	7.5, -2.0	41.7	2.06±0.08	$[0.49,1.02]^{v}$	
1824–1514	16.4, -1.2	35.2	2.19±0.18	[0.00,0.51]	
1836–4933	345.9,-18.3	9.8	2.14±0.35	[0.00,0.64]	
2016+3657	74.8, 1.0	34.7	2.09±0.11	[0.08,0.75]	
2020+4017	78.0, 2.1	123.7	2.08±0.04	[0.00,0.18]	γ-Cygni
2035+4441	83.2, 2.5	29.2	2.08±0.26	[0.00,0.42]	W63
2248+1745	86.0,-36.2	12.9	2.11±0.39	$[0.43,3.98]^{v}$	

(1): * indicates VP$\neq$P1234 but 68% confidence LowLim(τ) $\leq$ 0.20.

(2): in units of 10^{-8} cm^{-2} s^{-1}.

(3): ^{v} = variable; ^{lkv} = likely variable

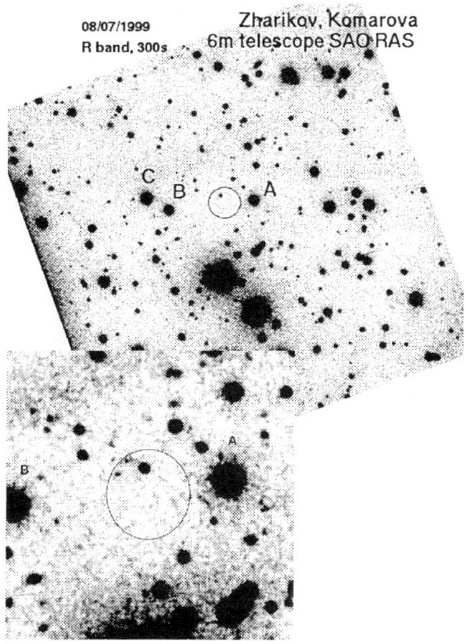

Figure 7 R image of the error box of RX J0007.0+7302 taken with the SAO 6 m telescope. The insert figure shows an 10" circle representing the *ROSAT* error box.

al. (1998). The γ-ray source is coincident with the supernova remnant G119.5+10.2 (CTA 1), of kinematical distance 1.4±0.3 kpc. *ROSAT* and *ASCA* observations (Slane et al. 1997) show the extended X-ray emission with a X-ray point source located in the central brighter part of the remnant, RX J0007.0+7302. The supportive *ASCA* data indicate that the central source is non-thermal and point to this object as a likely pulsar counterpart to 3EG J0010+7307. With the estimated distance of 1.4 kpc, the inferred luminosities $f_X d^2 \approx 3 \times 10^{30}$ erg s^{-1} and $f_\gamma d^2 \approx 4 \times 10^{33}$ erg s^{-1}. Slane et al. (1997) find the *ASCA* data consistent with a pulsar of period of about 0.1-0.2 seconds and $\dot{E}_{\rm rot} \sim 10^{36}$ erg s^{-1}. Present X-ray observations limit the pulsed fraction of X-rays to 75%.

Optical observations of RX J0007.0+7302 have been performed with the 2m telescope of Cananea (Brazier et al. 1998) and the 6m telescope of the Special Astrophysical Observatory (Carramiñana et al. 2000). Apart from a red object of magnitude $R \sim 22 - 23$, no likely counterpart is found to a limiting magnitude of ~ 25. Although an identification of the coincident object is still pending, its colours lend to believe that it is not related to the X-ray and γ-ray sources, supporting the pulsar hypothesis. Radio upper limits are at the level of 1 mJy at 370 MHz (Nice & Sayer 1997) implying $S_{400}d^2 \leq 2$, already in the "radio-quiet range" of Figure 1. Further multiwavelength studies of this object can provide valuable information about the X-ray source, its positional counterpart and the presence of pulsations.

4.2. 3EG J1835+5918

The most prominent outsider in the flux *versus* Galactic latitude relation (Fig. 5), this is the only steady, bright, high-latitude ($b = 25°$) unidentified γ-ray source in the sky. It also has the peculiarity of having the hardest spectral index within the 3EG Catalog: 1.69 ± 0.07. Using photons of energies above 1 GeV, Reimer et al. (2000) narrowed the uncertainty in position to ~ 8 arcminutes, allowing the error box to be covered by a single *ROSAT*-HRI pointing. While McLaughlin et al. (1996) estimated a variability index $V = 1.3$, marginally indicative of variability, Thompkins (1999) estimated a 68% confidence lower limit on τ of 0.16, consistent with a weakly variable source. Note that Thompkins (1999) consider this degree of variability as possible in γ-ray pulsars. A careful analysis, considering the statistical and systematic uncertainties the EGRET data, was made by Reimer et al. (2001) who showed the data to be compatible with a non-variable source. Taken together with the hard spectral index and the evidence of a break above 4 GeV, the evidence points to a Geminga-like pulsar, rather than a blazar. Also, the high ratio F_γ/F_{radio} practically rules out a blazar origin.

The *ROSAT*-HRI X-ray image of the 3EG J1835+5918 indicates the presence of ten point sources within the γ-ray error box. From respective optical follow-up studies, both Reimer et al. (2001) and Mirabal et al. (2000), who also included supportive *ASCA* data, point to RX J1836.2+5925, which has no optical counterpart was found down to a limiting magnitude of $V \gtrsim 23 - 25$, as the most intriguing and promising candidate counterpart of the EGRET source.

The high Galactic latitude of 3EG J1835+5918 practically bounds the distance to $\lesssim 1$ kpc. As the X-ray flux is rather low, a distance smaller than that of Geminga would indicate a rather old neutron star and, according to the correlation shown in Figure 4, its efficiency should be rather high. According to theoretical models, after the efficiency has reached a maximum level γ-ray emission ceases, probably at an age of the order of 10^6 years. Mirabal & Halpern (2001) found the *ROSAT* source to be ultrasoft source, with most of its estimated X-ray flux (3×10^{-13} erg cm^{-2} s^{-1}, without considering absorption) below $E <$ 0.3 keV. The interpretation as black-body emission is direct and provides

Table 4 Two fictitious set of plausible parameters for a Geminga-like pulsar associated to 3EG J1835+5918. Input assumption are a dynamical age of $P/2\dot{P} = 10^6$ years and $L_\gamma = 5 \times 10^{33}$ $(d/kpc)^2$ erg s^{-1} with 1 sr solid angle beam.

	Fictitious 3EG J1835+5918		Geminga
	Far (d = 1 kpc)	Nearby (d = 250 pc)	Pulsar
P (s)	0.009	0.475	0.237
$\dot{P}$	1.4×10^{-16}	7.5×10^{-15}	1.1×10^{-14}
$\dot{E}_{\rm rot}$ (erg s^{-1})	7.6×10^{36}	2.8×10^{33}	3.3×10^{34}
L_γ (erg s^{-1})	5.0×10^{33}	3.1×10^{32}	4.9×10^{32}
η_γ	6.5×10^{-4}	0.11	0.015
$B_{\rm dip}$ (G)	3.6×10^{10}	1.9×10^{12}	1.6×10^{12}

further evidence of a neutron star as the counterpart of the EGRET source. The black-body temperature is $\leq 5\times10^5$ °K which, together with an emitting area $A \leq 4\pi R_*^2$, would set the distance to the object to $d \approx$ 1.1 kpc $(A/4\pi R_*^2)^{1/2}$ $(T/5 \times 10^5$ °K$)^2$. This is probably an upper limit, as the emitting surface might be smaller, the star cooler and interstellar absorption has not been taken into account. The upper-limit on radio emission, 0.5 mJy at 370 MHz (Nice & Sayer 1997), would put the radio luminosity at $S_{400}d^2 < 0.5$ mJy kpc^2, in the "radio-quiet" range. Neutron star cooling models indicate that this temperature corresponds to an age of $\gtrsim 3 \times 10^5$ years (Page & Applegate 1992). This relatively old age is also consistent with the γ-ray spectral index (Fig. 4).

Pushing somewhat further the arguments, I constructed two possible set of parameters for a Geminga pulsar scenario, assuming the pulsar follows the $\eta_\gamma \dot{P}^{-1.3}$ fit of Figure 4 with an age of 10^6 years. If the assumed pulsar is located at 1 kpc then it should be much faster than the Crab, highly energetic and with a low magnetic field. In fact such a pulsar would look more like a ms-pulsars than a Crab-like or Vela-like pulsar. If the distance is assumed to be smaller, say 250 pc, then the pulsar would look like an older version of Geminga, with similar period but higher efficiency, due to the age. According to present theoretical ideas, such a pulsar would be in its last phases as γ-ray emitter This is a more conservative scenario than that of a ms-pulsar, albeit the reported EGRET detection of PSR J0218+4232 (Kuiper et al. 2000).

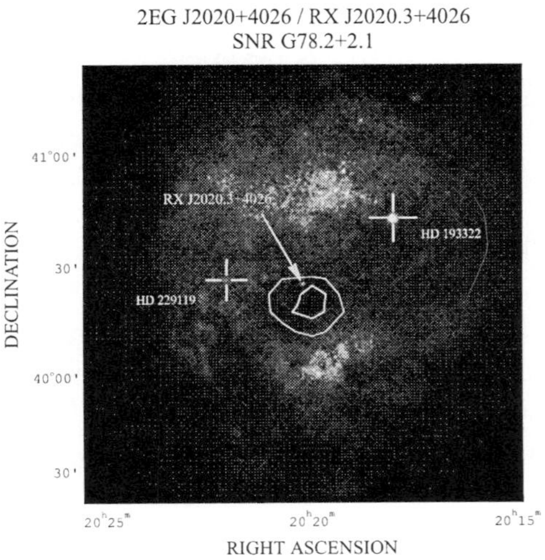

Figure 8 *ROSAT*-PSPC $2^\circ \times 2^\circ$ image of the γ-Cygni SNR. The EGRET 95% error box, made from the $E > 1$ GeV data, is shown with the X-ray proposed counterpart RX J2020.2+4026 clearly visible. The extended SNR emission is also evident.

4.3. 3EG J2020+4017

This is the brightest of the steady unidentified EGRET sources and possesses the basic properties of a Geminga-like γ-ray pulsar. The source was first discovered by the *COS-B* satellite, appearing in the Second *COS-B* catalog as 2CG 078+2 (Swanenburg et al. 1981). It is also included in the GeV catalog as GEV J2020+4023. It has a pulsar-like spectral index (2.08 ± 0.04) and Merck et al. (1996) showed evidence of a spectral break at about 4 GeV. Thompkins (1999) calculated $\tau = 0.07$, with a 68% confidence *upper* limit of 0.18, and explicitly quoted this as a low variability source. Given the systematic 10% of EGRET data, the emission is up to now consistent with a constant flux source.

As noted by Sturner & Dermer (1995), the EGRET source is located inside the SNR G78.2+2.1, also denoted as γ-Cygni because of its proximity to the 2nd magnitude star of the same name. Sturner & Dermer (1995) estimated an age for the remnant of 5400 yr. The *ROSAT* PSPC X-ray image of Brazier et al. (1996) (Fig. 8) clearly shows the extended diffuse emission of the SNR, with a single point source inside the 95% EGRET contour, RX J2020.2+4026. Assuming a power-law of index -2, the X-ray energy flux is about $\sim 4 \times 10^{-14}$ eg cm^{-2} s^{-1}. The detection of the source in the 0.5-0.9 keV range indicates a low column density.

Palomar Observatory Sky-Survey plates show a relatively bright star ($V \sim 14 - 15$) inside the *ROSAT* error box. This was spectroscopically identified as a K0V, making its association with the X-ray source unlikely. Furthermore, the distance of the star, assuming it to be in the main-sequence, is about 600 pc and does not match up the kinematical distance to the SNR (1.5 kpc). Because if its properties, the direct identification of the star with the γ-ray source is practically discarded. Still, a scenario of a Geminga pulsar around a normal star could be considered and in September 1997, the star was observed with higher spectral resolution (0.48 Å/pixel) for a week in search for orbital motions. No velocity change was found with a 3σ upper limit of $\sim$ 100 km/s in timescales of a few days. The star impairs setting limits fainter than $\sim 20^{\rm th}$ magnitude on other objects.

The definitive proof that 3EG J2020+4017 is a Geminga-like pulsar would be the detection of pulsations in the γ-ray data. There are enough γ-ray photons to make a significant light-curve if the pulsed fraction related to source + background photons is above $\sim$ 25%, but not for a direct period search (§5.2.1). The period (if there is such) needs to be found in other data, either high count rate X-rays or with future generation γ-ray satellites like *GLAST*.

4.4. 3EG J2227+6122

3EG J2227+6122 is an EGRET source with a soft Crab-like spectrum, of index 2.24±0.14, and no indication of a high energy break (Merck et al. 1996). The EGRET data are consistent with no variability: Thompkins (1999) calculated τ = 0.10, with a 68% lower limit equal to zero. The γ-ray source appears also in the 1st COMPTEL Catalogue (Schönfelder et al. 2000) and, with very low significance, in the GeV Catalog of Lamb & Macomb (1997).

Halpern et al. (2001a) performed a *ROSAT*-HRI study of the γ-ray error box, finding six possible X-ray counterparts and an extended diffuse X-ray flux. Using VLA they also found a radio nebula with the morphology of an incomplete shell. From the lack of an optical identification, they argued that the X-ray source RX J2229.0+6114 could be a rotation powered pulsar feeding electrons to an associated synchrotron nebula. More recently, Halpern et al. (2001b) performed *Chandra* and Jodrell Bank observations of the X-ray source, confirming the presence of the X-ray nebula and discovering a 51.6 ms radio pulsar, PSR J2229+6114. The radio emission is very weak: $\sim$ 0.25 mJy at 1412 MHz. Together with the assumed distance of 3 kpc, one gets $S_{1400}d^2 \sim 2.2$ mJy kpc, on the weak side of what is observed for radio pulsars (Fig 1; note the non detection at 400 MHz). Pulsations are also present in the *ASCA* data, with two peaks separated by $\sim$ 0.5 in phase in the light curve.

The evidence for an association between the pulsar-synchrotron nebula system and the EGRET source is very compelling, albeit the issue of whether the γ-ray source is the pulsar or the nebula. The identification of PSR J2229+6114 with 3EG J2227+6122 is pending, as the pulsations are not evident in single EGRET observations and, as Halpern et al. (2001b) pointed out, the present ephemerides might not be stable enough to coherently add all the available γ-ray data. Still, exploring the timing parameter space in the data can be a rewarding exercise.

4.5. RELATED OBJECTS

There are other sources which might be related to radio-quiet pulsars, although the existence of a plerionic environment opens the alternative of cosmic-ray gas interactions has the potential source of the γ-ray emission. X-ray synchrotron nebulae associated to large scale radio emission have been detected in 2EG J1811–2339 (3EG J1809–2328; Oka et al. 1999), and in 2EGS J1418–6049 (3EG J1420–6038; Roberts & Romani 1998; Roberts et al. 1999), as part of the *ASCA* survey of γ-ray sources (Roberts, Romani & Kawaii 2001). Note that both objects have 68% confidence lower limits on τ of $\sim$ 0.50 and are almost certainly variable. Roberts et al. (in these proceedings) present X-ray and radio observations of GeV sources, most of them found to be related to X-ray remnants. They quote three GeV sources as probable pulsars: GEV J2020+4023 (i.e. γ-Cygni), GEV 1835+5921 (i.e. 3EG J1835+5918) and GEV J1837–0610. Other GeV sources appear to be indirectly related to pulsars, through the synchrotron nebulae.

An interesting related source is 3EG J0634+0521, associated to a 34 ms X-ray pulsar around a Be star (Kaaret et al. 1999; Cusumano et al. 2000). This source seems to be similar to 2CG 135+01, tentatively associated to LSI +61°303 (Strickman et al. 1998), with the open possibility that both might be accretion powered. An alternative interpretation, supported by the orbital motion of the 34-ms pulsar, is that the system is physically similar to the binary pulsar PSR J1259–63, with a radio-quiet pulsar of just 1400 years (Kaaret et al. 2000). The identification of pulsations in the γ-ray data has not been made and the case is still open. Note that the EGRET source is highly variable: $\tau = 72.05$ with a 68% confidence lower limit of 5.15. If, as suggested by the variability, the source is not a rotation powered pulsar finding a suitable timing solution for the EGRET data might not be possible. Kaaret (in these proceedings) discuss in more detail this interesting object.

The evidence for a Geminga pulsar in these systems is not as strong as in the four cases discussed previously. A blind systematic search for pulsations on all EGRET sources detected during phases 1 and 2 EGRET data was performed by Mattox et al. (1996) with null results, pointing to Geminga as the only Geminga-like pulsar with a pulse strong enough to be detected in the EGRET data.

5. PERSPECTIVES

The major advance in our understanding of Geminga pulsars will come with the next generation of γ-ray telescopes, mainly *GLAST*. Still, some EGRET sources with similar properties are known and their identification as Geminga pulsars (or else) might well arrive before *GLAST* goes into orbit. Furthermore, multiwavelength astronomy is likely to provide fresh insight into more Geminga candidates in the next years. I briefly mention the perspectives before *GLAST* and with *GLAST*.

5.1. PRESENT MULTIWAVELENGTH

The main strategy to seek the identification of Geminga pulsars is well set: after selecting suitable γ-ray candidates one performs X-ray imaging of EGRET error boxes and turns to optical identification of X-ray point sources. If the candidate conforms with the expected characteristics of a γ-ray pulsar one should try to find a periodicity in any set of data and confirm it in the EGRET database. Going step by step:

- **selection of γ-ray sources:** as illustrated by the different tables in this article, some tens of suitable candidates from the Third EGRET catalog are already known. These are basically the steady, low variability and hard spectrum unidentified sources. Further studies might help to refine this selections. The next step forward in the direct study of γ-ray pulsars will come with the *AGILE* γ-ray satellite (Mereghetti et al., in these proceedings). *AGILE* is a small and light instrument with sensitivity and source location comparable to EGRET, and a larger field of view but poorer energy resolution. *AGILE* will provide a much needed restart to variability studies.

- **X-ray imaging:** several X-ray telescopes, with ever-increasing capabilities, have been coming online recently. The two most renamed examples are *Chandra-AXAF* and *Newton-XMM*. The main power of *Chandra* comes in its high resolution camera with subarcsecond spatial resolution and a nominal limiting sensitivity just below 10^{-15} erg cm^{-2} s^{-1}, with maximum effective area at about 1 keV. Such combination of sensitivity and refinement in positions can reduce ambiguities in optical observations of possible counterparts of X-ray sources. The most relevant instrument in *Newton*, to our purpose, is the EPIC camera. The spatial resolution is about

6-8", somewhat better than *ROSAT*-HRI but not comparable to *Chandra*. The sensitivity limit is $\sim 3.5 \times 10^{-15}$ erg cm^{-2} s^{-1} in the 0.4-4.5 keV band. The strength of EPIC for EGRET counterparts studies resides in its timing mode. Because of its larger effective area, *Newton* has a higher count rate -although with higher background. *Chandra* can also provide data in timing mode, with lower but cleaner count rates. Although the focus is in the imaging and photon counting capacities of both X-ray telescopes, X-ray spectroscopy is important in distinguishing thermal and power-law spectra. Finally, the increasing absorption towards the Galactic plane make the use of hard X-ray telescopes also attractive, specially for young pulsars with hard-spectra power-law spectra.

- **optical observations:** isolated neutron stars are some of the most difficult objects to study in the optical. Most of the effort so far has been in the spectroscopic or colour based rejection of potential candidates. In this respect it is important for imaging to reach magnitude 24 or 25, to constraint the nature of (undetected) objects (pressumed) present. In the absence of interstellar absorption space telescopes, namely the *HST*, can be an important tool for detecting very faint objects with UV excess. Finally, future studies aimed at sources close to the Galactic plane may have to deal with reddening and be limited if not performed in the infrared.

- **periodicity searches:** searching for unknown periodicities in astronomical data is not an easy task, specially when dealing with sources where the γ-ray dominates the rest of the spectrum by factors of thousands. Given that X-ray counterparts are a prerequisite in these studies they are the most promising objects for timing studies. The lowest count rates expected may correspond to thermal emission from $10^5 - 10^6$ year old neutron stars, like what may correspond to 3EG J1835+5918 or to the high latitude faint γ-ray sources. Closer to the Galactic plane one may expect younger pulsars, still showing magnetospheric hard X-ray emission, and probably better suited for X-ray periodicity searches. Although no candidate has yet been identified, optical objects of magnitude brighter than 22nd-24th may be good candidates for period searches using fast timing photometers on large optical telescopes.

- **periodicity confirmation:** EGRET data are handy and testing ephemerides on γ-ray time series does not involve any major difficulty. Near the Galactic plane one can aim to detect pulsations for pulsars with photon fluxes comparable to those of the faintest pulsars, $\sim 10^{-7}$ cm^{-2} s^{-1}. As time goes by, timing studies will be affected by the need to include the frequency derivative, $\dot{f}$, maybe already a limiting factor to detect pulsations in 3EG J2227+6122.

5.2. FROM EGRET TO *GLAST*

In the γ-ray front, present day (EGRET) data can be used mainly to confirm pulsations found at other wavelengths. The FITS photon lists (rdf files) can be easily read and exported to perform timing studies either with the standard EGRET tools or with own software. One steep further in the future, the *AGILE* γ-ray satellite is expected to be operational during 2003-2006 (Mereghetti et al. these proceedings). As mentioned above *AGILE* will have sensitivity and location accuracy comparable to EGRET, but its (three times) larger field of view together with improved off-axis sensitivity are bound to result in larger exposures for each source. *AGILE* will certainly be a valuable tool for γ-ray astronomers, ensuring coverage of the γ-ray sky until new telescopes go into orbit. The most promising γ-ray telescope in the forthcoming years is *GLAST*, the *Gamma-ray Large Area Space Telescope*, described in Gehrels (2001, these proceedings). *GLAST* will have a peak effective area $\gtrsim 8000$ cm^2 or larger, i.e. more than seven times larger that of EGRET. In addition, strong background from regions in the Galactic plane will be reduced by a factor of ~ 3.6 as the single photon angular resolution will be $\sim 3.5°$ at 100 MeV, compared to $5.8°(E/100 \text{ MeV})^{-0.534}$ for EGRET. The major improvements in the study of Geminga pulsars will come from its higher and cleaner count rate, allowing direct periodicity searches. The much larger field-of-view will probably result in longer exposure times, also helpful in searching for pulsations in fainter sources.

GLAST will locate γ-ray sources with arc-minute accuracy. Still, an arc-minute error box is generally too large to allow optical follow-up observations, specially when aiming at very faint (or undetectable) objects. X-ray observations will still be needed to precisely locate potential pulsars. *GLAST* might be able to find pulsations in the data, but the ultimate task of finding the object will still require multiwavelength observations. In fact, the positional accuracy of *GLAST* might be marginally enough for direct period searches. A positional error $\delta\theta$ can introduce a timing phase shift over a time T of the order of:

$$\delta\phi \simeq 2\pi f_p \, T \, (v_\oplus/c) \, \sin\theta \, \delta\theta \,, \tag{9.1}$$

where f_p is the frequency of the pulsar, $\hat{v}_\oplus \cdot \hat{n} = (v_\oplus/c)\cos\theta$ is the orbital velocity of the Earth projected in the direction of the source, of relative angle θ. When $\vec{v}_\oplus \cdot \hat{n}$ is maximum the shift is constant and is manifested

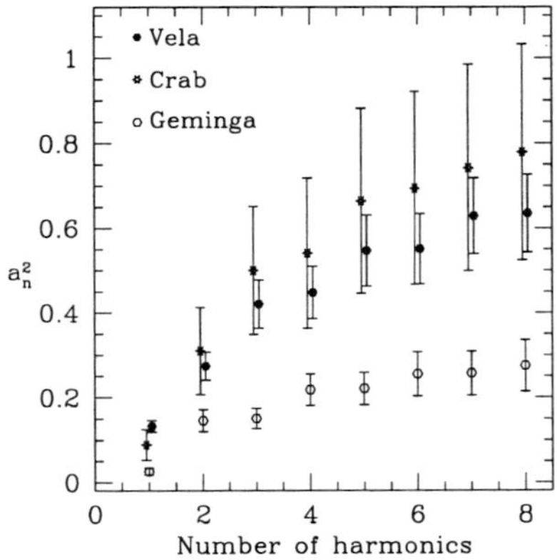

Figure 9 Inferred harmonic content of the Vela, Crab and Geminga light curves. Note that Geminga has a lower harmonic content than Crab and Vela and that it is distributed in the even numbered harmonics.

as a spurious Doppler shift in f. The main problem for period searches is when the *rate of change* of $\vec{v}_\oplus \cdot \hat{n}$ is maximum, which happens when it changes sign (i.e. around $\vec{v}_\oplus \cdot \hat{n} = 0$). In such a case the expected error in phase is about $\varepsilon(\phi) \simeq 2 \times 10^{-3}\ (f_p/1\ \mathrm{Hz})\ (T/1\ \mathrm{week})^2\ \cos\theta\ (\delta\theta/1')$, or about 0.6 for $f_p = 30$ Hz and $T = 3$ weeks. Still, the data are likely to have a count rate which will permit either searches on one week time intervals or with high enough signal-to-noise ratio to show even with a bad position. In the worst of the cases timing analysis can be repeated after multiwavelength observations have provided a more accurate position.

5.2.1 Direct period searches. It is worth considering what can be expected for direct period searches in *GLAST* data. Although the $\mathcal{H}$ test is the most sensitive of common periodicity tests (de Jager, Raubenheimer & Swanepoel 1989), one usually wants to implement a more practical approach which is to fix or limit the number of harmonics and directly estimate the Z_n^2 statistic (Buccheri & Sacco 1985). From the known light curves of Vela, Crab and Geminga (Fig 9) Z_4^2 or Z_6^2 might be the two best compromises between sensitivity and a practical search.

As a study case, the Z_4^2 test applied on a random sample produces numbers distributed according a χ^2 function with 8 degrees of freedom, which has the cumulative distribution:

$$P(\chi^2 \geq w) = \left\{1 + \frac{w}{2} + \frac{1}{2!}\left(\frac{w}{2}\right)^2 + \frac{1}{3!}\left(\frac{w}{2}\right)^3\right\} e^{-w/2}. \qquad (9.2)$$

In the presence of N_p pulsed events out of N total (pulsed + background) events the Z_n^2 statistic has a different expectation value than with pure noise:

$$\left\langle Z_n^2 \right\rangle = 2a_n^2\, N_p^2/N + 2n(1 - a_n N_p/N)\ , \qquad (9.3)$$

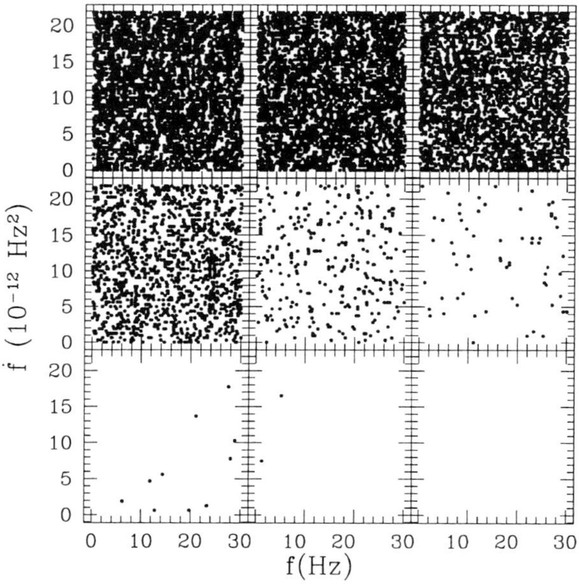

Figure 10 Z_4^2 periodicity search for 3EG J2020+4017, observation 203. The data consisted of 1129 photon events with $E > 200$ MeV spanning 20.75 days. The nine frames correspond to points with Z_4^2 values in the following ranges: *(top-left)* 32-36; *(top-middle)* 36-40; *(top-right)* 40-44; *(mid-left)* 44-48; *(mid-middle)* 48-52; *(mid-right)* 52-56; *(bottom-left)* 56-60; *(bottom-middle)* 60-64; *(bottom-right)* ≥ 64. An onverall 1% random probability event requires $Z_4^2 \geq 71.35$ and would appear in the *bottom-right* (empty) frame.

where the right-hand side term is expected for pure random data and a_n^2 is the added power of the n first harmonic components of the light curve ("pulsed fraction" up to that harmonic). Fig 9 shows the values for Vela, Crab and Geminga, estimated as $a_n^2 = Z_n^2\,(N/2N_\gamma^2)$ with N the number of events within the EGRET single photon cone of the pulsar position and N_γ the number of photons quoted in the 3EG Catalog.

In order to detect a periodic signal hidden in the data, it needs to be strong enough to satisfy $P\,(\geq w) \leq \varepsilon/3n_t$, where n_t is the number of independent frequencies tested, ε is a probability detection threshold typically set to 1%. The factor of 3 takes into account tries between independent frequencies, which should be introduced to boost the sensitivity of the search. A datafile of duration T has a set of $f_k = k/T$ independent frequencies (k an integer) and $\dot{f}_q = q/T^2$ independent frequency derivatives (q also integer). If the data sample has a duration of a day or less there is no need to introduce $\dot{f}$ and the number of frequencies one needs to test to arrive to a minimum frequency $f_{\rm m}$ is $n_t = f_{\rm m}T \approx 2.6 \times 10^6\,(f_{\rm m}/30\text{ Hz})\,(T/\text{days})$. For longer stretches of data one needs to test $n_t = f_{\rm m}\dot{f}_{\rm m}T^3$ combinations of frequencies and derivatives. To reach useful values, say $f_{\rm m} \lesssim 30$ Hz and $\dot{f}_{\rm m} \lesssim 2 \times 10^{-11}$ Hz^{-2}, one needs about $n_t \approx 3.8 \times 10^9$ independent tries.

As an example I consider the case of 3EG J2020+4017. As an exercise 1129 photons ($E \geq 200$ MeV) within the standrad EGRET photon

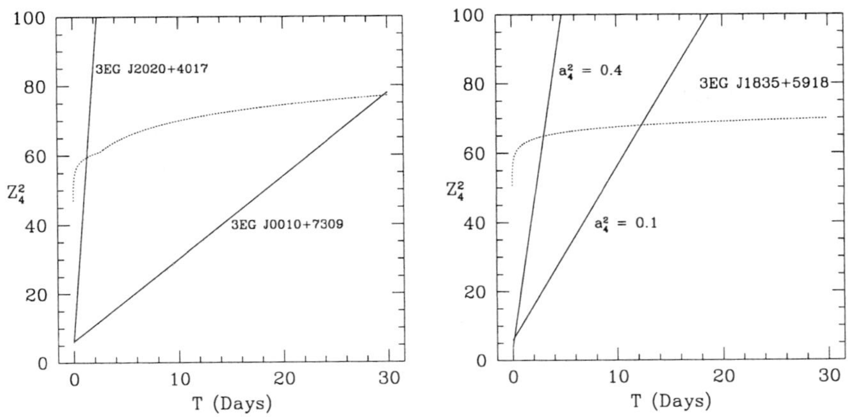

Figure 11 Periodicity searches with *GLAST*. *left:* estimated Z_4^2 versus integration time for a *GLAST* observation of 3EG J2020+4017, assuming $a_4^2 = 0.4$ and for 3EG J0010+7309 assuming $a_4^2 = 0.1$. The dotted line indicates the time dependency of the 1% overall random probability detection threshold, for a search down to $f \leq 50$ Hz and $\dot{f} \leq 2 \times 10^{-11}$ Hz^2. *right:* estimated Z_4^2 as a function of integrated time for a *GLAST* observation of 3EG J1835+5918, assuming $a_4^2 = 0.4$ and $a_4^2 = 0.1$. In this case the dotted line indicates the 1% overall random probability detection for a search down to $f \leq 250$ Hz, without considering $\dot{f}$, as it would be the case for a search of a ms-pulsar version of Geminga.

acceptance cone of the X-ray counterpart position were selected. The data span about 20.75 days and to test down to $f \gtrsim 30.13$ Hz and $|\dot{f}| \gtrsim 2.18 \times 10^{-11}$ Hz^2, I searched 3.78×10^9 independent $\{f, \dot{f}\}$ pairs, using 3 trial per Fourier interval. The search was conducted using an optimised direct Raleigh-test code on a Silicon Graphics Power Challenge computer running for 9 days. No coherent pulsations of sufficient statistical significance were found above the Z_4^2 detection threshold of 71.35, with the largest observed value been 62.2 (Fig 10). The non detection to the threshold level, set by $3n_t/\varepsilon$, indicates $a_n N_p < 288$ photons. On the other hand, the background counts in the 3EG Catalog are about 3.3 times larger than the source counts, so that in 1129 photons we expect to have $1129/4.3 \simeq 262$ source events. Given the high background in the source region the pulsed fraction of the γ-ray flux is not constrained. Only a 100% pulsed signal, all contained in the first four harmonics, had any chance of detection within these EGRET data.

Considering the scaling in effective area and single photon acceptance, one can estimate *GLAST* potential to detect pulsars among EGRET sources. Figure 11 illustrates plaussible searches for pulsars with parameters of classical young pulsars and of more rapid and older ms-pulsars. These Z_4^2 predictions are extrapolated from EGRET data, scaling the count rates by a factor of seven (effective areas) and reducing the underlying background by 0.364 (improved single photon cone) and, therefore should be close to what *GLAST* data will be like. If 100% pulsed, 3EG J2020+4017 might have been detected by EGRET in 23 days of observation. In a one month observation *GLAST* can be expected to detect pulsations if the (pulsed) fluxes are above 14×10^{-8} cm^2 s^{-1} in the Galactic plane (same background as for 3EG J2020+4017) or 6×10^{-8} cm^2 s^{-1} above the plane, as in the case of 3EG J1835+5918. Note that these values, similar to those of the faintest sources in the 3EG Catalog, refer to *pulsed* fluxes.

6. SUMMARY

The EGRET telescope helped to establish not only the nature of Geminga, but moreover pointed to the existence of a full population of radio-quiet, gamma-loud pulsars, objects that one hopes will eventually be studied primarily with γ-ray telescopes. At least four Geminga-like γ-ray candidates, with pulsar like X-ray counterparts, have been identified, one of them a radio-weak pulsar. And although *GLAST* will take upon the opportunity of studying directly and in detail radio-quiet γ-ray loud pulsars, there is a fair possibility of finding more Geminga candidates in the nearby future, including their pulses. With time Geminga will become the prototype of a long-time unique class of pulsars.

References

Bertsch, D.L., et al., 1992, Nat 357, 306

Bignami, G.F., Caraveo, P.A., & Mereghetti, S., 1993, Nat. 361, 704

Bignami, G.F., Caraveo, P.A., Lamb, R.C., 1983, ApJ 272, L9

Bloom, S.D., Hartman, R.C., Teräsranta, H., Tornikoski, M., & Valtaoja, E., 1997, ApJ 488, L23

Brazier, K.T.S., & Johnston, S., 1999, MNRAS 305, 671

Brazier, K.T.S, Kanbach, G., Carramiñana, A., Guichard, J., & Merck, M., 1996, MNRAS 281, 1033

Brazier, K.T.S, Reimer, O. Kanbach, G., & Carramiñana, A., 1998, MNRAS 295, 819

Buccheri, R. & Sacco, B. 1985, in *Data Analysis in Astronomy*, eds. Scarsi, Di Gesú, Crane, Levialdi, Plenum Press. p. 15

Burderi, L., Fauci, F., & Boriakoff, V., 1999, ApJ 512, L59

Caraveo, P.A., Bignami, G.F., Mignami, R., Taff, L.G., 1996, ApJ 461, L91

Caraveo, P.A., Bignami, G.F., Trümper, J.E., 1996, A&A Rev 7, 209

Carramiñana, A., Chavushyan, V., Zharikov, S., Reimer, O. & Brazier, K.T.S., 2000, AIP Conf Proc 510, 495

Cheng K.S., Ho C. & Ruderman M. 1986, ApJ 300, 500
Cusumano, G., Maccarone, M.C., Nicastro, L., Sacco, B., Kaaret, P., 2000, ApJ 528, L25
Daugherty, J.K. & Harding A.J. 1982, ApJ 252, 337
de Jager, O.C., Raubenheimer, B.C, & Swanepoel, J.W.H., 1989, A&A 221, 180
Dermer C.D. & Sturner S.J. 1994, ApJ 420, L75
Gehrels, N., Macomb, D.J., Berstch, D.L., Thompson, D.J., Hartman, R.C., 2000, Nat 404, 363
Grenier, I.A., 2000, A&A 364, L93
Halpern, J.P., Holt, S.S., 1992, Nat 357, 222
Halpern, J.P. & Tytler, D., 1988, ApJ 330, 201
Halpern, J.P., Gotthelf, E.V., Leighly, K.M. & Helfand, D.J., 2001, ApJ 547, 323
Halpern, J.P., et al., 2001, ApJ 552, L125
Hartman, R.C., et al., 1999, ApJS 123, 79
Hermsen, W., et al., 1992, IAU Circ 5541
Kaaret, P., Piraino, S., Halpern, J., Eracleous, M., 1999, ApJ 523, 197
Kaaret, P., Cusumano, G., Sacco, B., 2000, ApJ 542, L41
Kanbach, G., et al. 1994, A&A 289, 855
Kaspi, V.M., Lackey, J.R., Mattox, J.R., Manchester, R.N., Nailes, M., & Pace, R., 2000, ApJ 528, 445
Kassim, N.E. & Lazio, T.J.W., 1999, ApJ 527, L101
Kuiper, L., et al. 1999, A&A 337, 421
Kuiper, L., et al., 1999, A&A 351, 119
Kuiper, L., et al., 2000, A&A 359, 615
Kuz'min, A.D. & Losovskii, B.Ya., 1997, Astron. Let. 23, 283
Kwok, Cheng, & Lau, 1991, ApJ 379, 653
Lamb, R.C., Macomb, D.J., 1997, ApJ 488, 872
Malofeev, V.M., & Malov, O.I., 1997, Nat 389, 697
Mattox, J.R., et al., 1992, ApJ 401, L23
Mattox, J.R., Koh, D.T., Lamb, R.C., Macomb, D.J., Prince, T.A., Ray, P.S., 1996, A&A Suppl Ser 120, 95
McLaughlin, M.A., Mattox, J.R., Cordes, J.M., Thompson, D.J., 1996, ApJ 473, 763
McLaughlin, M.A., Cordes, J.M., Hankins, T.H., Moffett, D.A., 1999, ApJ 512, 929
Merck, M., et al., 1996, A&AS 120, 465
Mirabal, N., Halpern, J.P., Eracleous, M., Becker, R.H., 2000, ApJ 541, 180
Mirabal, N., & Halpern, J.P., 2001, ApJ 547, L137
Mukherjee, R., et al., 1995, ApJ 441, L61
Nel, H.I., et al., 1996, ApJ 465, 898

Nice, D.J., Sayer, R.W., 1997, ApJ 476, 261
Nolan, P.L., et al., 1993, ApJ 409, 697
Oka, T., et al., 1999, ApJ 526, 764
Page, D., & Applegate, J.H., 1992, ApJ 394, L17
Paredes, J.M., Martí, J., Ribó, M., & Massi, M., 2000, Science 288, 2340
Ramanamurthy, P.V, et al., 1995, ApJ 447, L109
Ramanamurthy, P.V, Fichtel, C.E., Kniffen, D.A., Sreekumar, P., Thompson, D.J., 1996, ApJ 458, 755
Reimer, O., Brazier, K.T.S, Carramiñana, A., Kanbach, G., Nolan, P.L., & Thompson, D.J. 2000, AIP Conf Proc 510, 489
Reimer, O., Brazier, K.T.S, Carramiñana, A., Kanbach, G., Nolan, P.L., & Thompson, D.J., 2001, MNRAS accepted (astro-ph/0102150)
Roberts, M.S.E., & Romani, R.W., 1998, ApJ 496, 827
Roberts, M.S.E., Romani, R.W., Johnston, S., & Green, A.J., 1999, ApJ 515, 712
Roberts, M.S.E, Romani, R.W., Kawaii, N., 2001, ApJS 133, 451
Romani, R.W. 1996, ApJ 470, 469
Romani, R.W., & Yadigaroglu, I.A., 1995, ApJ 438, 314
Ruderman, M.A., & Sutherland, P.G., 1975, ApJ 196, 51
Schönfelder, V., et al., 2000, A&A Suppl. Ser. 143, 145
Seiradakis, J.H., 1992, IAU Circ 5532
Shearer, A., et al., 1998, A&A 335, L21
Slane, P., Seward, F.D., Bandiera, R., Torii, K., Tsunemi, H., 1997, ApJ 485, 221
Shitov, Yu. P, Pugachev, V.D., 1997, New Astron 3, 101
Strickman, M.S., et al., 1998, ApJ 497, 419
Sturner, S.J., & Dermer, C.D., 1995, A&A 293, L17
Sturrock, P.A., 1971, ApJ 164, 529
Swanenburg, B.N., et al., 1981, ApJ 243, L69
Taylor, J.H., Manchester, R.N., & Lyne, A.G., 1993, ApJS 88, 529
Thompkins, W., 1999, PhD thesis, Stanford University
Thompson, D.J., Fichtel, C.E., Hartman, R.C., Kniffen, D.A., Lamb, R.C., 1977, ApJ 213, 252
Yadigaroglu, I.A., & Romani, R.W., 1995, ApJ 449, 211
Yadigaroglu, I.A., & Romani, R.W., 1997, ApJ 476, 347
Zhang, L., Cheng, K.S., 1998, A&A 335, 234

X-RAY AND RADIO OBSERVATIONS OF BRIGHT GEV SOURCES

Mallory S. E. Roberts
Dept of Physics, McGill University
3600 University St., Montréal, Québec H3A 2T8, Canada
roberts@physics.mcgill.ca

Roger W. Romani
Dept. of Physics, Stanford University
Stanford, Calif. 94305, U.S.A.

Nobuyuki Kawai
Cosmic Radiation Laboratory, The Institute of Physical and Chemical Research (RIKEN)
2-1 Hirosawa, Wako, Saitama 351-0198, Japan

Bryan M. Gaensler
Center for Space Research, Massachusetts Institute of Technology
70 Vassar Street, Cambridge, Mass. 02139, U.S.A.

Simon Johnston
Research Center for Theoretical Astrophysics
University of Sydney, NSW 2006, Australia

Keywords: Pulsars, Pulsar Wind Nebulae, Supernova Remnants, Variable gamma-Ray Emission, Wolf-Rayet Stars

Abstract We present X-ray and radio studies of sources which are bright above 1 GeV ($F_{>1\mathrm{GeV}} \geq 4 \times 10^{-8} \mathrm{ph\,cm^{-2}\,s^{-1}}$). Only 11 out of $\sim$ 30 of these γ−ray sources have been identified with lower energy counterparts: 5 blazars and 6 pulsars. Three of these pulsars are surrounded by radio pulsar wind nebulae (PWN), two of which are also seen as bright, extended X-ray synchrotron nebulae. The *ASCA* X-ray telescope has observed 28 of the bright GeV sources, revealing an excess of

A. Carramiñana et al. (eds.),
The Nature of Unidentified Galactic High-energy Gamma-ray Sources, 135–152.

$F_{2-10\mathrm{keV}} > 10^{-12} \mathrm{ergs\,cm^{-2}\,s^{-1}}$ sources within the EGRET error contours of the unidentified sources. Although several supernova remnants are positionally coincident with these sources, we find no X-ray evidence of high energy particle production in SNR shell shocks consistent with the GeV positions. We also present initial results from follow on radio imaging studies of several fields containing unidentified sources. We have discovered new X-ray/radio nebulae in three of these fields which are strong candidates for PWN. These sources, along with a the PWN around PSR B1853+01 in W44, are all positionally coincident with variable EGRET sources. This suggests a class of variable $\gamma-$ray sources associated with synchrotron emitting regions powered by the winds of young pulsars.

1. INTRODUCTION

Sources that are bright above 1 GeV are an important subset of the sources detected by EGRET. Lamb and Macomb (1997) first noted that the $\sim$ 30 sources with GeV fluxes $F_{>1\mathrm{GeV}} \geq 4 \times 10^{-8} \mathrm{ph\,cm^{-2}\,s^{-1}}$ are tightly clustered along the Galactic plane (Figure 1), making up the majority of low-latitude EGRET sources. 6 of the $\sim$ 8 identified $\gamma-$ray pulsars are in this set but only a small fraction (5/$\sim$ 80) of the known blazars are. The unidentified bright GeV sources are therefore good candidates for further pulsar identifications. Virtually none of them are likely to be background blazars seen through the Galaxy. Of special importance to our understanding of pulsar emission mechanisms is the number of these sources which are radio-quiet pulsars. Outer-gap models (Romani 1996) predict the majority of these sources should be pulsars whose $\gamma-$ray emission is beamed towards us, but whose radio emission is not. Polar-cap models (Harding 2001) predict relatively few such pulsars, making other source classes necessary. One such proposed class is supernova remnant (SNR) shocks, generally assumed to produce the majority of observed cosmic rays (eg. Berezhko & Völk 2000). Models of SNR predict hard, generically flat spectra in the GeV range (Baring et al. 1999). Other proposed source classes, such as colliding winds in massive binary systems (?) and isolated accreting black holes (Punsly 1999), have more tentative predictions of their $\gamma-$ray luminosity at GeV energies.

There are practical observational reasons to focus on the bright GeV sources among the Galactic source population of $\gamma-$ray sources. In general, the point spread function (PSF) of high energy $\gamma-$ray telescopes is roughly proportional to $E_\gamma^{-1/2}$. Observations of Galactic plane sources at energies below a few hundred MeV are plagued by source confusion

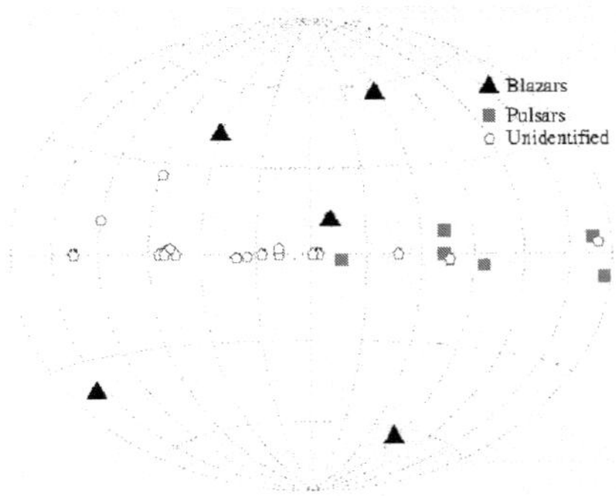

Figure 1 The Galactic distribution of bright GeV sources on the sky (Lamb & Macomb 1997).

and low signal-to-noise and will continue to be so in the *GLAST* era. If the energy dependence of the PSF is included in the analysis, sources with significant GeV emission can be more reliably localized and are less likely to be confused with background sources. A general steepening of the diffuse Galactic $\gamma-$ray emission (Hunter et al. 1997) above 1 GeV further helps the signal-to-noise ratio.

In this article, we present results of X-ray and radio studies of the bright GeV sources. We will first discuss the lower energy properties of the known and proposed source classes. We will then summarize the results of a recently completed 2–10 keV *ASCA* survey of nearly the entire sample (Roberts, Romani & Kawai 2001, hereafter RRK) and report on analysis of VLA and ATCA radio images of selected fields.

2. OBSERVATIONAL CHARACTERISTICS OF POTENTIAL LOWER ENERGY COUNTERPARTS

In general, when trying to find evidence of particle acceleration in the Galactic plane, one looks for evidence of non-thermal emission. In X-rays, it is necessary to observe at energies above $\sim$ 2 keV, both to see through the high absorption in the plane and to look for a hard power-law component of any emission. At radio wavelengths, one looks for a steep spectrum and high polarization. A high radio to infrared ratio is also a good indication of a non-thermal origin of radio emission. Blazars, young pulsars, and SNR can all show these characteristics allowing identification as potential $\gamma-$ray sources through a combination of their X-ray and radio emission.

Blazars are distinguished from other active galactic nuclei by being radio loud ($F_R \gtrsim 0.5$ Jy). They also have moderate X-ray emission

($F_X \gtrsim 10^{-13}$ ergs $cm^{-2}\,s^{-1}$, Sambruna 1997) with hard power-law spectra ($\Gamma \sim 1.7$). It is therefore relatively easy to rule out a blazar identification for most of the unidentified sources by looking at single dish radio survey data. If there is no bright, point-like radio source, then the source is probably not a blazar. Blazars are quite variable at most wavelengths, including γ−rays, unlike the known pulsars and what is expected from SNR shells. Therefore, any variable γ−ray source that is not a blazar is likely to belong to a new class of objects.

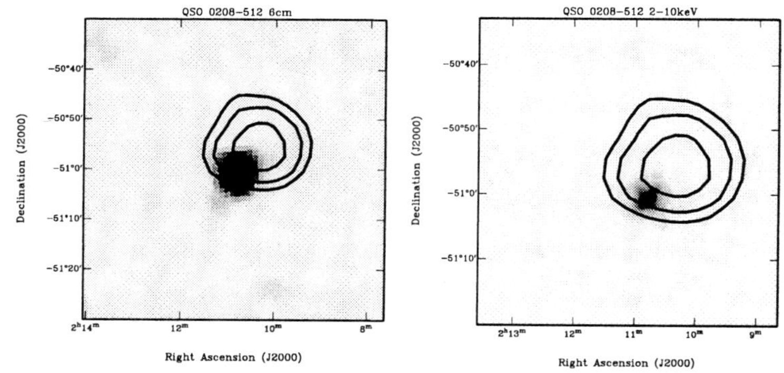

Figure 2 **Left:** 4850 MHz radio image of the GeV blazar QSO 0208−512 from the PMN survey (Condon et al. 1991). The contours mark the *EGRET* 68%, 95%, and 99% confidence position regions. **Right:** 2-10 keV *ASCA* image of QSO 0208−512.

The known γ−ray pulsars can be placed in two rough categories. The youngest ones (Crab and Vela) have bright synchrotron X-ray and radio wind nebulae surrounding them which dominate the emission at lower energies. These pulsar wind nebulae (PWN) are a common feature of the youngest pulsars ($\tau \lesssim 10^4$ years old). Older γ−ray pulsars, such as PSR 1055−52 and Geminga, appear as point sources in X-rays with a spectrum consisting of a soft thermal component (often highly absorbed) and a flat power law component (photon spectral index $\Gamma \lesssim 2$, Halpern & Wang 1997). These have weak or no PWN, and will be referred to here as isolated pulsars. The known γ−ray pulsars have not been observed to be variable in studies of $E \geq 100$ MeV EGRET emission (McLaughlin et al. 1996; Tompkins 1999). However, the Crab PWN appears variable to EGRET in the soft 70-150 MeV energy band (de Jager et al. 1996).

Supernova remnants are identified at radio wavelengths by having a shell type morphology, a steep radio spectrum, and significant polarization. They can also be distinguished from thermal radio sources, such as HII regions, by a high radio to infrared emission ratio (Whiteoak & Green 1996). PWN are distinguished in radio from SNR by having somewhat flatter spectra (energy spectral index $\alpha \sim -0.3 - -0.1$), somewhat higher polarization fraction ($\sim 10 - 30\%$), and amorphous morphologies (Frail & Scharringhausen 1997). In X-rays, SNR can have either a shell or filled center morphology with thermal plasma spectra having strong emission lines. PWN tend to have featureless, hard power-law X-ray spectra with toroidal, bow-shaped, or jet-like morphologies. Since SNR are generally several parsecs across, they are not expected to vary on timescales observable by EGRET, unless the $\gamma-$ray emission is highly localized.

3. ASCA X-RAY SURVEY

The images in this paper (unless otherwise noted) were made with the GIS instrument on the *ASCA* satellite. They come from the *ASCA* catalog of GeV sources (RRK) which covers about 85% of the EGRET 95% error contours for 19 of the 20 bright unidentified GeV sources listed in LM97. In order to screen out stellar sources, which tend to be quite bright in soft X-rays but faint above 2 keV, 2–10 keV images that have been exposure corrected and smoothed are shown. Some images are composites from multiple pointings, and sometimes field edge artifacts appear from scattered light. The details of the image reduction and analysis can be found in RRK. A log N–log S comparison in the 2–10 keV band of the unidentified GeV source fields to the relation derived from the *ASCA* Galactic Plane Survey ($|l| < 45°, |b| < 0.4°$, Sugizaki et al. 2001) shows a clear excess of sources with $F_{2-10\mathrm{keV}} \gtrsim 10^{-12}$ ergs cm^{-2}s^{-1} (Figure 3).

3.1. PREVIOUSLY KNOWN SUPERNOVA REMNANTS

There are five well observed SNR coincident with bright unidentified GeV sources. The X-ray images tend to be dominated by thermal plasma emission which falls off rapidly above 2 keV. Therefore, 4–10 keV contours on the 2–10 keV image can reveal a hard, power-law spectral component which is evidence of particle acceleration. W44 (GeV J1856+0115) and CTA 1 (GeV J0008+7304) both contain hard emission coincident with the GeV source (Figure 4). However, the hard source in W44 is a PWN associated with the young pulsar PSR B1853+01 (?), while the emission in CTA 1 is a suspected PWN associated with the X-ray pulsar candidate RXJ0007.0+7302 (Slane et al. 1997; Brazier et al. 1998). This suggests the GeV emission is related to the pulsars and not the SNR shell.

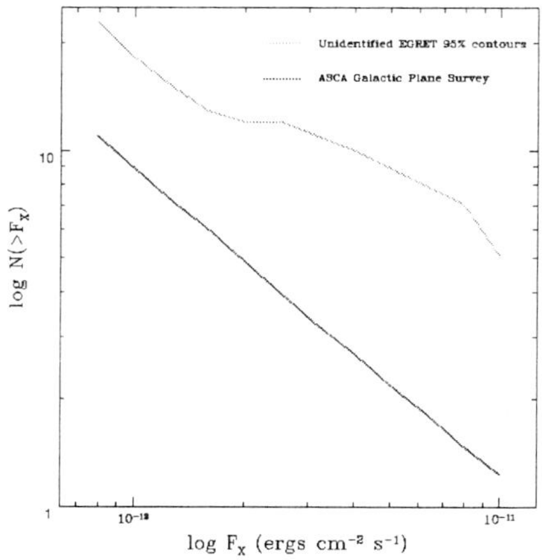

Figure 3 The *ASCA* 2–10 keV log N–log S distribution of X-ray sources found in the 95% error contours of unidentified GeV sources (RRK) compared to the *ASCA* Galactic Plane Survey (Sugizaki et al. 2001).

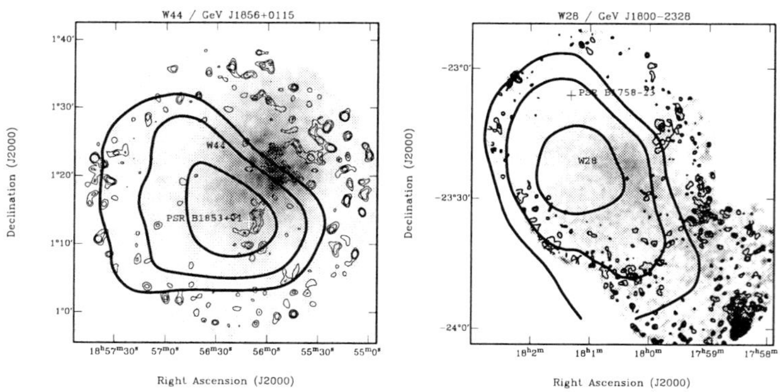

Figure 4 **Left:** The SNR W44 in 2-10 keV X-rays, with 4-10 keV light contours. The heavy contours are the *EGRET* 68%, 95%, and 99% confidence positions of GeV J1856+0114. **Right:** SNR CTA 1 X-ray image with hard X-ray and GeV J0008+7304 contours.

IC443 (GeV J0617+2237) and W28 (GeV J1800−2328) also have regions of hard X-ray emission, but in these cases the emission is not well correlated with the EGRET positions derived from GeV photons (Figure 5). The two regions of hard X-ray emission in the southern portion of the IC443 *ASCA* field, one of which now appears to be a pulsar with PWN (Keohane et al. 2001), have previously been suggested as sites of shock acceleration producing GeV emission (Keohane et al. 1997). The newer GeV based position seems to exclude this. W28 is coincident with the young pulsar PSR B1758−23. However, the high dispersion measure distance of 13.5 kpc (Kaspi et al. 1993), although admittedly unreliable, suggests the pulsar is not associated with the remnant or GeV J1800−2328.

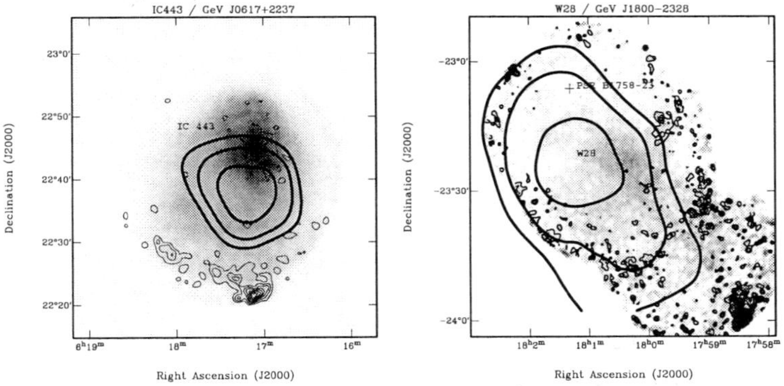

Figure 5 **Left:** SNR IC443 / GeV J0617+2237. **Right:** SNR W28 / GeV J1800−2328.

GeV J2020+4023 is well located near the center of the γ Cygni SNR shell. Despite very extensive observations with *ASCA*, there is no obvious X-ray emission associated with the γ-ray source. Brazier et al. (1996) found only a single X-ray point source in *ROSAT* PSPC data, possibly associated with a K star, and on the basis of the apparently very high L_γ/L_X ratio suggest an isolated pulsar origin for this source. Therefore, none of the SNR associated with GeV sources show evidence of X-ray shock emission consistent with the GeV position which can not be attributed to a young pulsar.

3.2. MASSIVE BINARIES

Colliding winds in massive binary systems are potential sources of $\gamma-$ray emission (Mücke & Pohl 2001). Statistical studies Romero, Benaglia, & Torres 1999 show marginal evidence for the association of Wolf-Rayet stars with unidentified $\gamma-$ray sources. Therefore it is interesting to note any new candidates among the bright GeV sources. There are two Wolf-Rayet + O star binary systems: 1E1024.0−5742 consistent with GeV J1025−5809 and WR141 consistent with GeV J2020+3658. Neither can be considered a very strong candidate. 1E1024.0−5742 is outside the somewhat large 95% positional contour. WR141 is well within the positional contour, but so is a second, equally bright hard point source. These two sources in GeV J2020+3658 are embedded in diffuse X-ray emission with a somewhat steep (photon index $\Gamma \sim 2.5$)

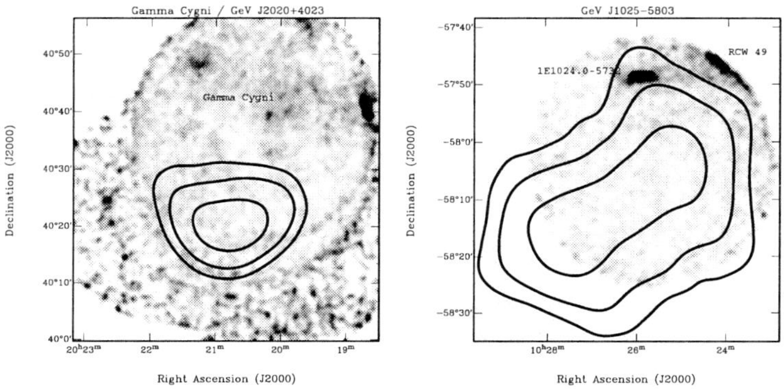

Figure 6 **Left:** 2-10 keV image of SNR $\gamma-$Cygni / GeV J2020+4023. **Right:** GeV J1025$-$5803 with the WN5+0 binary star 1E1024.0$-$5742. At the edge of the image is a colliding wind system in the RCW49 HII region.

spectrum, suggestive of a thermal SNR. However, a 20 cm observation of the field with the VLA in D array shows no bright radio nebula (Figure 7).

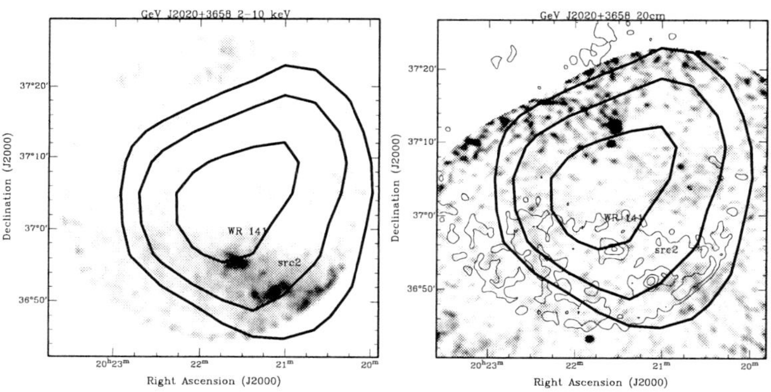

Figure 7 **Left:** GeV J2020+3658 in X-rays, showing WR 141 and the second hard source embedded in diffuse X-rays. **Right:** 20 cm VLA image of GeV J2020+3658 field with X-ray as light contours.

GeV J0241+6102 is consistent with the peculiar radio emitting Be-star/X-ray binary LSI+61 303. This source has been the subject of much study (cf. Strickman et al. 1998), but the nature of the compact object in the system is still uncertain. Although the $\gamma-$ray source is moderately variable, the variations have not been correlated with any known time scales of LSI+61 303. With the latest EGRET positional determination, the source now lies just outside the 95% confidence contour.

4. MULTI-WAVELENGTH STUDIES OF GEV SELECTED FIELDS

A crucial component in further categorizing the sources is $\gamma-$ray variability. Tompkins (1999) has measured the variability of all the sources listed in the third EGRET catalog (Hartman et al. 1999) using the $\tau \equiv \sigma_F/\mu_F$ statistic, where σ_F is the standard deviation of the measured fluxes from different viewing periods and μ_F is the mean flux from all viewing periods. Note this is a measure of how variable a source is rather than the usual χ^2 test of how inconsistent the source is with a constant model (cf. McLaughlin et al. 1996). In Figure 8 we plot all 28 sources in RRK the τ value versus $-\alpha_{X\gamma} \equiv -(1 + \log(F_\gamma/A_X)/6)$, a broadband energy 'spectral index' where F_γ is the photon flux above 1 GeV (equivalent to the power-law normalization at 1 GeV for a source with photon index $\Gamma = 2$) and A_X is the 1 keV X-ray power law normalization in photons $\mathrm{GeV}^{-1}\,\mathrm{cm}^{-2}$. In fields with more than one X-ray source, the $-\alpha_{X\gamma}$ values are derived from the brightest source consistent with being the GeV source, and can therefore be considered an upper limit. The plot is split into regions of high and low variability, and, from left to right, X-ray faint, moderate, and bright. The dotted line represents the systematic uncertainty in the τ measurement, which is consistent with no variability.

The only X-ray bright sources are the Crab nebula and possibly the source near the Galactic center. Vela is the only X-ray moderate pulsar, due to its PWN. The isolated pulsars are X-ray faint. All the identified GeV pulsars are consistent with no variability ($\tau = 0.1$, the systematic variability of EGRET), and the blazars are significantly variable. Four GeV sources in this sample are not in the 3EG catalog, and so were not measured for variability by Tompkins. These are shown along the bottom.

4.1. ISOLATED PULSAR CANDIDATES

There are three unidentified sources in the X-ray faint, low variability category which also contains the isolated pulsars. One is the source in SNR $\gamma-$Cygni. Since there are only weak upper limits on the non-thermal emission in W28 and IC443, the $\gamma-$ray sources positionally coincident with these SNR may also fall in this category. The other two sources are GeV J1835+5921 and GeV J1837−0610. GeV J1835+5921 is the only high latitude source among the bright unidentified GeV sources, and has been extensively studied (Mirabal et al. 2000; Mirabal & Halpern 2001; Reimer et al. 2001). Through X-ray, optical and radio studies, every source in the field has been identified and a

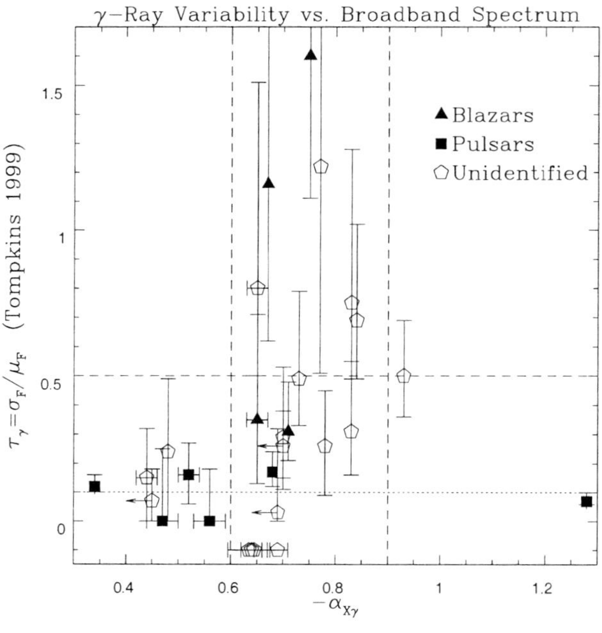

Figure 8 $E \geq 100$ MeV variability index $\tau \equiv \sigma_F/F$ versus the X-ray to γ−ray "spectral index" $-\alpha_{X\gamma}$ assuming the brightest non-thermal X-ray source consistent with the GeV source as the counterpart. Dashed lines denote regions of high and low variability, and from left to right X-ray faint, moderate, and bright. The dotted line is the level of variability due to systematic errors.

single faint X-ray source whose ratio of X-ray to optical luminosity is consistent with a neutron star identification has been discovered.

The field of GeV J1837−0610 contains one hard X-ray point source in the *ASCA* image (Figure 9) and faint diffuse emission. A 20 cm VLA D array image shows a faint ring surrounding the X-ray emission with a very bright HII region along one edge. This is suggestive of an old SNR interacting with the molecular cloud. However, the dynamic range requirements caused by the HII region make accurate spectral measurements of the ring difficult. Also in the field is the young (characteristic age of 34,000 yr) pulsar PSR J1837−0604 recently discovered by the Parkes Multibeam Survey (D'Amico et al. 2001). The dispersion measure distance to the pulsar of 6.2 kpc is consistent with the distance to the molecular cloud.

GeV J2035+4214 is in the X-ray moderate category and was not in the 3EG catalog and so was not measured by Tompkins for variability. The X-ray image (Figure 10) shows three sources: two hard point sources and a softer diffuse source. In the VLA 20 cm image it can be

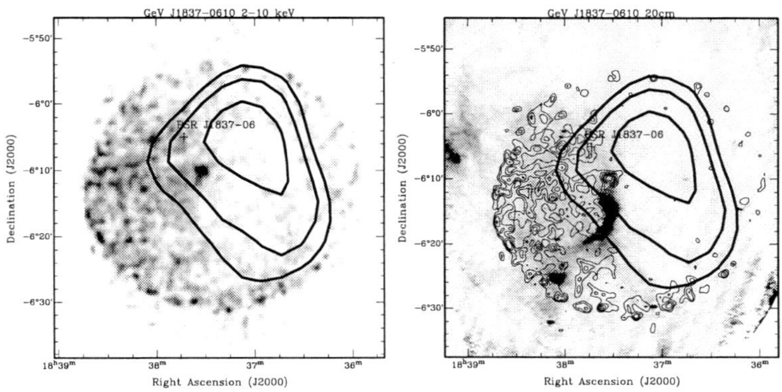

Figure 9 **Left:** GeV J1837−0610 in 2-10 keV X-rays, with the position of the young pulsar marked. **Right:** 20 cm VLA image of GeV J1837−0610 field with X-ray contours.

seen that one of the point sources is associated with a double-lobed radio galaxy. The other is embedded in the bright, nearby ($\lesssim 1$ kpc) molecular cloud DR17. The intensity of the brightest features of the molecular cloud make it difficult to determine if the dim region near the X-ray source has a different spectrum from the rest of the cloud. A somewhat speculative possibility for the emission from GeV J2035+4214 and GeV J1837−0610 is that shock accelerated particles from an old SNR (whose diffuse thermal X-ray emission we still see) are interacting with the molecular clouds, causing the GeV emission. However, the interpretation as isolated pulsars may be more conservative.

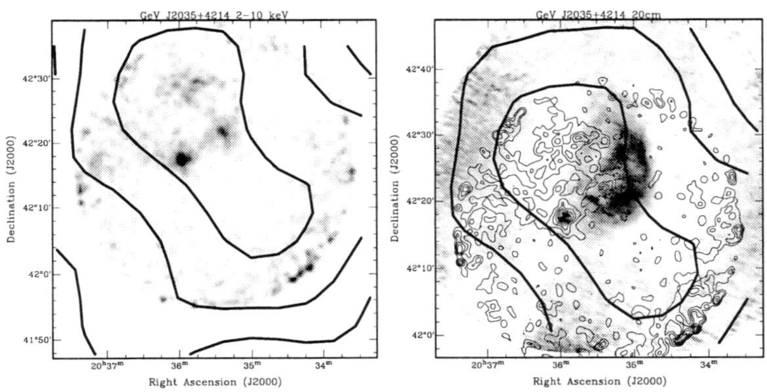

Figure 10 **Left:** GeV J2035+4214 in X-rays. **Right:** 20 cm VLA image of GeV J2035+4214 field showing the bright molecular cloud DR17.

4.2. CANDIDATE PULSAR WIND NEBULAE

The most intriguing sources are the X-ray moderate, high variability sources. There are four in this category, not including blazars. In addition, the source containing LSI+61 303 is on the high variability border, and the source inside CTA 1 is possibly variable. One of the four high variability sources is the PWN in W44 associated with PSR B1853+01. A second, GeV J1825−1310, is near the young pulsar PSR B1823−13. However, the *ASCA* image of the field (Figure 11) reveals a previously unknown, hard spectrum ($\Gamma \sim 2.2$) X-ray nebula. The other two variable sources also contain X-ray nebula, and are discussed in detail in the following sections.

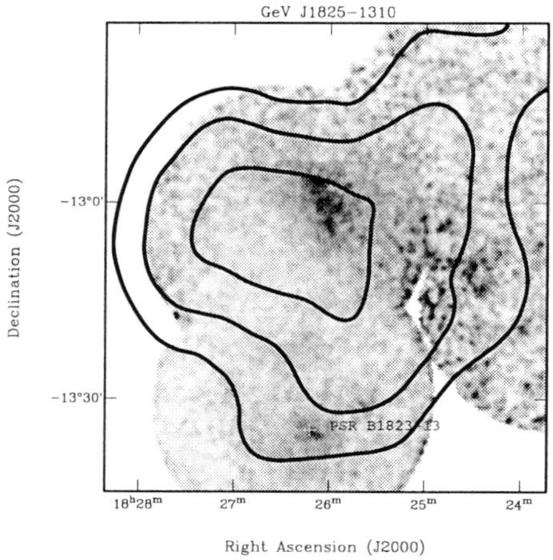

Figure 11 GeV J1825−1310 at 2-10 keV. The X-ray nebula is well centered with the young pulsar PSR B1823−13 to the south. The excess emission near where the four pointings join is from G18.1-0.2, a probable SNR in the Sharpless 53 HII cluster (Kassim et al. 1989), which is also consistent in position with the nearby softer γ−ray source 3EG J1823−1314.

4.2.1 The Kookaburra and the Rabbit.

GeV J1417−6100 is another X-ray moderate high variability source that has been the subject of much recent study (Roberts & Romani 1998; Roberts et al. 1999; Case & Bhattacharya 1999; Roberts 2000; D'Amico et al. 2001). The region contains the Kookaburra radio complex, within which are two extended hard X-ray sources and two hard point sources (Figure 12). One of the point sources is coincident with a weak, variable radio source, and is likely to be a Seyfert galaxy. The other point source appears variable, and may also be a Seyfert, although there is no radio counterpart. The Kookaburra complex itself consists of a mostly thermal shell, with two wings extending to the north and south which, from comparison to infrared images and some excess polarization, may be non-thermal. At the edge of the shell, near the southern wing, is a moderately bright radio nebula known as the Rabbit.

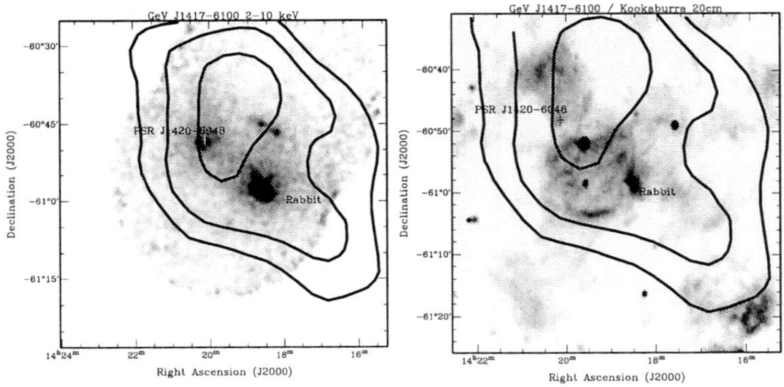

Figure 12 **Left:** GeV J1417−6100 in X-rays. **Right:** 20 cm ATCA image of GeV J1417−6100 field showing the Kookaburra Nebula and the locations of the Rabbit Nebula and PSR J1420-6049.

The Rabbit is spectrally distinct from the shell, and a significant source of polarized flux (Roberts et al. 1999). The morphology of the polarized flux is doubly peaked (Figure 13). Spectral tomography of the Rabbit suggests the polarized region has a spectral index of $\alpha_r \sim -0.3$, typical of a PWN. The brighter extended X-ray source is coincident with the Rabbit, and in the SIS image seems to contain a point source at the location of the upper polarization peak. The X-ray spectrum and morphology is also suggestive of a PWN, although detailed spatial structure is impossible to determine with the broad PSF of *ASCA*.

The Parkes Multibeam Survey recently discovered a very energetic young ($\dot{E} \sim 10^{37}$ ergs/s) pulsar, PSR J1420−6048, coincident with the second extended X-ray source in the upper wing of the Kookaburra (D'Amico et al. 2001). Folding the X-ray data at the radio period yields a marginal detection of an X-ray pulse. The detection is somewhat supported by the pulsed flux being consistent with the *ASCA* PSF, while the unpulsed flux has a significantly broader radial profile (Roberts et al. 2001). While several of the young objects in the region are at a distance of $\sim$ 2kpc, the dispersion measure and X-ray absorption towards the pulsar suggest a greater distance of $\sim$ 8 kpc. Such a distance may be problematic for an ID as the GeV source, although it should be noted that dispersion measure distances can be quite unreliable in complex regions such as this one.

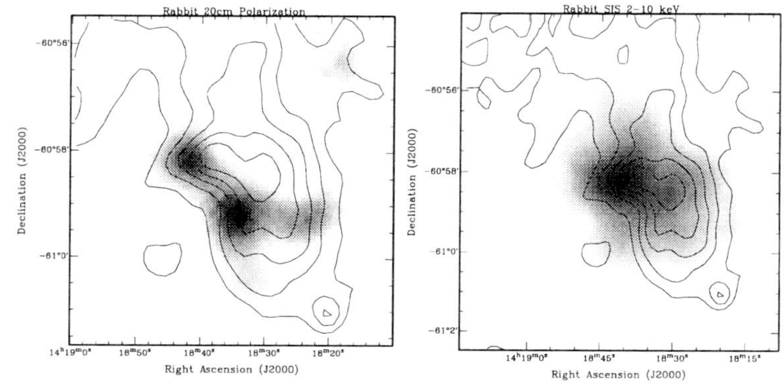

Figure 13 **Left:** 20 cm polarization image of Rabbit nebula, with radio continuum contours. **Right:** *ASCA* SIS image of Rabbit.

The upper wing of the Kookaburra has a strange rectangular shape reminiscent of SNR 3C 397 (Dyer & Reynolds 1999). There is no associated infrared flux seen by *IRAS* at 60 microns or the MSX mission at 8.3 microns. This, along with a suggestion of excess polarization in the ATCA radio maps, suggests a SNR ID. However, the radio spectral index measurements of the entire wing are currently too ambiguous to be certain, although the region immediately around PSR J1420−6048 is consistent with there being a small radio PWN (Roberts et al. 1999).

4.2.2 GeV J1809−2327; A New PWN. The final X-ray moderate, high variability GeV source in the sample is GeV J1809−2327. The *ASCA* image (Figure 14) shows an extended nebula centrally peaked with several smaller peaks to the south. The nebula has a power-law spectrum ($\Gamma \sim 2.2$), with the central peak being slightly harder. The smaller peaks are softer and are coincident with massive young stars. Oka et al. (1999) mapped the CO emission in the region and noted that the X-ray emission is surrounded by molecular gas in the Lynds 227 dark nebula. 20cm and 6cm VLA imaging of the region shows two nebulae (Roberts, Gaensler, & Romani 2001) consistent with the GeV source. The southern one seems to be a thermal molecular cloud which is also seen in the MSX 8.3 micron image (Figure 15). The northern source, coincident with the central peak of the X-ray nebula, has a non-thermal spectrum (energy index $\alpha \sim -0.4$) and is significantly polarized. These properties, along with its shape, suggest a PWN identification.

A short *Chandra* ACIS image (Figure 15) resolves the various point sources (Romani et al. 2001), and shows that the stellar sources are coincident with the middle, thermal cloud. The hard central peak of the

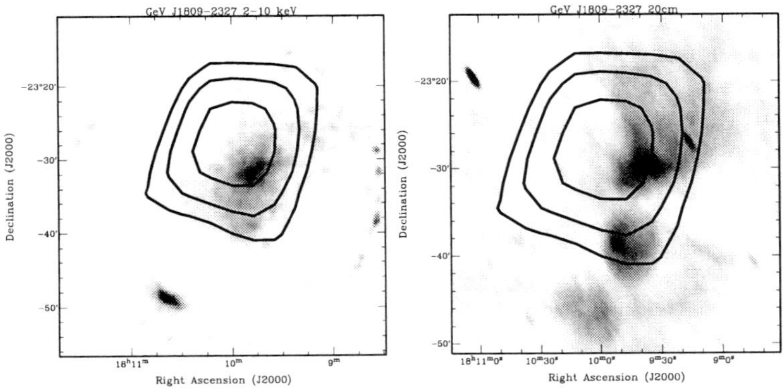

Figure 14 **Left:** *ASCA* 2–10 keV image of GeV J1809−2327. **Right:** 20 cm VLA image of GeV J1809−2327. The funnel shaped nebula is the likely PWN.

X-ray nebula is coincident with a point source at the edge of the PWN candidate. There is a small trail of emission leading back towards the center of the radio nebula. This strongly supports the identification of the X-ray/radio nebula as a PWN.

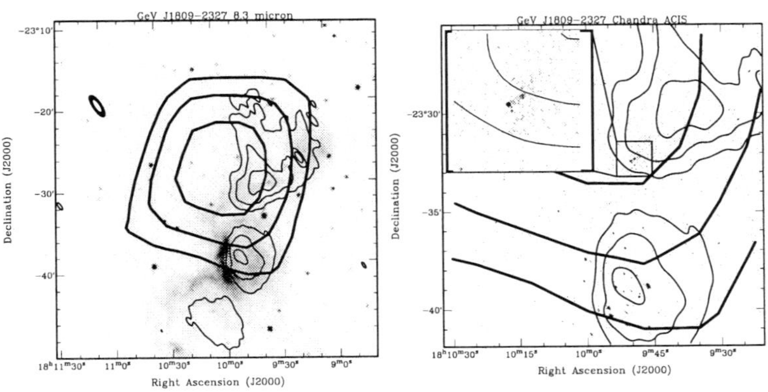

Figure 15 **Left:** GeV J1809−2327 at 8.3 microns, showing the thermal nature of the southern radio nebula. The light contours are of the 20cm radio emission. **Right:** *Chandra* ACIS image of GeV J1809−2327. Inset shows pulsar candidate.

5. SUMMARY

X-ray and radio studies of EGRET error contours have proven to be quite fruitful. There are now strong candidate low-energy counterparts for the majority of bright Galactic sources of GeV emission. *GLAST* should be able to make positive identifications of most of these sources. In the particular case of the two X-ray nebulae in the Kookaburra, *GLAST* would be able to distinguish between them even if both were sources of GeV emission (Roberts 2000). Although there are several massive colliding wind binary systems and SNR coincident with GeV sources, neither source class makes a convincing case for copious emission above a GeV. However, it should be noted that 100 MeV and above error ellipses around the SNR in this sample are much larger and often shifted from the GeV position. There are several other SNR coincident with sources in the 3rd EGRET catalog that have significantly steeper spectra than the sources in this sample. If SNR are, as a class, steep spectrum (photon index $\Gamma > 2$) $\gamma-$ray sources, then current models of SNR emission and cosmic-ray production may have to be revised.

Isolated $\gamma-$ray pulsars (without bright X-ray PWN) seem to be non-variable, and several new candidates for this older population of pulsars have been found. Looking at the total numbers of isolated pulsars and candidates, roughly half seem to be radio pulsars. X-ray pulse searches of the radio-quiet candidates with *Newton-XMM* could prove successful.

With a strong radio PWN candidate counterpart for GeV J1809$-$2327, 3 of the 4 X-ray moderate, high variability GeV sources in this flux-limited sample have now been shown to contain likely PWN which emit in both radio and X-rays (the other two being the Kookaburra/Rabbit and PSR B1853+01). The fourth, GeV J1825$-$1310, contains a candidate X-ray PWN which has yet to be carefully imaged in radio. In addition, the source coincident with the likely PWN in CTA 1 shows a hint of variability. Taken together, there is a strong case for synchrotron nebulae being able to produce high-energy ($E > 100$ MeV) $\gamma-$rays. Noting the apparent soft $\gamma-$ray variability of the Crab nebula, only the Vela pulsar exists as a counterexample. The variability studies of Tompkins (1999) were sensitive to timescales on the order of a few months. This is roughly the synchrotron cooling timescale of the X-ray emission. The presence of a strongly emitting $\gamma-$ray nebula would hinder pulse searches of these sources, since it would significantly reduce the pulse fraction in these sources. Therefore, even with the greater sensitivity of *GLAST*, it may be necessary to find pulsations at X-ray or radio wavelengths before $\gamma-$ray pulsations can be detected in these sources.

Acknowledgments

We would like to thank V. Kaspi for useful discussions and comments. MSER acknowledges support from the Québec Merit Fellowship Program. BMG acknowledges the support of NASA through Hubble Fellowship grant HF-01107.01-98A. The National Radio Astronomy Observatory Very Large Array is a facility of the National Science Foundation operated under cooperative agreement by Associated Universities, Inc. The Australia Telescope is funded by the Commonwealth of Australia for operation as a National Facility managed by CSIRO. This work made use of several on-line resources, including Skyview and W3Browse from the High Energy Astrophysics Science Archive Research Center a service of the Laboratory for High Energy Astrophysics at NASA/GSFC and the High Energy Astrophysics Division of the SAO, as well as the NASA/IPAC Infrared Science Archive.

References

Baring, M.G., Ellison, D.C., Reynolds, S.P., Grenier, I.A., Goret, P. (1999) *ApJ*,**513**,311

Berezhko, E.G., Völk, H.J. (2000) *ApJ*,**540**,923

Brazier, K.T.S., Kanbach, G., Carramiñana, A., Guichard, J., Merck, M. (1996) *MNRAS*, **281**, 1033

Brazier, K.T.S., Reimer, O., Kanbach, G., Carramiñana, A. (1998) *MNRAS*,**295**,819

Case, G. Bhattacharya, D. (1999), *ApJ*,**521**, 246

Condon, J.J. et al. (1991) *AJ*,**102**,2041

D'Amico, N. et al. (2001) in preparation

de Jager, O.C., Harding, A.K., Michelson, P.F., Nel, H.I., Nolan, P.L., Sreekumar, P., Thompson, D.J. (1996) *ApJ*,**457**,253

Dyer, K.K. Reynolds, S.P. (1999) *ApJ*,**526**,365

Frail, D.A. Scharringhausen, B.R. (1997) *ApJ*,**480**,364

Halpern, J.P. Wang, F.Y.-H. (1997) *ApJ*,**477**,905

Harding, A. K. (2001), *Proc. Int. Symp. on High Energy Gamma-Ray Astro. (Heidelberg)*, AIP Conf. Proc. 558, ed. F.A. Aharonian, H.J. Volk, p. 115.

Hartman, R.C. et al. (1999) *ApJS*,**123**,79

Hunter, S.D. et al. (1997) *ApJ*,**481**,205

Kaspi, V.M., Lyne, A.G., Manchester, R.N., Johnston, S., D'Amico, N., Shemar, S.L. (1993) *ApJL*,**409**,L57

Kassim, N.E., Weiler, K.W., Erickson, W.C., Wilson, T.L. (1989) *ApJ*,**338**,152

Keohane, J.W., Petre, R., Gotthelf, E.V., Ozaki, M., Koyama, K. (1997) *ApJ*,**484**,350

Keohane, J.W., Olbert, C.M., Clearfield, C.R., Williams, N.E., Frail, D.A. (2001) *BAAS*, Late Session 197th AAS Meeting,130.06

Lamb, D.Q., Macomb,D.J. (1997) *ApJ*,**488**,872

McLaughlin, M.A., Mattox, J.R., Cordes,J.M., Thompson, D.J. (1996) *ApJ*,**473**,763

Mirabal, N., Halpern, J.P., Eracleous, M., Becker, R.H. (2000) *ApJ*,**541**,180

Mirabal, N., Halpern, J.P. (2001) *ApJL*,**547**,L137

Mücke, A. Pohl, M. (2001) *Workshop on Interacting Winds From Massive Stars*, ASP Conference Series, Ed. Moffat, A.F.J. & St-Louis, N. (San Francisco)

Oka, T., Kawai, N., Naito, T., Horiuchi, T., Namiki, M., Saito, Y., Romani, R.W., Kifune, T. (1999), *ApJ*, **526**,764

Punsly, B. (1999) *ApJ*,**516**,141

Reimer, O., Brazier, K.T.S., Carramiñana, A., Kanbach, G., Nolan, P.L., Thompson, D.J. (2001) *MNRAS*, in press

Roberts, M.S.E., Gaensler, B.M., Romani, R.W. (2001) *in preparation*

Roberts, M.S.E., Romani, R.W. (1998) *ApJ*,**496**,827

Roberts, M.S.E., Romani, R.W., Johnston, S., Green, A.J. (1999) *ApJ*,**515**,[illegible]

Roberts, M.S.E. (2000) *Ph.D. Thesis, Stanford University*

Roberts, M.S.E., Romani, R.W., Kawai, N. (2001) *ApJS accepted*

Roberts, M.S.E. et al. (2001) *in preparation*

Romani, R.W. (1996) *ApJ*,**470**,469

Romani, R.W. et al. (2001) *in preparation*

Romero, G.E., Benaglia, P., Torres, D.F. 1999, *A&A*, **348**, 868

Sambruna, R. (1997) *ApJ*,**487**,536

Slane, P., Seward, F.D., Bandiera, R., Torii, K., Tsunemi, H. (1997) *ApJ*,**485**,221

Strickman, M.S., Tavani, M., Coe, M.J., Steele, I.A., Fabregat, J., Marti, J., Paredes, J.M., Ray, P.S. (1998) *ApJ*,**497**,419

Sugizaki, M., Mitsuda, K., Kaneda, H., Matsuzaki, K., Yamauchi, S. Koyama, K. (2001) *ApJS accepted*, astro-ph/0101093

Tompkins, W. (1999) *Ph.D. Thesis, Stanford University*

Whiteoak, J.B.Z. Green, A.J. (1996) *A&AS*,**118**,329

GAMMA-RAY PULSARS IN THE UNIDENTIFIED POPULATION: THE 'OUTER GAP' PERSPECTIVE

Roger W. Romani
Dept. of Physics, MC 4060
Stanford University, CA 94305
rwr@astro.stanford.edu

Keywords: Pulsars, Gamma rays

Abstract Given that pulsars and AGN are the only identified GeV point sources, it isn't too bold to infer a substantial pulsar contribution to the Unidentified population. Specific models however predict differently the **types** of pulsars present in the observed sample. The dominant factors are the efficiency of spindown-gamma conversion, the high energy beam shape and the gamma ray spectrum. I'll review briefly the main features of the 'outer gap' picture. Consequences for the relative numbers and Galactic distributions of the objects expected in the EGRET survey are then discussed, including young pulsars, 'Geminga's, binaries and millisecond pulsars. With recent counterpart work, we can already test and refine these predictions. The final, and main, point is that what we actually see teaches us in return important lessons about the pulsar physics. With up to a third of the spindown luminosity appearing as GeV gamma rays, these lessons give powerful constraints on the nature of the pulsar machine.

1. INTRODUCTION

Rotation-powered pulsars are attractive as High Energy Gamma Ray sources because the strong magnetic field couples their spin-down luminosity $\dot{E}_{SD} = I\Omega\dot{\Omega}$ into the low density exterior. This low-entropy deposition of a high luminosity provides a natural environment for γ-ray production. I will focus here mainly on magnetospheric γ-ray sources and will concentrate on the particular scenarios that pair produce, accelerate and radiate in the outer magnetosphere ('Outer Gap' models).

A. Carramiñana et al. (eds.),
The Nature of Unidentified Galactic High-energy Gamma-ray Sources, 153–166.

The basic picture is very generic: the rotationally-induced EMF gives rise to a potential drop across the open zone of the pulsar magnetosphere

$$\Delta\Phi_{open} \approx \frac{1}{2}\left(\frac{\Omega R_*}{c}\right)^2 R_* B_s \approx 10^{13}\text{V}\ B_{12}/P^2$$

Particle acceleration can occur when this voltage is spent in charge-starved gaps. The two traditional contenders are acceleration zones near the polar cap (PC) surface (cf. Baring, these proceedings) and vacuum regions bounded by the null charge surface, the closed zone and the light cylinder called outer gaps (Holloway 1973; Cheng, Ho & Ruderman 1986). Of course, seed charges must be produced in either case for a current to tap the potential drop. The main distinction is that for the polar cap zone of young pulsars fields are $\sim 10^{12}$G and $\gamma - B \longrightarrow e^{\pm}$ can be invoked, while in the outer magnetosphere $\gamma - \gamma \longrightarrow e^{\pm}$ provides the seed pairs. It is, in fact, the production of these pairs which limits the gap growth in the latter case (Romani 1996, Hirotani 2000). In turn this means that for the traditional IC-controlled PC picture, the potential drop is characteristically several$\times 10^{12}$V and the dominant escaping photon has an energy of several$\times$100 MeV (Daugherty & Harding 1996). Pair production is much more difficult in the outer gap case and so potentials grow to $\sim 10^{14}$V and the characteristic (curvature) photon has an energy of a few GeV (Romani 1996).

Using an open zone of cross section A, the maximum luminosity that a gap can produce is set by running a Goldreich-Julian current $J_{GJ} = \rho_{GJ} \cdot A \cdot c$ down the potential drop Φ_{gap}. In general a gap spans a portion f_{open} of the open field lines, so this is

$$J_{GJ} \approx \frac{\Omega B}{2\pi c}\left[\pi R_*^2 \left(\frac{\Omega R_*}{c}\right)\right] c f_{open},$$

giving a gap luminosity

$$L_{gap} \approx \frac{f_{open}}{2}\frac{\Omega^2 B_s R_*^3}{c}\Delta\Phi_{gap}.$$

Note that $\dot{E}_{SD} \sim \Omega^4 B_s^2$ so any gap radiator with a fixed 'break-down' voltage produces $L_{gap} \propto \dot{E}_{SD}^{1/2}$. Such a law must, of course, saturate at low $\dot{E}_{SD}$ and so gaps have a 'Death Line' where the required potential drop is barely achieved using the full open zone. For a 10^{14}V breakdown (the OG case) this gives a gap efficiency $\eta_{gap} \approx \left(\dot{E}_{SD}/10^{34}\text{erg/s}\right)^{-1/2} f_{open}$. For a $10^{12.5}$V polar cap, gap saturation and high efficiency are achieved later in the pulsar spin-down with $\eta_{gap} \approx \left(\dot{E}_{SD}/10^{31}\text{erg/s}\right)^{-1/2} f_{open}$.

This pattern is quite robust. More modern versions of the polar cap picture, for example, obtain larger surface potential drops when the GR-induced vacuum fields are included and when CR-limited pair cascades are assumed to control gap closure at higher altitudes (Harding & Muslimov 1998). Correspondingly, the efficiency saturates at higher $\dot{E}_{SD}$ in these models.

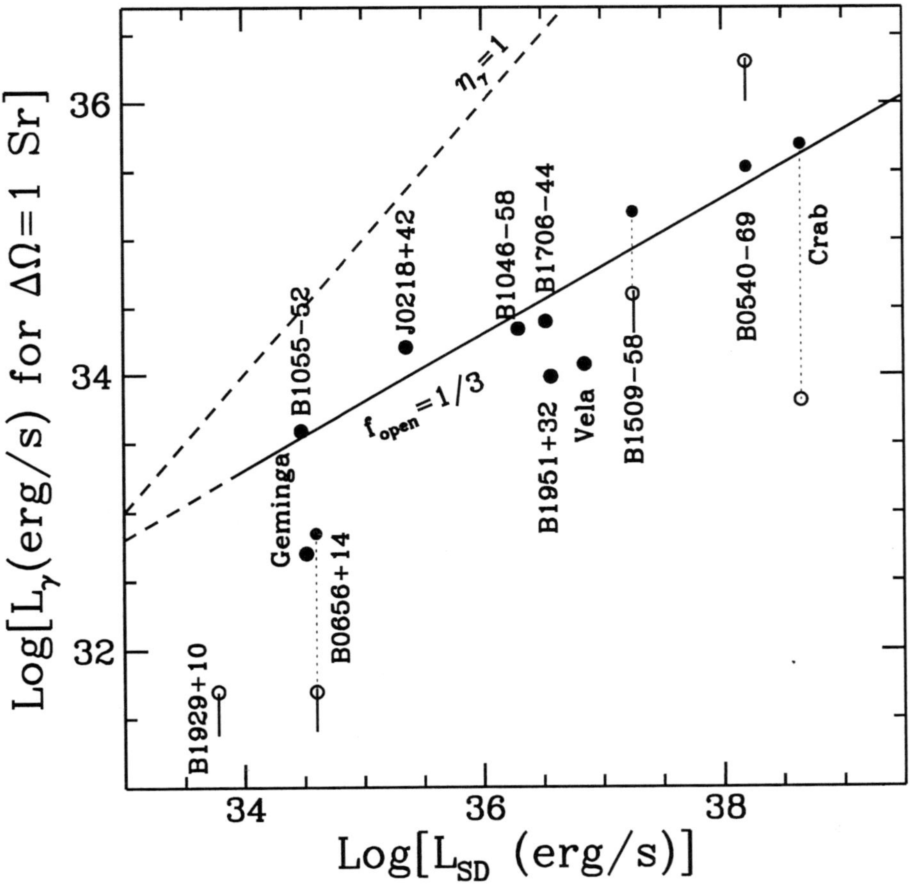

Figure 1 High energy photon luminosity *vs.* spin-down luminosity. Data points are from many observers.

Not surprisingly, observed pulsar γ-ray luminosities do crudely follow this scaling (Figure 1) which shows the high energy emission of the Gamma Ray Pulsars for an *assumed* beaming of 1 steradian on the sky. For the three youngest pulsars Crab, PSR B0540–69 and PSR B1509–58, most of this flux is from X-ray energies; for the other pulsars the EGRET range dominates, except for PSR B0656+14 where X-ray dominates and the $E > 100$MeV detection is marginal (Ramanamurthy, et al. 1996). For this source and for others where the EGRET flux or its upper limit are substantially less than the bolometric photon luminosity, the $E > 100$MeV flux or upper limit is shown by the open circle. Note that these luminosities generally have a factor ~ 2 uncertainty due to distance ambiguities alone. The diagonal solid line gives the predicted relationship for $\Delta\Phi = 10^{14}$V and $f_\gamma = 1/3$. The dashed line gives the limiting 100% γ-ray efficiency for a 1Sr beam.

2. DETAILS, PROFILES AND BEAMING

Most correlations look good on a Log-Log plot, so we should not be too impressed by Figure 1. PSR 1951+32 and Geminga, for example, lie a factor of 4 below the simple extrapolation. The EGRET fluxes of the Crab, PSR B1509–58, PSRs B1929+10 and PSR B0656+14 are also well below the prediction. So while a simple $\dot{E}_{SD}^{1/2}$ law looks good for the bolometric luminosity, clearly some additional factors determine the observed high energy *GeV* flux.

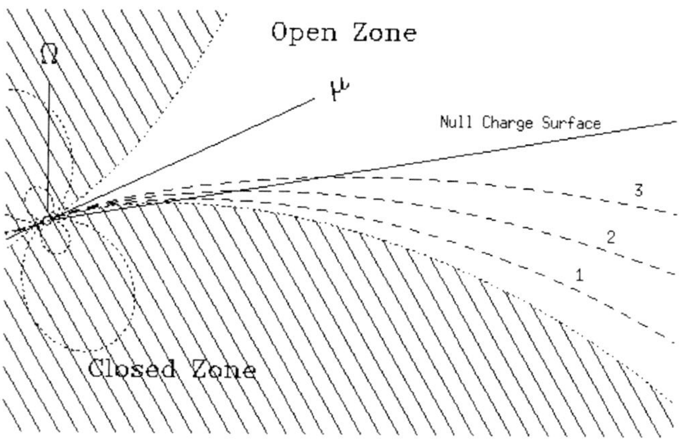

Figure 2 Outer Gap Beaming. The radio beam is confined the the magnetic axis μ. The γ-ray beam is wide, stretching from the null charge surface's projection at the spin equator to larger ζ at higher r. As the pulsar ages and gap closure becomes more difficult, the gap widens from dashed lines 1 to 3, and the γ-ray beam narrows.

One possibility is significant variation in the beaming. Another factor could be the fraction f_{open} of the open field lines participating in the high energy gap. Also dramatic variation in the spectrum can affect observability in the high energy band. In the OG picture, for example, the fraction of the sky covered by the pulsar beam does vary by substantial factors (Romani 1996, Zhang, Zhang & Cheng 2000). We expect that when the gap is narrow (f_{open} small) that it hugs the last closed field line

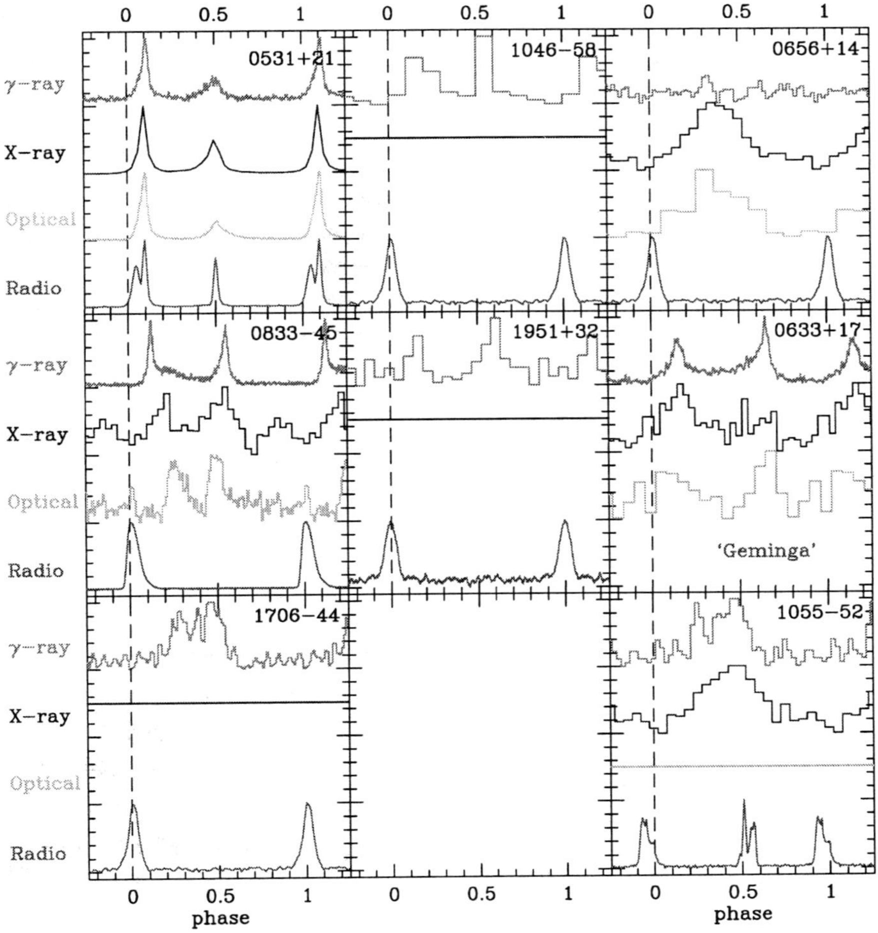

Figure 3 Compilation of multiwavelength lightcurves for γ-ray pulsars, using data from many observers. The closest approach of the magnetic axis μ is at phase 0. One and a half pulse periods are plotted.

surface and the GeV curvature gamma ray beam covers most of the sky. So in the case of the Crab and Vela the beam is large $\Delta\Omega \sim 10$Sr making the true GeV luminosity higher than that plotted. For Geminga, a larger f_{open} and $\Delta\Omega \sim 1$Sr are indicated. This narrower beam is confined to the spin equator (Figure 2). For some objects, e.g. PSR B0656+14 the magnetic inclination α is small and an observer line-of-sight intercepting the radio beam will miss the bulk of the GeV emission. Other interpretations exist. In the polar cap picture, for example, the low EGRET fluxes of PSR B0656+14 and PSR B1509–58 stem from a sharp (super-exponential) cutoff of the spectrum at E<100Mev induced by the large B field of these pulsars.

Clearly the extent and orientation of the radio, X-ray and γ-ray beams will influence the detectability of pulsars in these models. One important clue comes from the light curves of the detected objects (Figure 3). This compilation shows that the relationship between the various wavebands is complex, but a few clear patterns emerge. First, with the exception of the Crab pulsar, the radio emission does not line up with the high energy components. Second, at the highest (γ-ray) energies a basic pattern of a double pulse with the second component lagging the radio by $\sim$ 0.5 in phase seems to dominate. If we associate the radio emission with a hollow cone above the magnetic polar cap, the widely adopted radio picture which explains a wealth of phenomenology (e.g. Lyne & Graham-Smith 1998), then the high energy emission must have a different spatial origin. As Romani & Yadigaroglu (1995) showed, the light curves of the γ-ray pulsars are a natural consequence of the OG model and this geometry quantitatively predicts the correlation between γ-ray pulse separation and the lag from the radio, the polarization behaviour of the Crab and a number of other phenomena (see also Cheng, Ruderman & Zhang 2000). This success in modeling the light curves is not duplicated by the polar cap pictures, where one would expect the radio and high energy beams to have a similar direction and extent (cf. Figure 2).

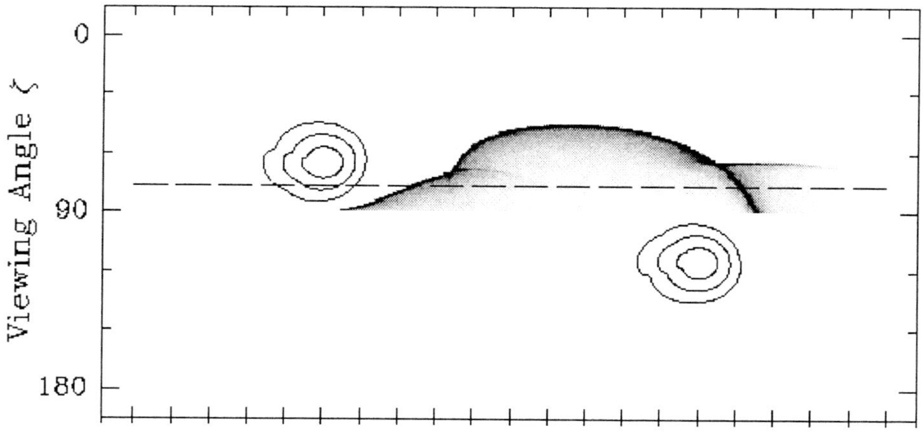

Figure 4 Sky map of radio (countours) and γ-ray (grey scale) beams (Vela-like $\alpha \approx 75^\circ$, producing Fig. 3 light curves). For Geminga ζ misses the radio polar cap; for PSR B0656+14-like objects with small α the Earth line of sight intersects the radio cap at small ζ, but misses most of the γ-ray beam near the spin equator.

3. NUMBER OF 'GEMINGA'S – CLASSIC RADIO PULSARS AND THE UNIDENTIFIED POPULATION

At this meeting our primary concern is the contribution of these pulsars to the unidentified source population. I believe that few would contest that there are several more classical spin-powered pulsars in the EGRET Unidentified source sample. It is also quite likely that isolated Vela-like pulsars do not account for all the observed sources (next section), but it is interesting to see if the classical pulsar population can account for the bulk of the sources. Different pulsar models make rather discrepant predictions. The pulsar beaming dominates the story, with the luminosity (perforce pinned to match the observed sample) having a smaller effect.

If one adopts the standard radio pulsar model (polar caps of no more than a few $\times(\Omega r_{NS}/c)^{1/2}$, isotropic α at birth), then for γ-ray sources the classical PC model is hopeless; one expects only a fraction of a pulsar visible to EGRET (Table 1, D&S luminosity model). In contrast, for the same assumptions the OG picture implies a couple of dozen young pulsars should be present in the EGRET catalog, with $\sim 1/3 - 1/2$ detected as radio pulsars, depending on the depth of the radio survey (YR, CZ). The remainder are the 'Geminga's (γ Only), with the radio beam missing the Earth. These are particularly intriguing candidates for the Unidentified population.

Table 1 Gamma Ray Pulsar Population Estimates

Model	γ+R	γ Only
Dermer&Sturner PC	0.04	0.03
Harding&Zhang PC	??	Aligned α
R&Y 95/97 OG	11(deep R)	11
Zhang, Zhang&Cheng OG	10	22

More modern polar cap models (e.g. Harding & Musslimov 1998, Zhang & Harding 2000) find that revised polar cap potentials can allow the polar gap to extend to $\sim 2r_{NS}$, with a concomitant widening of the polar cap beam. By additionally assuming that young pulsars all have small $\alpha \sim 20°$, they can substantially boost the detection fraction of observed radio pulsars, although detailed population estimates have not yet been made. However, the relatively small central γ-ray beams still ensure that the number of γ-ray sources expected is small for a standard type II supernova rate; the radio pulse offsets and the absence of radio emission for Geminga are also not explained in this model. Harding & Zhang (2001) have further speculated that the remnant curvature radiation above the polar cap at heights an appreciable fraction of r_{LC} might increase pulsar detectability through soft, faint 'Off Beam' emission. While an attractive scenario for making older, near-by pulsars visible, the luminosity in this picture is too low to add appreciably to the bright Galactic population.

To summarize the situation for the bright low-latitude Galactic γ-ray sources: OG beaming models give a reasonable interpretation of the γ-ray faint pulsars, such as PSR B0656+14 and the radio-faint pulsars such as Geminga. The overall numbers expected in the EGRET sample, $\sim 15 - 30$ for a canonical supernova rate of 1/100y and the luminosity distributions and correlations with regions of high mass star formation are in good agreement with the data (Yadigaroglu & Romani 1997; Zhang, Zhang & Cheng 2000). However, some aspects are less satisfactory. For example, the fraction of γ-ray sources that are radio-quiet like Geminga is expected to be $\sim 1/2 - 2/3$. With 8 classical radio pulsars already identified in the EGRET sample (and only one identified Geminga) this suggests that most of the remaining sources should be radio quiet, leaving uncomfortably little room for other source classes. Exacerbating this are studies that show that the spectra (Merck et al. 1996) and variability (McLaughlin, et al. 1996; Tompkins 1999) of many of these low latitude sources differ appreciably from the hard-spectrum, constant flux observed from the isolated pulsars. This leaves uncomfortably little room for the Gemingas expected in OG models, suggesting that the γ-ray beams are somewhat narrower or that the radio beams are somewhat broader than in the initial OG studies.

At intermediate latitudes $|b| \geq 10°$ there is of course additional evidence for a more complicated story. Models of the young isolated pulsar population, do predict a handful of objects in this range (e.g. Yadigaroglu & Romani 1997) and some of the intermediate latitude objects are very likely to be classical pulsars (e.g. the strong cases for CTA 1, Brazier, et al 1998 and 3EG J1835+5918, Mirabel & Halpern 2001) However, Grenier & Perot (1999 see, also, these proceedings) and Gehrels, et al. (2000) have clearly shown a further Galactic population at intermediate latitudes distinct from that in the disk. So, happily, the story is undoubtably more complex than a simple inference of more Velas and more Gemingas.

4. OTHER PULSAR GAMMA-RAYS

In the spirit of Occam's Razor, however, I would argue that it is still worth extending the pulsar story as far as possible in explaining the unidentified sources. I will comment briefly on some other realizations of pulsar γ-ray sources, the connection with OG models and the present evidence for such sources in the EGRET population. First it is worth remembering that the circumstantial evidence implicating high mass stars and their pulsar progeny is very strong (Kaaret & Cottam 1996, Yadigaroglu & Romani 1997), so it is important to consider how other components of the pulsar γ-ray emission might contribute to the population.

The *pulsed* γ-rays from the objects in Figure 3 represent less than $\sim 20\%$ of $\dot{E}$ even for very wide beams. Much of the rest of the energy must be emitted as a $e^{\pm}/B$ wind and we know from the Crab that with a suitable circumstellar environment, shocks in this flow can produce appreciable EGRET emission. Moreover, this flux is variable (e.g. de Jager, et al. 1996). So it is plausible that a significant, or even dominant, fraction of the EGRET emission from an unidentified source might come from a variable plerionic component. Roberts, et al. (these proceedings and references therein) describe the discovery of several non-thermal X-ray/radio nebulae (e.g. 2EG J1811–2339, 3EG J1420–6038) associated with variable GeV sources, which are good candidates for such plerions. OG beaming computations (e.g. RY95) predict that 25% or more of young pulsars should be visible in neither the radio or pulsed γ-rays, leaving only the isotropic plerion emissions. Depending on the fraction of L_γ in the plerion, up to a half dozen such sources should be present in the 3EG catalog. Other sources, such as PSR B1046–58 (Kaspi et al 2000) may be composite, showing a pulsed component buried in stronger unpulsed γ-ray emission.

In the survey of Galactic GeV sources, Roberts, et al. also found several several plausible associations with high mass binaries. Similar γ-ray association has long been made for the peculiar object LSI+61 303. More recently, the very interesting case of 3EG J0634+0521 (Cusumano, et al 2000; Kaaret, et al 2000 and these proceedings) has associated an X-ray pulsar+Be star binary with the γ-ray source. Both OG and PC models would predict that fraction of the neutron star/OB-star binaries are active rotation-powered pulsars capable of γ-ray emission. Radio detection of such pulsars is generally difficult (eg PSR B1259–63 Johnston et al 1992) due to scattering and DM smearing but γ-ray pulsations should be unaffected for a range of mass loss rates allowing the pulsar wind to hold off accretion. For a Galactic birthrate of HMXB of $\sim 2.5 \times 10^{-3}$/yr

(Iben, Tutukov & Yungelson 1995), OG beaming, a typical γ-ray pulsar age of 10^5y and a logarithmic distribution of semi-major axes, one would expect $\sim 3-5$ active pulsar/B-star binaries within the $\sim$ 3kpc depth of the EGRET survey volume. Eclipses and intermittent accretion poisoning might well make such pulsars γ-ray variable, even if magnetospheric emission dominates the flux.

The light cylinder B and potential drop $\Delta\Phi$ of the millisecond pulsars (MSP) with the highest $\dot{E}$ spindown luminosity are surprisingly similar to those of the young γ-ray pulsars. In the young pulsars (except for the $\tau \sim 10^3$y objects) we expect thermal surface emission to supply the bulk of the target X-rays for $\gamma - \gamma$ pair production. In contrast, MSP surfaces are old and cold. However, these high $\dot{E}$ MSP have recently been found to produce strong magnetospheric X-ray emission (Becker & Truemper 1999), which provides an additional plausible source for the required photons. The likely γ-ray detection of PSR B0218+4232 (Kuiper et al 2000) suggests that a number of MSP must be hiding in the Unidentified EGRET population.

Finally, there has been much interest in associating a portion of the intermediate latitude population noted above with Gould's belt. The inference is that the high mass stars of this structure, being born at $\sim$ 25/Myr, produce pulsars that somehow generate low luminosity, long-lived γ-ray sources, perhaps from 'Off-Beam' emission covering a large fraction of the sky (Harding & Zhang 2001).

5. THE FULL POPULATION

All of these scenarios ultimately trace their γ-ray luminosity to the the spin-down energetics of the underlying pulsar population (§1). So the sort of population synthesis comparison with the unidentified source catalog pursued in YR97 can test consistency with a range of pulsar models. With either specific luminosity and beaming functions or more generic η_γ efficiency laws, one can predict the γ-ray sky for known pulsar populations. Several groups are now pursuing such tests of the unidentified population. We have recently extended the Monte Carlo population syntheses of YR97 to include the nearby Gould Belt stars and disk and bulge populations of Galactic millisecond pulsars (Elteto & Romani, in preparation). We compare the model populations with the 3EG catalog, after incorporating the detailed variation of sensitivity on the sky, by use of a 3-D K-S test (Gosset 1987, see also Merghetti, these proceedings).

Examples of best fit matches to the observed sample of 'persistent' (Grenier and Perrot 1999) sources are shown in Figure 5. The grey scale shows the expected probability of EGRET source detection for the given

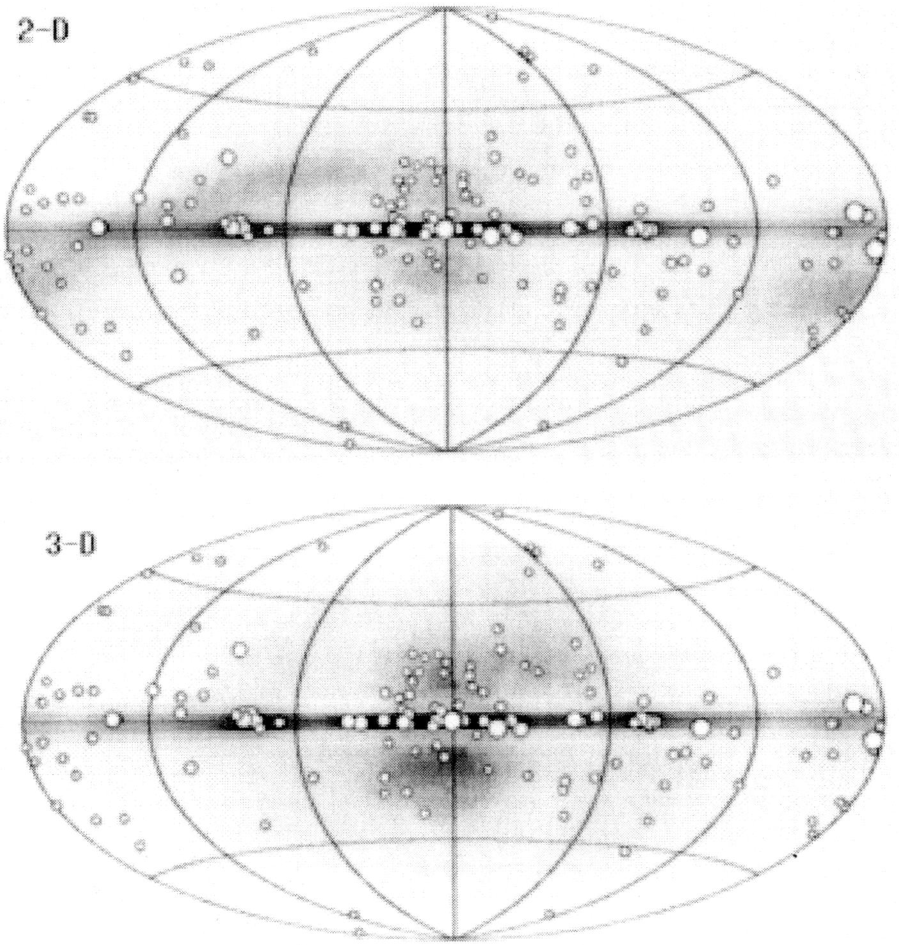

Figure 5 Monte-Carlo models of the pulsar and unidentified 'Persistent' source population. Above: Young+Gould belt, Below: Young+MSP

fit model, while the dots give the observed sources with size proportional to the Log of the EGRET flux. In the upper panel, we have used only the spatial information (2-D KS test), while in the lower panel we include the match to a model luminosity function at each sky location (3-D KS test). A young disk population is, of course, required in each case, with parameters similar to those found in YR97. If only the positions of the sources are considered then a Gould belt distribution of sources does provide a plausible source distribution, as found by Grenier, Gehrels, and colleagues. However, for reasonable pulsar population parameters and γ-ray luminosity law (e.g. $\eta_\gamma \propto \dot{E}_{SD}^{-1/2}$), the observed luminosity function and spatial distribution is a very poor match to the model predictions. In addition, even with the η_γ law tuned to maximum efficiency and distributing the γ-rays isotropically, it is difficult to produce more than a few detected EGRET sources for the Gould belt supernova rate of $\sim$ 25/Myr (Poeppel 1997). More generally, changes to the pulsar emission model dramatic enough to produce the large number $\sim 30-40$ of observed intermediate latitude sources appear to ruin the match to the young disk population, as well. So while a Gould belt association implicates high mass stars, the connection with young pulsars of the sort detected with EGRET is not clear.

An alternative source population, shown in the lower panel of figure 5, provides a substantially better match to the spatial and luminosity distributions. This is the Galactic millisecond pulsar population which is inferred to have both a disk and bulge component (Cordes & Chernoff 1997). The fit shown has about half of the observed intermediate latitude sources from each. While a detailed luminosity/beaming model for MSP is not unfortunately available, interestingly for plausible gap efficiency models, the shortest period MSP dominate the model EGRET sources. These are often in short period binaries and will be rather difficult to detect in the radio, so perhaps the the lack of many radio MSP associations is not too worrisome. If a strong bulge component is really present, however, it is more worrisome that the EGRET sources are not strongly associated with the globular clusters (Michelson et al 1994). Moreover, neither Gould belt or MSP pulsar models (or a combination of these) provide a fully satisfactory fit to the intermediate latitude sources sample at present – it is likely that some aspect of the population, beaming, or luminosity evolution needs to be changed.

6. SUMMARY

Rotation-powered pulsars are undoubtably present in the Unidentified EGRET source population and we can expect that careful counterpart studies can produce several further identifications over the next few years. Not all of these objects will be the classic, isolated Vela-like pulsars dominating the brighter pulsed detections. There are intriguing hints that other pulsar γ-ray manifestations such as plerion shocks, pulsars with high mass companions and MSP make a substantial contribution to the source sample. OG models for the magnetospheric emission, with their wide beams and high Γ $e^{\pm}$ winds, are an attractive framework with which to understand such pulsar sources. However, the range of variability and spectrum properties seen, as well as the large number of sources in the intermediate latitude sample provides a significant challenge to all present pulsar models.

It goes without saying that in the GLAST era, many of the present puzzles should be swiftly resolved, with accurate source positions and direct pulsation searches of the EGRET-detected sources. Detailed tests of the OG beam picture, particularly through population studies, will become routine. We can also expect a number of the alternative manifestations of OG pulsar emission, eg. MSP and binaries, to join the population. However, the existing pulsar/Unidentified source story is surely incomplete. With a luminosity threshold allowing several 100 to 10^3 pulsar detections, new object classes (and a new window on the Galactic neutron star population) will undoubtably be found. And even if rotation-powered pulsars are shown to dominate the Galactic sample, there will certainly be objects not fitting the basic pulsar classes above. The Mystery of the Unidentified Gamma-Ray Sources is alive and well after thirty years and it is a pleasure to anticipate that it will not be fully solved any time soon.

Acknowledgments

I thank Alberto Carramiñana and the LOC for hosting a delightful workshop. Work on the Unidentified source counterparts has been pursued with a number of colleagues, especially Mallory Roberts, Nobu Kawai and Simon Johnston. I also thank the ATNF of CSIRO for hospitality while this contribution was being completed.

References

Becker, W. & Truemper, J 1999, *AA*, **341**, 803
Brazier, K.T.S., Reimer, O., Kanbach, G. & Carramiñana, A. 1998, *MNRaS*, **295**, 819
Cheng, K.S., Ho. C. & Ruderman, M.A. 1986, *ApJ*, **300**, 522
Cheng, K.S., Ruderman, M. A. & Zhang, L. 2000, *ApJ*, **537**, 964
Cordes, J.M & Chernoff, D. F. 1997, *ApJ*, **482**, 971
Cusumano, G. et al 2000, *ApJ*, **528**, L25
Daugherty, J.K. & Harding, A.K. 1996, *ApJ*, **458**, 278
de Jager, O.C. et al. 1996, *ApJ*, **457**, 253
Gehrels, N., et al 2000, *Nature*, **404**, 363
Gosset, E. 1987, *AA*, **188**, 258
Grenier, I.A. & Perrot, C. 1999 in *Proc. XXVI Int'l Cosmic Ray Conf*, **3**, 476
Harding, A.K. & Musslimov, A.G. 1998, *ApJ* **508**, 328
Harding, A.K. & Zhang, B. 2001, *ApJ* in press
Hirotani, K. 2000, *MNRaS*, **317**, 225
Holloway, N.J. 1973, *Nature*, **246**, 6

Iben, I., Tutukov, A.V. & Yungelson, L.R. 1995, *ApJS*, **100**, 217
Kaaret, P., Cusumano, G. & Sacco, B. 2000, *ApJ*, **542**, L41
Kaaret, P. & Cottam, J. 1996, *ApJ*, **462**, L35
Kaspi, V. M. et al. 2000, *ApJ*, **528**, 445
Kuiper, L. et al. 2000, *AA*, **359**, 615
Lyne, A.G, & Graham-Smith, F. 1998 *Pulsar Astronomy* (Cambridge:Cambr
McLaughlin, M.A., et al 1996, *ApJ*, **473**, 763
Merck, M. et al. 1996, *AASupl*, **120**, 465
Michelson, P.F., et al. 1994, *ApJ*, **435**, 218
Mirabel, N. & Halpern, J.P. 2001, *ApJ*, **547**, L137
Poeppel, W.G.L. 1997. *Fund. Cos. phys.*, **18**, 1
Ramanamurthy, P.V., et al. 1996, *ApJ*, **458**, 755
Romani, R.W. 1996, *ApJ*, **470**, 469
Romani, R.W. & Yadigaroglu 1995, *ApJ*, **438**, 314 (RY95)
Tompkins, W. 1999, *PhD Thesis, Stanford University*
Yadigaroglu, I.-A. & Romani, R.W. 1997, *ApJ*, **476**, 347 (YR97)
Zhang, B. & Harding, A.K. 2000, *ApJ* **532**, 1150
Zhang, L., Zhang, Y.J. and Cheng, K.S. 2000, *AA*, **357**, 957

IDENTIFYING THE MYSTERIOUS EGRET SOURCES: SIGNATURES OF POLAR CAP PULSAR MODELS

Matthew G. Baring

Department of Physics and Astronomy,
Rice University, MS-108,
P. O. Box 1892,
Houston, TX 77251-1892,
U. S. A.
baring@rice.edu

Keywords: gamma-ray pulsars, unidentified EGRET sources

Abstract The advent of the next generation of gamma-ray experiments, led by *GLAST*, *AGILE*, *INTEGRAL* and a host of atmospheric Čerenkov telescopes coming on line in the next few years, will enable ground-breaking discoveries relating to the presently enigmatic set of EGRET/CGRO UID galactic sources that have yet to find definitive identifications. Pulsars are principal candidates for such sources, and many are expected to be detected by *GLAST*, some that are radio-selected, like most of the present EGRET/Comptel pulsars, and perhaps even more that are detected via independent pulsation searches. At this juncture, it is salient to outline the principal predictions of pulsar models that might aid identification of gamma-ray sources, and moreover propel subsequent interpretation of their properties. This review summarizes relevant characteristics of the polar cap model, emphasizing where possible distinctions from the competing outer gap model. Foremost among these considerations are the hard X-ray to gamma-ray spectral shape, high energy cutoffs and pulse profiles, and how these characteristics generally depend on pulsar period and period derivative, as well as observational viewing angle. The polar cap model exhibits definitive signatures that will be readily tested by the detections of *GLAST* and other experiments, thereby establishing cogent observational diagnostics. The paper focuses on different classes of pulsars that might define agendas and parameter regimes for blind gamma-ray pulsation searches; examples include the highly-magnetized ones that are currently quite topical in astrophysics.

A. Carramiñana et al. (eds.),
The Nature of Unidentified Galactic High-energy Gamma-ray Sources, 167–188.

1. INTRODUCTION

Pulsars are a central part of any discussion of candidates for the putatively galactic population of EGRET unidentified (UID) sources. A major factor in this is their inherent nature in being both among the brightest galactic sources (for observational summaries, see Kanbach, these proceedings, Thompson 2001), and moreover being distinctive via their pulsations. These features have fed the historical evolution of the field of unidentified gamma-ray sources, with pulsars leading the *post-facto* galactic identifications (e.g. see Thompson, these proceedings). There is a pervasive feeling in the gamma-ray community, as embodied in the course of this Workshop, that such a situation will persist. This perception is driven by the expectation that the *Gamma-Ray Large Area Space Telescope* (*GLAST*: `http://www-glast.stanford.edu`) will detect pulsars in profusion, some that are radio-selected, like most of the present EGRET/Comptel pulsars, and perhaps even more that are detected via independent (blind: i.e. not radio or X-ray selected) pulsation searches. Current estimates of the anticipated *GLAST* pool of pulsars range from dozens to several hundred (e.g. Harding, 2001b), depending on whether an outer gap or polar cap model is preferred, and on specific assumptions pertaining to each model. This population should account for a significant fraction of the present EGRET UID galactic sources. Furthermore, it should provide an ample dataset for exploring pulsar physics in general, and enable discrimination between the outer gap and polar cap scenarios in particular.

Discussions of the generic features of each leading model for gamma-ray emission in pulsars are therefore timely. The outer gap (OG) case, where the accelerating potential is far removed from the stellar surface and more proximate to the light cylinder, is advocated by Romani (these proceedings). This paper presents the perspective of the polar cap (PC) model, where particle acceleration is effected either just above or within a stellar radius or two of the neutron star surface, and is presumed to be attached to the last open field line near the polar cap (e.g. Sturrock 1971, Ruderman & Sutherland 1975, Arons & Scharlemann 1979; see also the recent review by Harding 2001a). Motivations for preferring a polar cap scenario include, but are not limited to, (i) a difficulty in accepting a viewpoint that electrodynamic dissipation near the stellar surface is only a minor contributor to the pulsar's high energy radiative signals, (ii) the connection of gamma-rays to coherent radio emission in PC models, mediated by single photon pair creation in strong magnetic fields, (iii) the relatively narrow range of ages permitted for gamma-ray pulsars according to the outer gap model, (iv) the fact that the polar cap prediction (Harding 1981) of pulsar luminosity scaling approximately with open line field voltage was confirmed by the EGRET/Comptel pulsar collection (e.g. see Fig. 4 of Thompson 2001, for the status quo of this correlation), and (v) the predictions (Cheng, Ho & Ruderman 1986; Cheng 1994; Yadigaroglu & Romani 1995; Romani 1996) from earlier inceptions of outer gap models of many radio-quiet gamma-ray pulsars and also inverse Compton emission at TeV energies, forecasts that have not been borne out by subsequent observations. While the first two of these are largely conceptual preferences, the last two connect to reality. Geminga stands alone is the only potentially radio-quiet "garden-variety" gamma-ray pulsar, and the constraining limits to TeV (pulsed) emission from pulsars obtained by Atmospheric Čerenkov Telescopes (e.g. Nel et al. 1993; Lessard et al. 2000) have forced the revision (e.g. Hirotani, 2000) of TeV flux estimates from outer gap models.

The features of the polar cap model are intimately connected to the extremely strong magnetic field that threads the emission region. It is this field that controls the maximum energy of emission, the nature of pair creation, and a host of physics that determines the pulsar spectrum and influences the pulse shape. This review will highlight some of these properties, and explore how (mostly) spectral and temporal characteristics generally depend on pulsar period and period derivative, as well as on observational viewing angle. The polar cap model exhibits definitive signatures that will be readily tested by the detections of *GLAST* and other experiments, thereby establishing palpable observational diagnostics. Discerning which of the outer gap and polar cap models is most appropriate for gamma-ray pulsars, or whether each has its own domain of applicability, is a central quest for pulsar astrophysicists. While preparatory analysis for the *GLAST* era is a worthwhile goal alone, perhaps most salient for the subject of this meeting is a definition of identification strategies for gamma-ray sources, based on the global characteristics of pulsar models and the assumption that pulsars constitute a sizeable fraction of the EGRET UID collection; this will form the focus at the end of this review.

2. POLAR CAP MODELS OF GAMMA-RAY PULSARS

The polar cap scenario is attractive from a physics perspective. Nevertheless, like its outer gap competitor, it remains unproven. Hence it is imperative to establish definitive/unambiguous properties and predictions that enable a determination of its applicability to gamma-ray pulsars. This can be achieved via two approaches. The first is to isolate individual pulsars for analysis, and explore the model behavior of phase-resolved spectra (and also polarization swing profiles in an ideal world where gamma-ray polarimetry is accessible: see Section 3) that can be compared to high quality temporal and spectral data. This has been the approach of Daugherty & Harding (1996), Dyks & Rudak (2000) and Romani (1996) using the Vela pulsar as a test case in supporting their competing perspectives. The major drawback of such isolated focuses is that the models have enough parameters to render model discrimination near impossible. Therefore, multiple objects need to be considered, a time-consuming task for pulsar-by-pulsar analyses. This leads to the second diagnostics method: population statistics. It is expedient to identify global characteristics of the models so as to define parametric trends that can be confirmed or disproven given a large database such as that to be afforded by *GLAST*. Given the disparity in emission region geometry and operable physics incorporated in the polar cap and outer gap scenarios, it is improbable that they will provide a collection of similar or coincident behavioral trends. Hence, the emphasis of this paper will be to address the more global signatures of polar cap models by first defining the relevant cascade and radiative properties.

2.1. BASIC PROPERTIES OF CASCADES

Polar cap models for pulsar high-energy emission are generally based on the idea, dating from the earliest pulsar models of Sturrock (1971) and Ruderman & Sutherland (1975; hereafter RS75), of particle acceleration and radiation near the neutron star surface at the magnetic poles. Within this broad class, there is a large variation, with the primary division being whether or not there is free emission of particles from the neutron star surface. This question hinges on whether the surface temperature T of the neutron star (many of which have now been measured in the range $T \sim 10^5 - 10^6$ K; Becker & Trümper 1997) exceeds the ion, $T_{\rm i}$ and electron, $T_{\rm e}$, thermal emission temperatures. If $T < T_{\rm i}$, a vacuum gap will develop at the surface, due to the trapping of ions in the neutron star crust (RS75, Usov & Melrose 1995). In this case, the particle acceleration and radiation will take place very near the neutron star surface. If $T > T_{\rm e}$, free emission of particles of either sign of charge will occur. The flow of particles is then limited only by space charge, and an accelerating potential will develop (Arons & Scharlemann 1979; Muslimov & Tsygan 1992) due to an inability of the particle flow all along each open field line to supply the corotation charge that is required to short out the electric field component $E_{||}$ along the magnetic field lines. In space charge-limited flow models, the accelerating $E_{||}$ is screened at a height where the particles radiate γ-rays that produce pairs. This so-called pair formation front (e.g. Arons 1983, Harding & Muslimov 1998) can occur at high altitudes above the polar cap, a property that may prove necessary to explain the spectral cutoffs in the some or most of the EGRET pulsars.

The acceleration of primary electrons is rapid and ceases when one of two types of radiative cooling becomes significant. This establishes the maximum Lorentz factor γ_e of these particles, and a quasi-monenergetic primary distribution is established prior to cascading. The two cooling mechanisms are curvature radiation induced by the magnetic field line curvature, the process that is more widely cited in pulsar literature as a primary emission mechanism, and resonant (magnetic) inverse Compton scattering of thermal X-rays from the stellar surface (e.g. Sturner and Dermer 1994), a relatively recent consideration. Both are strong functions of the magnetic field strength and either the electron's Lorentz factor or the field geometry. Curvature radiation-initiated cascades generally have $\gamma_e \sim 10^7$ (e.g. Daugherty & Harding 1989; see also Harding & Muslimov 1998), while inverse-Compton seeded pair cascades yield $\gamma_e \sim 3 \times 10^5 - 10^6$ (e.g. Sturner 1995; see also Harding & Muslimov 1998). Such photons propagate through the magnetosphere until they achieve sufficient angles $\theta_{\rm kB}$ with respect to the magnetic field to permit the creation of pairs via $\gamma \to e^+e^-$ above the threshold energy of $2m_ec^2/\sin\theta_{\rm kB}$. This propagation is influenced by general relativistic distortions of photon trajectories and field structure (e.g. Gonthier & Harding 1994; Harding, Baring & Gonthier 1997), as is the magnitude of the field in the local inertial frame. For small polar cap sizes, corresponding to longer pulsar periods, it is the failure of the primary photons to acquire sufficient angles $\theta_{\rm kB}$ at low to moderate altitudes (prior to dipole field decline) that is primarily responsible for the existence of a theoretical death line for radio pulsars (Sturrock, Baker & Turk 1976): pair creation is quenched at high altitudes since the rate is a strongly increasing function of $\mathbf{B}$ (e.g. Tsai & Erber, 1974).

The first generation of pair creation initiates the pair cascade, with pairs generally being created in excited transverse (to the field) momentum states, the so-called Landau levels. De-excitation via cyclotron and synchrotron radiation is then extremely rapid, on timescales of 10^{-16} sec or less for typical neutron star fields of $B_0 \gtrsim 10^{12}$ Gauss (n.b. subscripts zero denote polar surface fields). These secondary photons can then travel to higher altitudes and create further pairs and successive generations of photons. So proceeds the cascade, with a typical number of generations being around 3–4, and the total number of pairs per primary electron being of the order of $10^3 - 10^4$ (Daugherty & Harding

1982). The cumulative product is an emission spectrum that comprises a curvature/inverse Compton continuum that is cut off at hard gamma-ray energies by pair creation, with the addition of several synchrotron components at successively lower energies, terminating when the magnetosphere becomes transparent to $\gamma \to e^+e^-$ at significant altitudes. The details of such spectral formation are discussed below. A notable exception to this cascade scenario arises in highly-magnetized pulsars, PSR 1509–58 (with $B_0 \sim 3 \times 10^{13}$ Gauss) being the case in point. When the surface polar field B_0 exceeds around 6×10^{12} Gauss, pairs are produced in the zeroth (ground state) Landau level (Baring & Harding 2001, hereafter BH01), so that cyclotron/synchrotron emission is prohibited. Cascading is then effectively squelched and the pair yield diminished (BH01). A possible amelioration of this circumstance was posited by Zhang & Harding (2000a), namely the Landau level excitation of higher generation pairs via Compton scatterings with X-rays from the surface. Since such excitation can only arise above the cyclotron resonance (e.g. Gonthier et al. 2000), Baring & Harding (2001) determined that the population of excited Landau states relative to that in the ground state is small.

A more fascinating variation on the cascade theme involves the phenomenon of photon splitting, again applicable to high field pulsars. Magnetic photon splitting, $\gamma \to \gamma\gamma$, a third-order (in the fine structure constant $e^2/\hbar c$) quantum electrodynamical (QED) process in which a single photon splits into two lower-energy photons (Adler 1971, Baring & Harding 1997), operates efficiently and competes effectively with pulsar pair production only in magnetic fields above $\sim 10^{13}$ Gauss (Harding, Baring & Gonthier 1997, hereafter HBG97). This region of high magnetic field strength lies in the upper-right part of the $P - \dot{P}$ diagram. Splitting is forbidden in field-free regions by Furry's theorem, a symmetry property of QED. In regimes of weak or modest fields when vacuum dispersion effects are small, $\gamma \to \gamma\gamma$ is a collinear process, conserving both energy and momentum. The rate of splitting, like that of magnetic pair creation $\gamma \to e^+e^-$, is generally a rapidly increasing function of field strength (the exception being at fields $B \gtrsim 10^{14}$ Gauss), photon energy and photon propagation angle with respect to the field. However, splitting possesses no energy threshold, so that it can and does dominate the first order process of pair creation if $\mathbf{B}$ is sufficiently high. This leads to an alternative channel for cascade cessation, with gamma-rays being reprocessed without yielding pairs so that synchrotron generations are suppressed. The result is distinctive bumps and polarization signals in the EGRET/Comptel band (HBG97). The issue of splitting-influenced pulsar cascades is addressed below in Section 3 and in depth in HBG97 and Baring & Harding (2001).

3. PREDICTIONS OF POLAR CAP MODELS

Having assembled the basic ingredients of pulsar cascades, the task here is to identify the array of radiative signatures that are the hallmark of polar cap models. Since the cascade physics differs according to the period P and period-derivative $\dot{P}$ of the pulsar, for the purposes of this discussion a division is made between the canonical Crab-like/Vela-like young pulsars with moderately strong fields, and their more highly-magnetized cousins like PSR 1509–58. The focus in this section is mainly on pulsars whose beam sweeps across the line of sight to Earth, so-called *on-beam* pulsars; off-beam pulsars will also be mentioned briefly.

3.1. CRAB-LIKE AND VELA-LIKE PULSARS

Here we consider "standard" bright young pulsars like the Crab, Vela and Geminga with moderately high, but not extremely high, surface polar fields B_0. These constitute the majority of EGRET/Comptel pulsars, with PSR 1509–58 (addressed in the next subsection) and the millisecond pulsar PSR J0218+4232 (Kuiper et al. 2000) being the notable exceptions. Spectral properties will be the first focus. From the discussion of cascade properties above, this case corresponds to synchrotron-curvature cascades if the pulsar period is substantially shorter than a second. The curvature spectrum is generated by a quasi-monoenergetic electron injection that results from the rapid electrodynamic acceleration. Since the curvature mechanism is essentially identical to the synchrotron one, except that the relevant curvature scale is the radius of field curvature as opposed to the gyroradius, simple synchrotron formalism can be applied (e.g. see Jackson 1975) to yield a spectrum of $\varepsilon^{-2/3}$ below a maximum cutoff energy. This result holds provide that there is no cooling during emission, which is generally not the case. With curvature cooling operating (i.e. when $P \gtrsim 0.3\,\text{sec}$), the spectrum steepens to $\varepsilon^{-5/3}$ (e.g. Daugherty & Harding 1982), a power-law that extends down to energies at which cooling becomes inefficient, and then the flat $\varepsilon^{-2/3}$ form is again assumed.

Superposed on this are contributions from synchrotron emission from successive generations. Since these components result from pairs that are created by the photon spectrum belonging to a previous cascade generation, and the pair injection traces the input photon spectrum, it is straightforward (e.g. Wei, Song & Lu 1997; Harding & Daugherty 1998; Baring & Harding 2000) to determine that the spectral index α_i of the ith generation (i.e. for $dn_\gamma^{(i)}(\varepsilon)/d\varepsilon \propto \varepsilon^{-\alpha_i}$) satisfies the recurrence relation $\alpha_{i+1} = (\alpha_i+1)/2$. This result assumes that synchrotron cooling of the pairs rapidly depopulates excited Landau levels and steepens the pair spectrum by an index of unity. It then follows that (Harding & Daugherty 1998)

$$\alpha_n = 2 - \frac{2-\alpha_1}{2^{n-1}} \quad , \tag{12.1}$$

where the primary index is $\alpha_1 \approx 5/3$. Hence, each generation steepens in spectral index, so that, due to its preponderance of photons, the last generation determines the emergent cascade index. Notice that the asymptotic index for a large number of generations is 2. This automatically creates difficulties within this framework for pulsars with spectra steeper than ε^{-2}, the Crab being the notable case. However, it must be remembered that such simple analytic determinations are "one-zone" computations, and that the cascade spans a range of altitudes and field geometries, all of which modify this picture of steepening. In particular, the finite generational energy degradation $\chi = 0.5\max[B/B_{\rm cr},\, 0.1]$ (i.e. such that photon energies satisfy the recurrence relation $\varepsilon_{i+1} \sim \chi\varepsilon_i$) disrupts the idealized picture of power-laws generating power-laws, since the structure introduced by pair creation cutoffs (discussed below) influences successive generations to introduce additional steepening. Note that, hereafter, $B_{\rm cr} = 4.413\times 10^{13}$ G denotes the quantum critical field.

The spectral index is not a free parameter, but is determined roughly by noting that cascade cessation occurs when the mean free path for pair creation is comparable to the stellar radius R_0, i.e. the scalelength for field decline. This then establishes (e.g. Harding, private communication; Baring & Harding 2000) the effective maximum number of cascade generations (permitted to be a non-integer, following Lu, Wei & Song 1994)

$$n = 2 + \frac{1}{\log_e \chi}\,\log_e\left[\frac{4}{3\pi}\,\frac{B_{\rm cr}}{B_0\gamma_1^3}\left(\frac{r}{R_0}\right)^4 \frac{Pc}{\lambda\!\!\!^{-}_{\rm c}}\right] \tag{12.2}$$

for pulsar period P (in seconds) and electron primary Lorentz factor γ_1. Here $\lambda\!\!\!^{-}_{\rm c}$ is the Compton wavelength over 2π. Observe the appearance of a local field $B_0(r/R_0)^{-3}$ factor. This relation provides a closed system, establishing the spectral index as a function of spin-down parameters P and B_0 and the (unknown) altitude $r-R_0$ of typical emission above the surface. Variations on this *generation index* formulation can be found in Lu, Wei & Song (1994) and Wei, Lu & Song (1997), and also in Zhang & Harding (2000a) who treat (resonant) inverse Compton cascading as well to probe spectral signatures in the X-ray band. While details of "altitudinal smearing" will muddy this spectral index determination (as is evident in the comparison of such predictions with EGRET pulsar characteristics in Harding & Daugherty 1998), one expects the overall global trends with B_0 and pulsar period to be approximately true for polar cap models, providing a powerful observational diagnostic. Generally, the number of generations and the spectral index are *increasing functions of* B_0, *but decreasing functions of the altitude.*

It must be emphasized that the mean spectral index of EGRET UIDS that are associated with the Gould Belt is around 2.25 (Grenier et al. 2000; see also Gehrels et al. 2000), greater than that of most EGRET pulsars; resolution of this mismatch may be found in the discussion of off-beam pulsars below. A brief comment on the phase-dependence of spectra is also warranted, since pulse-phase spectroscopy is an objective of pulsar studies, attainable for a large number of pulsars in the *GLAST* era. Constraints on model phase space are narrowed considerably by exploration of the spectral variations with phase, though both polar cap (Daugherty & Harding 1996; Dyks & Rudak 2000) and outer gap (Romani 1996) models have successfully accounted for Vela's properties (see also Zhang & Cheng 2001 for an outer gap consideration of Geminga). A characteristic that has emerged in gamma-ray pulsars is that their trailing peak and interpulse spectra are flatter (e.g. Thompson 2001; see also Kanbach et al. 1994 for Vela details) than their leading peak spectra, features that need to be modelled. The interpulse emission is generally expected to be harder in the polar cap model, since it is primarily due to lower altitudes where synchrotron components contribute less substantially (Daugherty & Harding 1996), and the spectrum possesses mostly primary curvature radiation. The range of altitudes probed is a function of geometrical perspective, and hence also of pulse phase. Given the number of free parameters in models, it is critical to survey an array of pulsars in model/data comparisons of spectral and temporal properties.

The next major spectral feature is the maximum energy of emission, which is controlled by attenuation due to pair creation during photon propagation through the pulsar magnetosphere. This attenuation leads to the reprocessing that spawns lower energy photons, and provides a characteristic super-exponential turnover (e.g. Daugherty & Harding 1996) that contrasts that expected in outer gap models (e.g. see Thompson 2001 for a comparison). $\gamma \to e^{\pm}$ occurs at the threshold $\varepsilon \sin\theta_{\rm kB} = 2$ for $B \gtrsim 0.1 B_{\rm cr}$ and above threshold at $\varepsilon \sin\theta_{\rm kB} \sim 0.2 B_{\rm cr}/B$ for lower fields (e.g. see Daugherty & Harding 1983). Here, $\theta_{\rm kB}$ is the angle of photon propagation relative to $\mathbf{B}$, and hereafter photon energies ε are expressed in units of $m_e c^2$. Hence, the mean free path for photon attenuation in *curved* fields is $\lambda_{\rm pp} \sim \rho_c/\varepsilon \max\{2,\, 0.2/B\}$, i.e. when $\varepsilon \sin\theta_{\rm kB}$ crosses above threshold.

The radius of field curvature is $\rho_c = [Prc/2\pi]^{1/2}$ for a pulsar period P. Pair creation cutoff energies $\varepsilon_{\rm MAX}$, derived from the codes developed in HBG97 and Baring & Harding (2001), are plotted in Figure 1. While these are refined estimates, including the effects of general relativity on spacetime curvature, field enhancement and photon energy, their empirical dependence on B_0, R_0 and pulsar period P (in seconds) can be summarized in the relation (see also Harding, 2001a)

$$\varepsilon_{\rm MAX} \approx 0.4\sqrt{P}\left(\frac{r}{R_0}\right)^{1/2} \max\left\{1,\ \frac{0.1\,B_{\rm cr}}{B_0}\left(\frac{r}{R_0}\right)^3\right\} \text{GeV} . \qquad (12.3)$$

Refinements to this estimate to include the effects of photon splitting in higher fields (near $B_{\rm cr}$) are discussed in Baring & Harding (2001). The overall trend is clear: there is a strong anti-correlation between the maximum energy and the surface magnetic field, which seems to be augmented by an apparent decline of emission altitude with B_0. Such a trend is a distinctive characteristic that can be probed by *GLAST* and is unlikely to be reproduced by outer gap models. Note also, that the maximum energy is generally in the 1–10 GeV band for normal young pulsars, can be much lower (e.g. HBG97, BH01) for highly magnetized ones, and also much higher for millisecond pulsars (Bulik, Rudak & Dyks 2000) so that sub TeV-band signals are possible for polar cap models via synchrotron/curvature cascades if the field is low enough. It also should be remarked that the cutoff energy depends on pulse phase, with slightly greater values achieved between the pulse peaks in the case of Vela modelling (Daugherty & Harding 1996); such a property matches the EGRET observations (Kanbach et al. 1994).

The other major spectral feature in the gamma-ray band corresponds to the lower energy of the cascade, which from Monte Carlo simulations (e.g. Daugherty & Harding 1982) corresponds to Lorentz factors $\gamma_{\rm MIN}$ of around 50–100 for Vela-like pulsars. The synchrotron photon energy for this Lorentz factor is $\varepsilon_{\rm MIN} \sim \gamma_{\rm MIN} B/B_{\rm cr}$, noting that a factor of $1/\gamma_{\rm MIN}$ is introduced to account for the cascade beaming. This energy is typically in soft gamma-rays, and generally $\gamma_{\rm MIN}$ depends on B_0, P, etc. in more or less the same manner (Baring & Harding 2000) that $\varepsilon_{\rm MAX}$ does in Eq. (12.3), since the same pair creation physics applies to both. However, the reprocessing that leads to the establishment of $\varepsilon_{\rm MIN}$ occurs at lower altitudes (yielding possibly different parametric dependences) than the ultimate attenuation that defines $\varepsilon_{\rm MAX}$, so that in general $\varepsilon_{\rm MIN} \ll \varepsilon_{\rm MAX}$; it is difficult to be more specific than this inequality. In addition, since there is probably little altitudinal dependence in $\varepsilon_{\rm MIN}$, contrasting that for $\varepsilon_{\rm MAX}$ inferred from Figure 1, one might expect a spectral "narrowing" with increasing B_0. The spectrum possesses a

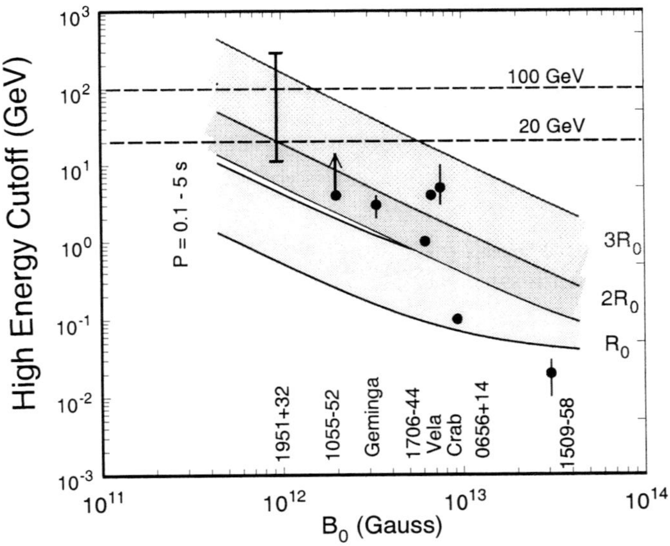

Figure 1 Maximum pulsar emission energies (from Baring & Harding 2000) imposed by pair creation attenuation at different altitudes, described empirically via Equation (12.3). For each altitude, a range of pulse periods (polar cap sizes) is represented by a shaded band. These energies are determined by the more involved photon propagation/attenuation code described in Baring & Harding (2001). Inferred cutoff energies (or ranges) for 8 gamma-ray pulsars of different B_0 are indicated, from which a trend of declining altitude of emission with increasing B_0 is suggested.

break at $\varepsilon_{\rm MIN}$, below which it assumes the flat $\varepsilon^{-2/3}$ form that signifies curvature or synchrotron emission from quasi-monoenergetic pairs. Such a slope is consistent with the broad-band optical/ hard X-ray non-thermal continuum of the Vela pulsar that lies underneath the thermal surface X-rays (e.g. see Pavlov et al. 2001 for a spectrum). Note that when the surface field becomes sufficiently high, the cyclotron scale imposes structure on the spectrum and redefines the variation of $\varepsilon_{\rm MIN}$ with B_0; this situation is sampled by the highly-magnetized pulsars.

3.2. PSR 1509–58 AND HIGH B PULSARS

Having established the generic properties of the more familiar gamma-ray pulsars, it is salient to move on to their more highly-magnetized siblings. The operative cascade physics is identified in Section 2. High fields of pulsars like PSR 1509–58 inhibit cascading (BH01) via the suppression of pair creation. This removes the complexity of the sequence of synchrotron components and leaves a bare curvature (or inverse Compton) primary electron spectrum, which is generically flat. The spectral index of PSR 1509–58, roughly 1.6 (e.g. see HBG97), is consistent with curvature emission from cooled primaries. Accordingly, it can be deduced that increasing the surface field must at some point flatten the pulsar continuum. This starkly contrasts the inference from Eq. (12.1) of steepening with increasing B_0 ; this trend reversal, probably at around $B_0 \sim 10^{13}$ Gauss, is a distinctive prediction of the polar cap model. Therefore, to address the focus of this paper, based on these spectral slope issues, it appears unlikely that *on-beam* high field pulsars are optimal candidates for EGRET UIDs, which appear to have spectra steeper on average than ε^{-2} , at least those correlated with the Gould Belt (Gehrels et al. 2000, Grenier et al. 2000; see also the reviews by Grenier and Gehrels in these proceedings). Note that the determination of UID spectral indices is subject to selection effects against detecting steep spectrum sources; this suggests that the true average may be even steeper.

This pessimism is reinforced by the expectation that high field pulsars have low maximum energies, according to the pair creation turnovers predicted and observed in Figure 1. This is dramatically emphasized by the lack of an EGRET detection for PSR 1509–58. Such an absence of emission is extremely constraining on spectral models, and was exploited by HBG97 to assert that photon splitting was acting in PSR 1509–58. Splitting attenuates photons at somewhat lower energies than does pair creation in such high B pulsars (see BH01 for details). This implies that the true $\varepsilon_{\rm MAX}$ curves in Figure 1 lie somewhat lower than the $\gamma \to e^{\pm}$ ones displayed at $B \gtrsim 10^{13}$ Gauss when $\gamma \to \gamma\gamma$ is taken into account. HBG97 observed that splitting naturally accounts (pair creation alone cannot, as is obvious from the figure) for the inferred turnover at around 10–30 MeV in PSR 1509–58 if a standard polar cap size is assumed and general relativity is incorporated in photon transport calculations. This study provided a nice piece of circumstantial evidence for the action of photon splitting, an interesting prospect for physics.

Returning to the EGRET UIDs, these opacity constraints indicate that high B sources can only be candidates if their emission regions are at relatively high altitudes, for which there is no confirmed observational evidence. Nevertheless, if the galactic UIDs turn out to be high B pulsars, they would then be expected to have steep spectra if the cutoff always matches the EGRET band, an improbable fine-tuning. The cutoffs would not be super-exponential, the signature of pair creation, but rather more gradual if splitting is operating. In such a case, there would be a segregation of flatter Crab-like and Vela-like pulsars in the galactic plane, and steep-spectrum higher field pulsars at low to moderate galactic latitudes. How such a correlation between B_0 and latitude (i.e. perhaps also kick velocity) would be attained is presently unclear.

Before closing this subsection it is desirable to make a small pitch for **gamma-ray polarimetry**. This is perhaps most relevant to high B pulsars, but is still quite important for the more common EGRET pulsars. Gamma-ray polarimetry is traditionally a haven for skeptics, though the mood of the high energy astrophysics community is rapidly changing given the prospects (Lei, Dean & Hills, 1997) that the *INTEGRAL* mission will detect polarization at the 10% level from the Crab pulsar (at 200–600 keV), and also in a handful of other sources. Hard gamma-ray experiments like *GLAST* are generally not afforded the opportunity to act as polarimeters, being limited by multiple scattering in trackers above 300 MeV. Despite early estimates of *GLAST*'s potential polarimetric capability (Yadigaroglu 1997), current design precludes a major focus on this observational goal. Medium energy gamma-ray experiments, on the other hand, are ideally suited to polarization studies (via their sampling of Compton scattering kinematics), and accordingly considerable emphasis has been placed in recent workshops on such new developments for next-generation advanced Compton telescopes (e.g. see Kanbach et al. 2000, and the Web pages for the MEGA [`http://www.gamma.mpe-garching.mpg.de/MEGA/mega.html`] and ACT [`http://gamma.nrl.navy.mil/ngram/`] consortia). Science motivations for polarimetry are obvious for pulsars. The presence of strong fields virtually guarantees a strong polarization signal in polar cap models, and when these couple with spectral structure and temporal information, particularly powerful observational diagnostics are achieved. This may be fruitful at the lower end of the cascade continuum in Vela-like objects, but it is an especially valuable tool for highly-magnetized pulsars since the attenuation cutoffs fall in the Comptel band, and should exhibit strong and distinctive polarization signatures. A concerted effort to realize the historically ambitious goal of gamma-ray polarimetry may yield dramatic science gains in the near future.

3.3. MAGNETARS: NOT RELEVANT FOR UIDS?

A natural step from these considerations of high field pulsars is to magnetars, specifically anomalous X-ray pulsars (AXPs) and soft gamma repeaters (SGRs). These can be quickly dismissed as potential candidates for EGRET UIDs unless we have been unlucky in the observational process. To date, identified emission from these sources does not exceed 1 MeV for the SGRs (e.g. see data for giant flares in Mazets et al. 1981 for the 5th March 1979 [SGR 0525-66] event, and Hurley et al 1999 for SGR 1900+14) and considerably less for AXPs. Moreover, their spectra are typically steep, at least in quiescent epochs, generally precluding detection in the EGRET band. The possible exceptions are the so-called "initial hard spikes" in the two cases of giant flares of SGR 0525-66 (in 1979) and SGR 1900+14 (in 1998), neither of which was seen by a hard gamma-ray mission. Hence, observationally there is still the possibility that this particular mode of activity may generate super-MeV emission, a question that *GLAST* might be able to answer. From the theoretical viewpoint, high energy emission is not readily expected from these sources, if the magnetar interpretation is adopted. This is because the magnetospheric opacity due to pair creation and photon splitting will inhibit escape of photons above around 50–200 MeV (e.g. see Figure 9 of BH01). This bound applies specifically to a "pulsar mode," where (e.g. giant flare) emission is strongly coupled to a beam near the polar cap; extending to equatorial regions enhances the opacity and can force the spectrum down into the hard X-ray/soft gamma-ray band (Baring 1995, Harding & Baring 1997), more commensurate with that seen in normal SGR outbursts and AXPs. The escape clause here is to reduce the strength of the ambient magnetic field, i.e. either to relinquish (untenable to some) the magnetar interpretation, or to move the giant flare emission region to higher altitudes (not necessarily outer gaps), a prospect that is difficult to reconcile with the enormous energy liberated in SGR giant flares.

3.4. RADIO QUIESCENCE AT HIGH B?

An issue that impacts the discussion of unidentified gamma-ray sources is the existence or otherwise of radio counterparts. This concerns highly-magnetized pulsars, if they can be radio quiet without dramatically inhibiting gamma-ray emission, as has been suggested by Baring and Harding (1998, see also Zhang & Harding 2000b). Since it is commonly assumed (e.g. Sturrock 1971; Ruderman & Sutherland 1975; Arons & Scharlemann 1979; for a dissenting view, see Weatherall & Eilek 1997) that a plentiful supply of pairs is a prerequisite for coherent radio emission at observable flux levels, any suppression of pair creation in pulsars implies that the emission of radio waves should be strongly inhibited. Baring & Harding (1998) posited the idea that photon splitting could effect a suppression of pair creation by providing a more competitive mode of photon attenuation for high polar fields. Accordingly, they predicted an approximate boundary in the P-$\dot{P}$ diagram that delineated pulsars of lower $\dot{P}$ (or B_0) that could be radio-loud, and those of unusually large period derivative, which where necessarily radio quiet due to the action of splitting. The boundary was:

$$\dot{P} \approx 7.9 \times 10^{-13} \left(\frac{P}{1\,\mathrm{sec}}\right)^{-11/15} \tag{12.4}$$

The fact that this *boundary of quiescence* neatly separated members of the 1995 version of the Princeton Pulsar Catalog (Taylor et al. 1993) from the small family of purportedly radio-quiet magnetars (i.e. AXPs and SGRs), was an enticing piece of support for the proposition. Yet this concept pre-dated results from the new Parkes Multi-Beam survey [`http://www.atnf.csiro.au/~pulsar/psr/pmsurv/pmwww/`] that has discovered a small number of pulsars of higher magnetization than previously known (e.g. Camilo et al. 2000), with three lying above this putative quiescence boundary (BH01). This development proves not to be unduly disturbing, since only a small change in the emission altitude can comfortably accommodate the new detections (BH01). Moreover, the "polarity" of the rotating magnetosphere can influence the nature of the acceleration gap, with significant consequences for the boundary of quiescence (Zhang & Harding 2000b). What is more telling, from an observational perspective, is that one of the radio pulsars recently discovered in the Parkes Multi-Beam Survey, PSR J1814–1744, lies very close to the anomalous X-ray (AXP) pulsar CTB 109, so that a single radio quiescence line cannot separate the radio pulsar and AXP/magnetar populations. This proximity coupled with the fact that PSR J1814–1744 has not been detected in X-rays (Pivovaroff, Kaspi & Camilo 2000) strongly suggests that a quantity other than P and B_0 has a profound influence on the properties of highly-magnetized radio pulsars and AXPs.

From a theoretical standpoint, the suppression of pair creation at high fields by photon splitting is not unequivocal. The extensive investigation by Baring & Harding (2001) of photon propagation and attenuation in general relativistic magnetospheres revealed that suppression was significant only if both polarization states (in the external magnetic field) of photons could split, rather than just one. This point addresses a subtlety of QED dispersion of the magnetized vacuum, principally in relation to selection rules derived by Adler (1971): only one polarization state is amenable to splitting in the limit of weak to moderate vacuum dispersion. While almost certainly applicable to typical pulsars, this contention, based on the leading order contribution to the dispersive properties provided by the vacuum, may or may not persist in supercritical ($B \gtrsim 4.41 \times 10^{13}$ Gauss) fields where higher order QED corrections become operative. Hence, whether or not splitting can act to inhibit pair creation critically depends on this unanswered question of physics, the mathematical solution of which is potentially difficult or intractable.

3.5. OFF-BEAM PULSARS

The situation concerning on-beam pulsars motivates an expansion of perspective. From the foregoing discussions, EGRET-type pulsars typically have spectra flatter than EGRET UIDs, and high field counterparts might match the UID slopes if a conspiracy establishes their turnovers at just the right energies to mimic the steeper UID spectra. In the absence of a comfortable explanation of UIDs within the context of on-beam polar cap pulsars, Harding & Zhang (2001) recently proposed *off-beam* pulsars as candidates for the unidentified sources. Effectively, the line of sight to Earth does not cut the rim of the polar cap in these sources, but rather samples a broader (spatial) wing corresponding to higher altitudes above the neutron star surface, from which the emission is typically at lower energies, but with a harder spectrum. The hardness originates in the curvature primary photons, with a simultaneous drop in the maximum energy due to the combination of pair creation attenuation and field geometry. The net effect is that the spectrum in the EGRET band is steeper for these sources, due to the influence of a cutoff in the near-GeV range, however the solid angle of emission increases from that of on-beam pulsars. Hence, pulsation searches will be biased towards on-beam pulsars despite off-beam ones constituting a larger percentage of the population. While an attractive proposition in several ways, this suggestion still mandates some fine-tuning of the observational perspective to generate spectra that match UID observations. This issue plagues the high-field pulsar explanation also, and in fact, these two alternatives pose a challenge: how can one discriminate between off-beam/moderate B and on-beam/high B scenarios given that they display similar spectral properties. The answer may be provided by population statistics.

4. GLOBAL PROPERTIES FOR POPULATION STUDIES

4.1. GAMMA-RAY LUMINOSITIES

As indicated in the Introduction, one of the principal successes of the polar cap model is its prediction (Harding 1981) of an almost linear correlation between the inferred luminosity of gamma-ray pulsars and B_0/P^2, i.e. the voltage on their last open field lines for standard polar caps. This correlation, while not exact, largely due to the uncertainty in determining source distance by folding (radio) dispersion measures into the Taylor-Cordes (1993) galactic electron model, is distinctly different from the canonical spin-down luminosity, $\propto B_0^2/P^4$, which is mirrored by the X-ray pulsar population (Becker & Trümper 1997). An enticing feature of this prediction was that only 2 gamma-ray pulsars were known at the time it was proposed, and subsequent predictions by competing analyses/models (e.g. Sturner & Dermer 1994; Romani & Yadigaroglu 1995; Cheng & Zhang 1998; Rudak & Dyks 1999) and revisions (Zhang & Harding 2000a) all post-dated the EGRET database. The current status is that the polar cap expectations (Sturner & Dermer 1994; Zhang & Harding 2000a) match the data more accurately than their outer gap counterparts (Romani & Yadigaroglu 1995; Cheng & Zhang 1998), with each group of researchers offering different B_0 and P dependences for the luminosity (see Harding 2001a for a review). To some extent, this situation is limited by small number statistics, an issue that will be irrelevant in the *GLAST* era, when such correlations will be established on a really firm basis.

Setting aside partisan theoretical justifications, this observational correlation motivates a revision of historical thinking. Traditionally, the EGRET community has used the spin-down luminosity B_0^2/P^4 as an indicator of a pulsar's observability. While theoretically motivated in some sense, this choice does not match the established trend, and can dictate periods that are selected in pulsation searches. While this has netted most pulsars high up on a $B_0^2/P^4/d_{\rm PSR}^2$ rank-ordered list (where $d_{\rm PSR}$ is the pulsar distance), certain gamma-ray pulsars (notably the longer period pulsars PSR 0656+14 and PSR 1055–52) are surprisingly low in spin-down luminosity, and millisecond pulsars have proven extraordinarily difficult to detect (up till PSR 0218+4232, see Kuiper et al. 2000) given their short periods. Clearly, a gamma-ray luminosity dependence $L_\gamma(P, \dot{P})$ that differs from the spin-down one will dramatically modify the observability criterion, particularly if the period dependence is substantially different. Furthermore, the spectral shape also influences the observability (Baring & Harding 2000), a more subtle influence. This is a consequence of how the luminosity is distributed in the gamma-ray band, specifically that portion that emerges above the threshold sensitivity for a specific gamma-ray detector. The driving parameters for such an apportionment are $\varepsilon_{\rm MAX}$, index α_n, and to a lesser extent $\varepsilon_{\rm MIN}$, since the spectra are generally flat enough for the bulk of the luminosity to emerge at the highest energies. These parameters control the normalization of the pulsar gamma-ray power-law.

An appropriate definition of a detector's observability $\mathcal{O}(\varepsilon_{\rm TH})$ is the *integral* flux above an effective instrumental energy threshold $\varepsilon_{\rm TH}$. For pulsars with $\varepsilon_{\rm MIN} \ll \varepsilon_{\rm TH}$, the usual case for *GLAST* considerations, this scales as the luminosity divided by the spectral normalization, yielding $\mathcal{O}(\varepsilon_{\rm TH}) \propto L_\gamma(P,\dot{P})\,\varepsilon_{\rm MAX}^{(\alpha_n-2)}/d_{\rm PSR}^2$ (Baring & Harding 2000). Modifications to this dependence are possible, in particular if $\varepsilon_{\rm MIN} \gtrsim \varepsilon_{\rm TH}$, in which case $\mathcal{O}(\varepsilon_{\rm TH}) \propto L_\gamma(P,\dot{P})/\varepsilon_{\rm MIN}/d_{\rm PSR}^2$. Either of these possibilities yields substantially different observabilities from the spin-down formula, assuming that $\varepsilon_{\rm MAX}$ and $\varepsilon_{\rm MIN}$ scale with B_0 and P approximately as the low field alternative offered in Eq. (12.3). This leads to the conclusion that observabilities predicted for *GLAST* pulsation searches should follow a dependence somewhere in between B_0/P^2 and $B_0^2/P^{5/2}$. Using the latter possibility, Baring & Harding (2000) generated a revised rank-ordered listing that indicated a dramatic rearrangement from the traditional EGRET ordering. Notable changes included the much higher ranking of the "outlier" longer period gamma-ray pulsars PSR 0656+14 and PSR 1055–52, and the marked lowering of millisecond pulsars (PSR 1939+2134, PSR 0437–4715, PSR 1744–1134, etc.) in the ranks, specifically out of the top 40. Both of these reflect the weaker dependence of the revised observability on P. The old and new rankings are listed in Table 1 for the confirmed and candidate (non-millisecond) gamma-ray pulsars; it becomes clear that such revisions mute questions of why PSR 1055–52 was seen by EGRET. Refinements of such rank orderings are in progress, an interesting preparatory step for the *GLAST* mission.

Table 1 CGRO Gamma-Ray Pulsars and rank-ordered listings

PSR	*P (sec)*	$\dot{P}$	$B_0^2/P^4/d_{\rm PSR}^2$ *rank*	$B_0^2/P^{5/2}/d_{\rm PSR}^2$ *rank*
Vela	0.089	1.25×10^{-13}	2	1
Crab	0.033	4.21×10^{-13}	1	2
Geminga	0.237	1.1×10^{-14}	4	3
1509–58	0.150	1.5×10^{-12}	5	5
1706–44	0.102	9.3×10^{-14}	7	6
0656+14	0.385	5.5×10^{-14}	20	13
1951+32	0.040	5.85×10^{-15}	6	14
1055–52	0.197	5.83×10^{-15}	33	23

4.2. GAMMA-RAY VS. RADIO OBSERVABILITY

Pulsation searches for EGRET UIDs that have spatial associations with known radio pulsars naturally bias the search phase space in P

and $\dot{P}$. Yet there is no guarantee that every gamma-ray pulsar (or EGRET UID) is a radio pulsar, and vice versa. The polar cap and outer gap models make distinctly different predictions of the correlation between observing gamma-ray and radio emission from pulsars, based largely on assumed emission region geometries. Outer gap models (e.g. Romani & Yadigaroglu 1995; Zhang et al. 2000) suggest an almost complete disconnect between emission in the two wavebands so that detected (normal as opposed to high B) gamma-ray pulsars should mostly be radio-quiet due to the much larger solid angle in the gamma-ray beam. The fact that Geminga is the only radio-quiet gamma-ray pulsar may be bothersome to outer gap proponents. Perhaps more disconcerting is that the best determinations of altitudes for the radio emission (e.g. Gil & Han 1996) may suggest more of a connection with polar caps than outer gaps. Even if the origin of both radio and gamma-ray emission is connected to pair creation in a polar cap, the solid angles of these components should be somewhat different. There are large uncertainties present in any prediction of the ratio of numbers of gamma-ray and radio pulsars, since models must incorporate details of the distribution of fields and periods at birth, the spatial and velocity distributions of pulsars in the galaxy, the influence of galactic gravitational potentials, and luminosity and solid angle geometry prescriptions for both the gamma-ray and radio emission. Assembling such ingredients, both Sturner and Dermer (1996) and the very recent analysis of Gonthier et al. (2001) find that only a minority of EGRET pulsars would be expected to be radio-quiet. This fraction increases substantially for *GLAST* to an almost 50/50 radio-quiet/radio-loud situation according to Gonthier et al. (2001); this result is a consequence of *GLAST*'s improved sensitivity enabling it to sample deeper than typical radio surveys. Hence blind pulsation searches shall be a much more salient tool for *GLAST*, and the gamma-ray UID community may well have to forgo attachments to radio pulsar counterparts.

5. CONCLUSION

This review has detailed some of the expectations for polar cap pulsar properties that will prove extremely useful subsequent to pulsar identification for gamma-ray sources. These include spectral trends with B_0 and P, identifying the science gains to be made given a considerable gamma-ray pulsar database in the *GLAST* era. Yet the discussion has also elucidated possible guidelines for UID pulsation searches so that pulsar science is not merely a *post-facto* consideration for the identification process. It is clear that standard gamma-ray pulsars like the

Crab, Geminga and Vela might not constitute the majority of the subset of UIDs that eventually turn out to be pulsars, at least if the mean *GLAST* gamma-ray source characteristics are similar to those of the EGRET UIDs. Turning to off-beam pulsars or highly-magnetized on-beam ones as candidates does not dramatically change the search phase space (P and $\dot{P}$) from current agendas. Lowering the instrumental threshold energy as much as possible is a worthwhile goal, provided that it does not compromise threshold sensitivity and angular resolution properties. While it has been argued that observability criteria need revision from historical preferences, and that there is probably less need to be biased against searching at longer periods, since magnetars appear to be unlikely candidates for UIDs, there is no compelling reason to search on supersecond periods and in the domain $\dot{P} \gtrsim 3 \times 10^{-12}$ sec/sec. Continued identification efforts with the EGRET database and refinements to theoretical models in the next half decade will help set the stage for the watershed of gamma-ray identifications to be attained by *GLAST*.

References

Adler, S. L. 1971, Ann. Phys. **67**, 599.

Arons, J. 1983, ApJ **266**, 215.

Arons, J. & Scharlemann, E. T. 1979 ApJ **854**, 879.

Baring, M. G. 1995, ApJ **440**, L69.

Baring, M. G. & Harding A. K. 1997, ApJ **482**, 372.

Baring, M. G. & Harding A. K. 1998, ApJ **507**, L55.

Baring, M. G. & Harding A. K. 2000, AAS HEAD Meeting, Honolulu, HI, Bull. AAS **32**, 12.43.

Baring, M. G. & Harding A. K. 2001, ApJ **547**, 929 (BH01).

Becker, W. & Trümper, J. 1997, Astron. Astr. **326**, 682.

Bulik, T., Rudak, B. & Dyks, J. 2000, MNRAS **317**, 97.

Camilo, F., et al. 2000, ApJ **541**, 367.

Cheng, K. S. 1994, in *Towards a Major Atmospheric Čerenkov Detector*, ed. T. Kifune (Tokyo, Universal Academy) p. 25.

Cheng, K. S., Ho, C. & Ruderman, M. A. 1986, ApJ **300**, 500.

Cheng, K. S. & Zhang, L. 1998, ApJ **493**, L35.

Daugherty, J. K. & Harding A. K. 1982, ApJ **252**, 337.

Daugherty, J. K. & Harding A. K. 1983, ApJ **273**, 761.

Daugherty, J. K. & Harding A. K. 1989, ApJ **336**, 861.

Daugherty, J. K. & Harding, A. K. 1996, ApJ **458**, 278.

Dyks, J. & Rudak, B. 2000, MNRAS **319**, 477.

Gehrels, N., et al. 2000, Nature **404**, 363.

Gil, J. A. & Han, J. L. 1996, ApJ **458**, 265.

Gonthier, P. L. & Harding, A. K. 1994, ApJ **425**, 767.

Gonthier, P. L., Harding, A. K., Baring, M. G., Costello, R. M. & Mercer, C. L. 2000, ApJ **540**, 907.

Gonthier, P. L., et al. 2001, ApJ submitted.

Grenier, I. A. 2000, Astron. Astr. (Lett.) **364**, L93.

Harding, A. K. 1981, ApJ **245**, 267.

Harding, A. K. 2001a, *High Energy Gamma-Ray Astronomy*, eds. F. A. Aharonian, H. Völk, AIP Conf. Proc. 558, p. 115.

Harding, A. K. 2001b, to appear in *Proc. of Gamma 2001 Symposium*, eds. N. Gehrels, C. Shrader & S. Ritz (AIP, New York)

Harding A. K. & Baring, M. G. 1996, Proc. Huntsville Gamma-Ray Burst Workshop, eds. Kouveliotou, C., Briggs, M. S., and Fishman, G. J., (AIP 384, New York) p. 941

Harding A. K., Baring, M. G. & Gonthier, P. L. 1997, ApJ **476**, 246 (HBG97).

Harding, A. K. & Daugherty, J. K. 1998, Adv. Space Res. **21(1/2)**, 251.

Harding, A. K. & Muslimov, A. G. 1998, ApJ **508**, 328.

Harding, A. K. & Zhang, B. 2001, ApJ **548**, L37.

Hirotani, K. 2000, ApJ **549**, 495.

Hurley, K. et al. 1999, ApJ **510**, L111.

Jackson, J. D. 1975, *Classical Electrodynamics,* (Wiley and Sons, New York)

Kanbach, G., et al. 1994, Astron. Astr. **289**, 855.

Kanbach, G., et al. 2000, AAS HEAD Meeting, Honolulu, HI, Bull. AAS **32**, 16.06.

Kuiper, L., et al. 2000, Astron. Astr. **359**, 615.

Lei, F., Dean, A. J. & Hills, G. L. 1997, Space Sci. Rev. **82**, 309.

Lessard, R. W. 2000, ApJ **531**, 942.

Lu, T., Wei, D. M. & Song, L. M. 1994, Astron. Astr. **290**, 815.

Mazets, E. P. et al. 1981, Adv. Space Sci. **80**, 3,85,119.

Muslimov, A. G. & Tsygan, A. I. 1992, MNRAS **255**, 61.

Nel, H. I., et al. 1993, ApJ **418**, 836.

Pavlov, G. G., et al. 2001, ApJ submitted [astro-ph/0103171].

Pivovaroff, M. J., Kaspi, V. M. & Camilo, F. 2000, ApJ **535**, 379.

Romani, R. W. 1996, ApJ **470**, 469.

Romani, R. W. & Yadigaroglu, I.-A. 1995, ApJ **438**, 314.

Rudak, B. & Dyks, J. 1999, MNRAS **303**, 477.

Ruderman, M. A. & Sutherland, P. G. 1975, ApJ **196**, 51 (RS75).

Sturner, S. J. 1995, ApJ **446**, 292.

Sturner, S. J. & Dermer, C. D., 1994, ApJ **420**, L79.

Sturrock, P. A. 1971, ApJ **164**, 529.

Sturrock, P. A., Baker, K. & Turk, J. S. 1976, ApJ **206**, 273.

Taylor, J. H. & Cordes, J. M. ApJ **411**, 674.
Taylor, J. H., Manchester, R. N. & Lyne, A. G. 1993, ApJ Supp. **88**, 529.
Thompson, D. J. 2001, *High Energy Gamma-Ray Astronomy*, eds. F. A. Aharonian, H. Völk, AIP Conf. Proc. 558, p. 103
Tsai, W.-Y. & Erber, T. 1974, Phys. Rev. D **10**, 492.
Usov, V. V. & Melrose, D. B. 1995, Aust. J. Phys. **48**, 571.
Weatherall, J. C. & Eilek, J. A. 1997, ApJ **474**, 407.
Wei, D. M., Song, L. M. & Lu, T. 1997, Astron. Astr. **323**, 98.
Yadigaroglu, I.-A. 1997, Exp. Astron. **7**, 221.
Yadigaroglu, I.-A. & Romani, R. W. 1995, ApJ **449**, 211.
Zhang, B. & Harding, A. K. 2000a, ApJ **532**, 1150.
Zhang, B. & Harding, A. K. 2000b, ApJ **535**, L51.
Zhang, L. & Cheng, K. S. 2001, MNRAS **320**, 477.
Zhang, L., Zhang, Y. J. & Cheng, K. S. 2000, Astron. Astr. **357**, 957.

IV

SUPERNOVA REMNANTS AND MASSIVE STARS

MASSIVE STARS AND GAMMA-RAY SOURCES

Philip Kaaret
Harvard-Smithsonian Center for Astrophysics
60 Garden Street
Cambridge, MA 02138
USA
pkaaret@cfa.harvard.edu

Keywords: gamma rays: observations, gamma rays: sources, pulsars: general, stars: individual: 2EG J0635+0521, 3EG J0634+0521, PSR J1259–63, SAX J0635+0533, stars: neutron

Abstract I describe statistical correlations between tracers of Population I objects and EGRET sources, present results from a multiwavelength search of the error box of the unidentified EGRET source in Monoceros, and speculate on a potential new class of gamma-ray emitting objects related to massive stars.

1. INTRODUCTION

Ever since their discovery more than 20 years ago, the nature of the unidentified gamma-ray point sources concentrated along the Galactic plane has been one of the outstanding puzzles in gamma-ray astronomy (Swanenburg et al. 1981). The chief impediment to progress in identification of the sources has been the large gamma-ray error boxes which invariably contain multiple optical and X-ray counterpart candidates. The only convincing gamma-ray source identifications for Galactic sources have been made through detection of a pulsed gamma-ray signal at a period known from radio or X-ray observations (Bignami & Hermsen 1983; Halpern & Holt 1992; Thompson et al. 1994; Ramanamurthy et al. 1995). However, only a small fraction of the total gamma-ray source population has been identified as pulsars and the question of the nature of the unidentified population remains.

A. Carramiñana et al. (eds.),
The Nature of Unidentified Galactic High-energy Gamma-ray Sources, 191–196.

2. STATISTICAL CORRELATIONS

Statistical analyses of correlations between the positions of the unidentified gamma-ray sources and candidate classes of counterpart objects have the potential to provide clues concerning the nature of the gamma-ray sources (e.g. Montemerle 1979). Studies carried out with EGRET data (Thompson et al. 1995) performed by Kaaret & Cottam (1996) showed that the EGRET sources along the Galactic plane are correlated with the position of OB associations, indicating that most of the Galactic EGRET sources are Population I objects associated with sites of massive star formation. As young, rotation-powered pulsars are the only known class of Galactic gamma-ray emitter and are progeny of population I objects, a tenable hypothesis is that most of the unidentified Galactic sources are rotation-powered pulsars (Kaaret & Cottam 1996; Yadigaroglu & Romani 1997).

However, there are several classes of potential gamma-ray emitters, namely supernova remnants and massive stars, which are also related to Population I objects and are well correlated with OB associations. In general, the EGRET sources can be shown to be well correlated with any tracer of Galactic Population I objects (see the review by Romero in this volume). Thus, while sufficient to establish that the majority of the Galactic gamma-ray sources are Population I objects, statistical correlations are insufficient to uniquely identify the gamma-ray sources with any particular class of Population I objects. Proof of source identifications requires detailed study of individual objects.

3. A YOUNG, BINARY PULSAR IN A GAMMA-RAY ERROR BOX

Much of the effort in identifying the Galactic gamma-ray sources discovered by EGRET has been directed towards finding potential X-ray counterparts. X-rays are the preferred band because the X-ray to gamma-ray flux ratios of the known gamma-ray sources are reasonably high, leading to the expectation that X-ray counterparts of the gamma-ray sources should exist. Given that the X-ray to gamma-ray flux ratios are higher than for the optical or radio bands, use of the X-ray band also helps to reduce the number of potential counterparts.

X-ray observations made using the BeppoSAX satellite of the error box of the gamma-ray source 2EG J0635+0521 (also 3EG J0634+0521) in Monoceros led to the discovery of a relatively bright X-ray source with a very hard X-ray spectrum, a power-law with photon index of 1.5 ± 0.08 with emission up to at least 40 keV (Kaaret et al. 1999). The source remains in the improved error box from the 3rd EGRET

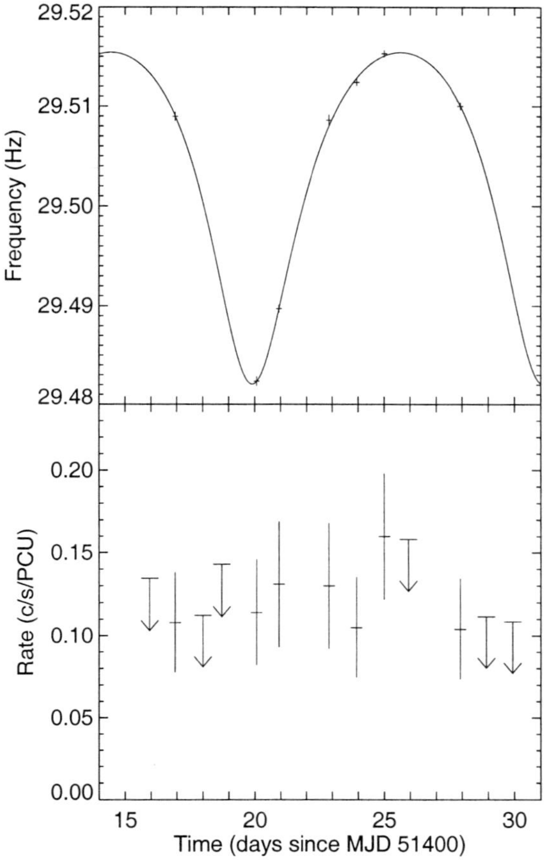

Figure 1 Pulsation frequency and flux versus time for SAX J0635+0533 (from Kaaret et al. 2000). The upper panel shows RXTE frequency measurements (crosses) and a fitted orbital profile (solid curve). The lower panel shows the pulsed count rate in the 4.4–23.6 keV band.

catalog for the corresponding source 3EG J0634+0521 (Hartman et al. 1999). Optical observations of the X-ray error box identified a Be-star optical counterpart, indicating that the compact object responsible for the X-ray emission was likely in a binary system (Kaaret et al. 1999). A pulsation search conducted by incoherently adding power spectra from relatively brief integration intervals in order to maintain good sensitivity even for pulsations with varying frequency due to orbital motion within the binary system led to the discovery of pulsations with a period of 34 ms (Cusumano et al. 1999).

Subsequent X-ray monitoring of SAX J0635+0533 using the Rossi X-Ray Timing Explorer led to measurement of the orbital parameters for

Table 1 Orbital Parameters of 2EG J0635+0521.

Parameter	Value
Orbital period	$11.2^{+0.5}_{-0.4}$ days
Projected semimajor axis	$a \sin i = 83^{+11}_{-8}$ lt s
Mass function	$5.0^{+2.8}_{-1.6}\, M_\odot$
Orbital eccentricity	$0.29^{+0.09}_{-0.06}$
Epoch of periastron	MJD $51419.8^{+0.5}_{-0.9}$
Longitude of periastron	$-4° \pm -24°$

the system, shown in Table 1, and also to an upper bound on the intrinsic frequency derivative of the system of $\dot{\nu} = -3.3 \times 10^{-10}\,\mathrm{Hz\,s^{-1}}$ equivalent to a period derivative $\dot{P} = 3.8 \times 10^{-13}$ (Kaaret et al. 2000). The high frequency derivative implies a very high torque on the system, which would require an accretion rate equal to 40 times the critical Eddington accretion rate. It is unclear how such a large rate could be present given the low X-ray luminosity observed from the system and the fact that pulsations (presumably originating from the neutron star surface) are observed. These considerations argue against the torque being due to accretion.

The torque and X-ray luminosity of SAX J0635+0533 are both consistent with those found for young, rotation-powered pulsars. In addition, the pulsed flux appears constant across all RXTE observations, as would be expected from a rotation-powered pulsar. In this case, the period derivative would imply a characteristic age $P/2\dot{P} \sim 1400$ years and a spin-down power $\dot{E} = 5 \times 10^{38}\,\mathrm{erg\,s^{-1}}$. These numbers are similar to those of the young, rotation-powered pulsars such as the Crab.

4. SPECULATION

Five of the top six radio pulsars ranked by spin-down power divided by distance squared are known gamma-ray sources. Added to the list, SAX J0635+0533 would have the third highest rank, and is, therefore, a likely gamma-ray source if it is, indeed, a rotation-powered pulsar. Improved X-ray timing, particularly if done simultaneously with future gamma-ray observations, may allow determination of an ephemeris sufficiently accurate to fold the gamma-ray data and potentially positively identify SAX J0635+0533 with 2EG J0635+0521.

Young pulsars in binaries may have sufficiently strong outflows to over-power accretion. This possibility was suggested by Illarionov &

Sunyaev (1975). In this case, the outflow from the rotation-powered pulsar would prevent the stellar wind from the companion from entering into the light cylinder of the pulsar, and the magnetospheric processes creating the pulsar emission would be unaffected by the presence of the companion. In essence, the pulsar would behave as an isolated rotation-powered pulsar. If so, then the X-ray and gamma-ray emission from such young pulsars in binaries may be similar to that from isolated pulsars. However, the radio emission would likely be absorbed in the stellar wind plasma.

The gamma-ray emitting phase would end if the stellar wind from the companion penetrates the light cylinder of the neutron star or if the pulsar slows sufficiently that the magnetospheric processes leading to gamma-ray emission turn off. The latter consideration implies that the life time will not exceed the $\sim 5\times10^5$ year life time of isolated gamma-ray pulsars. The time at which the stellar wind penetrates the light cylinder likely depends on the initial spin period and magnetic field of the neutron star and on the mass loss rate of the companion and the orbital geometry of the system. Radio pulsar/massive star binaries such PSR J1259–63 (Johnston et al. 1992) have life times of many millions of years, well in excess of the life times for isolated gamma-ray pulsars. However, the known radio pulsar/massive star binaries are highly eccentric and are unlikely to represent the true distribution of neutron star/massive star binaries (Lipunov et al. 1994). The orbital eccentricity of SAX J0635+0533 is much lower than that of any known radio pulsar/massive star binary.

The total number of binary gamma-ray pulsars can be estimated if the birth rate and the life time in the gamma-ray emitting phase is known. The birth rate of Be/X-ray binaries in the Galaxy is estimated to be of the order of $\sim 10^{-3}$/year. As these represent only one class of massive star/neutron star binary, the total birth rate may be higher. Using this birth rate and allowing gamma-ray life time estimates in the range of $10^3 - 10^5$ years leads to an estimate of the number of currently gamma-ray emitting young, binary pulsars in the Galaxy of 1-100. The fraction of gamma-ray emitting pulsars in binaries should be equal to the binary fraction for massive stars, estimated as 0.36 by Garmany, Conti, & Massey (1980), multiplied by the fraction of binary systems which survive their first supernova explosion leading to the birth of a neutron star, estimated as 0.43 by Bethe & Brown (1998). This leads to a fraction of 0.15 or roughly 1 in 7. The fraction of observable gamma-ray pulsars in binaries may be lower if the presence of the companion decreases the gamma-ray emitting lifetime as described above.

We emphasize that positive identification of SAX J0635+0533 with the gamma-ray source in Monoceros has not been obtained and thus, the comments in this section should be considered speculation. However, if the identification of SAX J0635+0533 with 2EG J0635+0521 is confirmed via detection of gamma-ray pulsations, then it would represent the first example of a new class of Galactic gamma-ray sources: young, rotation-powered pulsars in binaries.

Acknowledgments

I am grateful to Dr. Alberto Carramiñana and all members of INAOE for their hospitality during this conference.

References

Bethe, H.A. & Brown, G.E. 1998, ApJ, 506, 780

Bignami, G.F. & Hermsen, W. 1983, ARA&A, 21, 67

Cusumano, G., Maccarone, M.C., Nicastro, L., Sacco, B. & Kaaret, P. 2000, ApJ, 528, L25

Garmany, C.D., Conti, P.S. & Massey, P. 1980, ApJ, 242, 1063

Halpern, J.P. & Holt, S.S. 1992, Nature, 357, 222

Hartman, R.C. et al. 1999, ApJS, 123, 79

Illarionov, A. F., & Sunyaev, R. A. 1975, A&Ap, 39, 185

Johnston, S. et al. 1992, ApJ, 387, L41

Kaaret, P. & Cottam, J. 1996, ApJ, 462, L35

Kaaret, P., Piraino, S., Halpern, J. & Eracleous, M. 1999, ApJ, 523, 197

Kaaret, P., Cusumano, G. & Sacco, B. 2000, ApJ, 542, L41

Lipunov, V.M., Nazin, S.N., Osminkin, E.Yu. & Prokhorov, M.E. 1994, A&Ap, 282, 61

Montemerle, T. 1979, ApJ, 231, 95

Ramanamurthy, P.V. et al. 1995, ApJ, 447, L109

Swanenburg, B.N. et al. 1981, ApJ, 243, L69

Thompson, D.J. et al. 1994, ApJ, 436, 229

Thompson, D.J. et al. 1995, ApJS, 101, 259

Yadigaroglu, I.-A. & Romani, R.W. 1997, ApJ, 476, 356

DIFFUSE SOURCES OF HIGH-ENERGY GAMMA RAYS IN THE MILKY WAY

Seth W. Digel
Universities Space Research Association
Code 661, NASA Goddard Space Flight Center
Greenbelt, Maryland 20771
USA
digel@gsfc.nasa.gov

Keywords: Milky Way galaxy, gamma rays, cosmic rays

Abstract Cosmic-ray interactions with interstellar gas and low-energy photons make the Milky Way galaxy a gamma-ray source that spans the whole sky. The low instrumental background, large effective area, good angular resolution, and long mission life of the Energetic Gamma-Ray Experiment Telescope (EGRET) on the Compton Gamma Ray Observatory provided the first opportunity for comprehensive study of this diffuse gamma-ray emission. I will discuss what has been learned from EGRET about the cosmic rays and interstellar gas in the Milky Way, including the apparent excess of GeV emission from the interstellar medium, the potential Galactic halo at GeV energies, and the implications of EGRET observations for the nature of dark matter in the Milky Way. For many gamma-ray astronomers, the emission of the Milky Way is primarily a background to be removed in the study of point sources. I will illustrate the potential for diffuse emission to confuse or contaminate the EGRET source catalogs, and conclude with the prospects for advances with the *Gamma-ray Large Area Space Telescope (GLAST)*.

1. INTRODUCTION

The existence of diffuse, high-energy gamma-ray emission from the Milky Way was anticipated long before its detection. The processes relevant to the production of interstellar gamma rays – bremsstrahlung and inverse Compton scattering of cosmic-ray electrons, and decay of secondary pions from collisions of cosmic-ray nucleons with interstellar nuclei – have been understood for 50 years (see Morrison 1958). Cosmic rays themselves have been studied for 90 years, and the distribution of

A. Carramiñana et al. (eds.),
The Nature of Unidentified Galactic High-energy Gamma-ray Sources, 197–211.

the interstellar gas in the Milky Way was mapped in the 21-cm line of H I more than 40 years ago (van de Hulst, Muller, & Oort 1954). With all of these ingredients in hand, it was long understood that the Milky Way should be a diffuse source of high-energy gamma rays.

Diffuse gamma rays from the Milky Way were not definitely detected until *OSO-3* was launched in 1967. The satellite carried an instrument with sufficient capability to discriminate against the cosmic-ray background and which provided $\sim 20°$ angular resolution. The final map was based on 621 gamma rays with energies ≥ 50 MeV, and showed the spatially extended emission associated with the midplane of the Milky Way (Kraushaar et al. 1972). The picture was considerably refined by the SAS-2 (1972–1973) and COS-B (1975–1982) missions, which detected the extragalactic background of gamma rays and gamma-ray pulsars as bright sources in the Milky Way. The EGRET instrument on the Compton Gamma-Ray Observatory (1991–2000) completed the first all-sky survey at high-energies, detecting more than 1.4 million gamma rays (> 30 MeV), more than 60% of which are interstellar in origin (Fig. 1). The large volume of data, together with the good angular resolution and excellent bacgkround rejection of EGRET, made possible detailed studies of the interstellar emission. Discoveries with EGRET are discussed in Section 4. A brief discussion of the potential for further progress with *GLAST* follows.

Figure 1 Intensity of gamma rays (> 100 MeV) observed by EGRET through Phase 5. The broad, intense band near the equator is diffuse emission from the Milky Way. The intensity scale ranges from 1×10^{-5} cm^{-2} s^{-1} sr^{-1} to 5×10^{-4} cm^{-2} s^{-1} sr^{-1} in ten logarithmic steps. The data have been smoothed slightly by convolution with a gaussian of FWHM 1.5°.

2. GAMMA-RAY PRODUCTION MECHANISMS

The processes relevant to the production of interstellar gamma rays are summarized in Table 1, which compares the characteristic spectral shapes and energies of the gamma rays with those of the cosmic-ray electrons and protons. Bremsstrahlung scattering is described in detail by Koch & Motz (1959), inverse Compton scattering by Moskalenko & Strong (2000), and pion decay from proton interactions by Dermer (1986).

The different spectral characteristics of gamma rays produced by the three interaction processes and differences in relevant energy ranges of cosmic rays together suggest the potential for studying cosmic rays with remote observations of gamma rays. Interstellar gamma-ray spectra carry information about the cosmic-ray spectra that would otherwise be unavailable.

Table 1 Cosmic-Ray Interactions[a]

Process	*Photon Spectral Index*	*Photon Energy*
Bremsstrahlung	Γ_e	$\leq E_e$
Inverse Compton	$(\Gamma_e + 1)/2$	$\ll E_e$
Pion Decay	peak at $M_{\pi^0}/2$, then $\sim \Gamma_p$	$\ll E_p$

[a] These are the characteristic power-law spectral indicies and photon energies relative to the spectral indicies (Γ_p, Γ_e) and energies (E_p, E_e) of cosmic-ray protons and electrons.

3. COSMIC RAYS

The spectra of cosmic rays are directly measured only in the solar system, of course, and the local measurements are severely affected by solar modulation below $\sim$ 20 GeV. Measurement of non-thermal radio continuum emission can be used to infer the spectral shape for cosmic-ray electrons at remote distances, but not the absolute normalization, and the measurements are effectively averaged over entire lines of sight. Calculations of the interstellar gamma-ray emission must assume a distribution of cosmic rays and then test for consistency with gamma-ray observations (see Sec. 4.4). Model testing in this way is perhaps the most direct way to infer the distribution of cosmic rays across the Galaxy.

4. RESULTS FROM EGRET

4.1. COSMIC RAYS ARE GALACTIC

Confirmation that cosmic rays with energies up to $\sim$ 100 GeV are produced in galaxies, rather than metagalactic or universal in origin, came from EGRET observations of the Magellanic Clouds. The Large Magellanic Cloud was detected with a flux of $(1.9 \pm 0.4) \times 10^{-7}$ cm^{-2} s^{-1} ($>$ 100 MeV), which implied a cosmic-ray density similar to that in the Milky Way (Sreekumar et al. 1992). However, the Small Magellanic Cloud nearby was not detected, and the upper limit for the flux, 0.5×10^{-7} cm^{-2} s^{-1} ($>$ 100 MeV), is several times less than it would be if the cosmic-ray density were the same as in the Large Magellanic Cloud or the Milky Way (Sreekumar et al. 1993).

4.2. PION BUMP

The EGRET spectrum of the inner Milky Way (Fig. 2) indicates the presence of cosmic-ray protons via the characteristic shape of the π^0 decay spectrum. Note that the π^0 bump, at 68 MeV, is not detected directly; the broad pion decay continuum at higher energies, which changes the overall spectral shape, is evident. (The discrepancy between the observed and modelled spectrum above $\sim$ 1 GeV is discussed in Sec. 4.5.) Spectral measurements with EGRET provided the expected confirmation of cosmic-ray protons as a significant source of interstellar gamma rays at high energies.

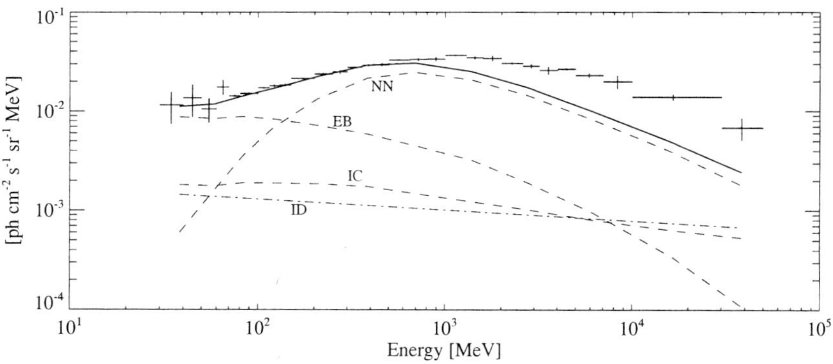

Figure 2 Spectrum of the inner Galaxy ($|l| < 60°$, $|b| < 10°$) observed by EGRET (Hunter et al. 1997). Point sources have been subtracted. Components of the interstellar emission, as derived from the model of Hunter et al. (1997) are labelled: EB for electron Bremsstrahlung; IC for inverse Compton, ID for isotropic, extragalactic emission; and NN for π^0 decay from cosmic-ray proton interactions. The contribution of the protons is evident. The apparent excess of emission for energies above 1 GeV is discussed in Section 4.5.

4.3. MOLECULAR MASS CALIBRATION

The interstellar medium is largely cold H I and H_2, the latter being predominant in the denser regions, in so-called molecular clouds. Although cold H I can be observed via its 21-cm spin-flip spectral line, H_2 is generally not directly detectable at interstellar conditions. The 2.6-mm $J = 1$–0 line of CO, the second most abundant interstellar molecule after H_2, has long been used as an indirect tracer; CO is rotationally excited by collisions with H_2. Empirically, the column density $N(H_2)$ of molecular hydrogen has been found to be proportional to W_{CO}, the integrated intensity of the 2.6-mm line (e.g., Dickman 1978). Initially, the proportionality could be calibrated only for translucent molecular clouds, which have optical extinctions low enough that galaxy counts can be used together with the dust-to-gas ratio to calibrate the proportionality.

Gamma rays have long been recognized as an effective tracer for calibrating $N(H_2)/W_{CO} \equiv X$ in massive, opaque molecular clouds. High-energy cosmic rays freely penetrate interstellar clouds and are smoothly distributed on scales larger than the clouds. In addition, the clouds are essentially transparent to gamma rays. Using EGRET data, values of X are found by fitting a simple, linear model in which the gamma-ray emission from an interstellar cloud is the sum of the emission from the atomic and molecular components. The X-ratio is then derived from the ratio of the emissivity of atomic hydrogen in the vicinity of the cloud to the effective emissivity per unit W_{CO} in the molecular parts of the cloud (e.g., Hunter et al. 1994).

EGRET spatially resolved several large, nearby interstellar clouds (essential for distinguishing the atomic and CO-correlated emissivities), making feasible the study of X on the scale of individual molecular clouds. Values of X from EGRET analyses are listed in Table 2 for four local complexes of interstellar clouds. The range of X is quite small, even between the Ophiuchus and Orion clouds, which differ by approximately a factor of 100 in mass ($\sim 10^4$ $M_\odot$ vs. $\sim 10^6$ $M_\odot$). Among the local clouds in Table 2, one trend may be evident: an increase of X with Galactocentric distance, which might be expected owing to the general decrease of metallicity and of temperatures of molecular clouds with Galactocentric distance (e.g., Elmegreen 1989). The evidence from gamma-ray analyses remains inconclusive, however, owing to the generally low column densities of molecular gas beyond the solar circle, and therefore the limited photon statistics available.

EGRET data have also been used to study the gamma-ray emission of the whole Milky Way, under the assumption that X is constant through-

Table 2 Determinations of $N(H_2)/W_{CO}$ with EGRET

Region	*Reference*	*X-ratio*[a]
Milky Way average	Strong & Mattox (1996)	1.9 ± 0.2
	Hunter et al. (1997)	1.56 ± 0.05
Ophiuchus ($R \sim 8.4$ kpc)	Hunter et al. (1994)	1.1 ± 0.2
Cepheus (8.7)	Digel et al. (1996)	0.92 ± 0.14
Orion (8.9)	Digel et al. (1999)	1.35 ± 0.15
Monoceros (9.2)	Digel et al. (2001)	1.87 ± 0.29

[a] units 10^{20} cm^{-2} K^{-1} km^{-1} s.

out (i.e., with only a single adjustable value for X), with results listed in Table 2. The models of Strong & Mattox (1996) and Hunter et al. (1997) are quite different (see Sec. 4.4), and the agreement between values of X derived from the two works may be considered to be rather close. In a whole-Galaxy study with a single value of X, the maximum likelihood or best-fit value represents a kind of emission-weighted (or really exposure-and-emission-weighted) average that may not apply to any specific region in particular.

4.4. INTERSTELLAR EMISSION MODELS

Three approaches have been used to model the interstellar emission of the entire Milky Way. The motivation for developing these models is to gain insights into cosmic rays and the interstellar medium as well as to distinguish between point sources and diffuse gamma-ray emission. The latter is necessary owing to the limited angular resolution of EGRET (typically several degrees, Thompson et al. 1993) and the low fluxes of gamma rays. The whole-Galaxy models may be distinguished primarily by how the spatial and spectral distribution of cosmic rays are modelled.

The EGRET team approximated the cosmic-ray density as proportional to the gas density on ~ 1.7 kpc scales (Hunter et al. 1997), motivated by an expectation that cosmic rays would be coupled to interstellar gas via the magnetic field threading the interstellar medium (Parker 1966). The smoothed surface density of gas, estimated from an inversion of H I and CO spectral line surveys, is used to set the overall intensities of the cosmic rays, and their spectra are assumed to have the same shape as measured in the solar system.

Strong & Mattox (1996) assumed that the cosmic-ray density is azimuthally symmetric in the Milky Way and fit the densities for a range of energies in each of a set of annular rings (typically 2 kpc wide) centered on the Galactic center. This approach has the advantage of not depending on the uncertainties inherent in the determination of the 3-dimensional distribution of gas (see Hunter et al. 1997), and does allow the spectrum of cosmic rays to vary. The tradeoff is the presupposition that the spectrum depends only on Galactocentric distance.

Both the Hunter et al. and Strong & Mattox approaches are remarkably good at reproducing the large-scale features of the diffuse emission observed by EGRET, especially on consideration of the difficulty of deriving three-dimensional distributions of gas from spectral line surveys. Discrepancies with the observations remain, generally at lower intensities.

Strong, Moskalenko & Reimer (2000) have undertaken a model of the cosmic-ray distribution in the Milky Way using realistic distributions of cosmic-ray sources and numerically solving the equations for cosmic-ray production and propagation of all species up to Ni. They constrain the model using locally-measured spectra of cosmic rays, radio continuum observations of synchrotron emission, and EGRET observations of diffuse emission. The injection spectra of cosmic-ray protons and electrons and the spatial distribution of cosmic-ray sources are adjustable to give agreement with the constraints. Approaches like this are promising, although in many aspects they necessarily are not well constrained by the data. Pohl & Esposito (1998) also numerically modelled cosmic-ray propagation to study gamma-ray emission.

Studies of EGRET data are useful for quantitative investigation of the assumptions about cosmic-ray densities in the whole-Galaxy models. For example, Digel et al. (2001) studied the gamma-ray emissivity across the outer Galaxy in Monoceros, a longitude range that contains a massive segment of the Perseus arm. The model of Hunter et al. (1997) predicts a corresponding large enhancement of cosmic-ray density in the Perseus arm (the peak in the profile of Fig. 3). The study of Digel et al., which allowed the cosmic-ray density to be a free parameter in the distance ranges indicated in the figure, did confirm the coupling of cosmic rays to the Perseus arm, but with a greater arm-interarm contrast and a smaller enhancement in the arm itself. These findings suggest that the coupling model of Hunter et al. should be modified in the outer Galaxy.

4.5. GEV EXCESS

Hunter et al. (1997) noted that the average spectrum of the inner Galaxy is harder than expected at energies > 1 GeV, having a significant excess at high energies relative to the prediction from their model of

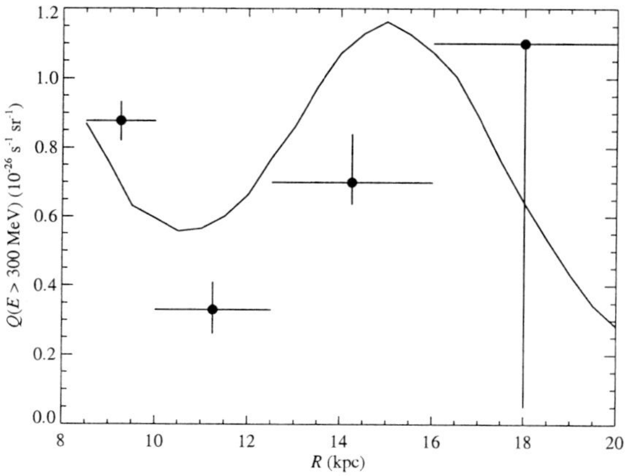

Figure 3 Profile of gamma-ray emissivity across the outer Galaxy in Monoceros (Digel et al. 2001) compared with the cosmic-ray density predicted by the model of Hunter et al. (1997) for the same longitude range (210–250°). For the energy range shown (> 300 MeV), most of the gamma-ray emission is from cosmic-ray proton interactions. The Hunter et al. profile is normalized to local measurements of the emissivity $Q(> 100$ MeV).

interstellar emission (Fig. 2). They ruled out calibration errors as the source of the discrepancy, and Mori (1997) concluded that it could not be attributed to errors in the production function for π^0 decay. One significant consideration may be that the interstellar emission model of Hunter et al. is based on the implicit assumption that the shapes of cosmic-ray spectra do not vary across the Milky Way.

The excess is clearly associated with the interstellar medium. The extragalactic isotropic spectrum does not harden similarly at high energies; it is well described by a single power law across the entire energy range of EGRET (Sreekumar et al. 1998). The excess may be a local (within 1 kpc) effect. Studies of individual nearby interstellar cloud complexes reveal the excess as an increased gamma-ray emissivity at high energies (Fig. 4*a*). However, the excess is not detected in the emissivity of more distant regions, such as the interstellar clouds in the Perseus arm of the outer Galaxy (Fig. 4*b*; Digel et al. 2001).

Pohl & Esposito (1998) point out that the distribution of cosmic-ray electrons in particular should be very non-uniform, owing to rapid energy losses relative to the timescale for diffusion away from the SNR sites of production. The spectra of cosmic-ray electrons should be harder near SNR, and Pohl & Esposito conclude that the GeV excess can be explained if the locally-measured electron spectrum is somewhat softer than representative for the local region.

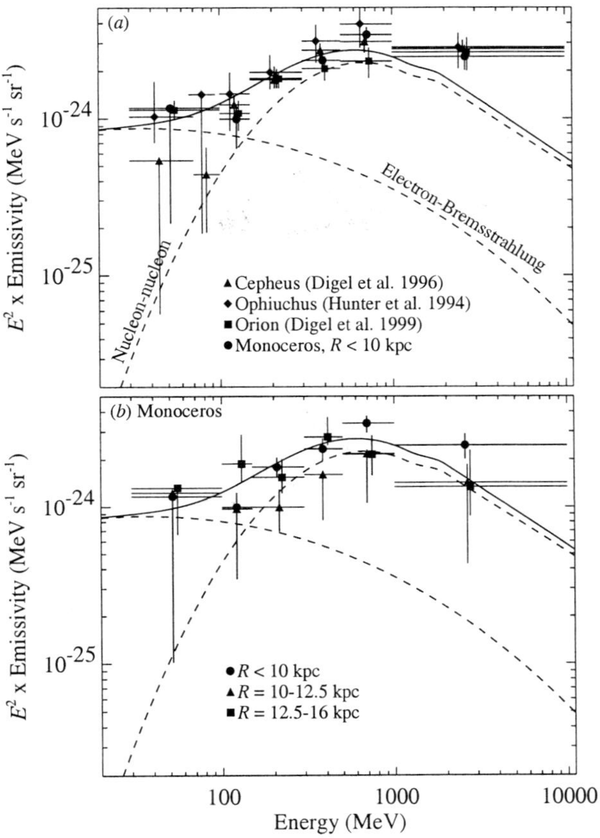

Figure 4 (*a*) Comparison of differential emissivities for local cloud complexes. (*b*) Differential emissivities in three Galactocentric distance ranges in Monoceros. Circles indicate the local ($R < 10$ kpc) range, triangles the interarm (R = 10–12.5 kpc) range, and diamonds the Perseus arm (R = 12.5–16 kpc) range. The horizontal error bars indicate the energy range and the vertical error bars the $\pm 1\sigma$ uncertainty of the emissivity. For upper limits, the 2-σ values are indicated. Also plotted are the nucleon-nucleon (π^0) and electron-bremsstrahlung emissivities, and their sum, for the solar vicinity from Bertsch et al. (1993). The emissivities for Monoceros are from Digel et al. (2001).

4.6. HIGH-LATITUDE EMISSIVITY EXCESS

Chen, Dwyer, & Kaaret (1995) pointed out that variations of cosmic-ray density on a very local scale are evident in the EGRET data. Figure 5 illustrates their finding of a significant excess of gamma-ray emissivity at high latitudes in the first quadrant (near $l, b = 75°, +45°$). In this large solid angle on the sky, either the cosmic-ray density is different by a factor ~ 2 from the average high-latitude sky, the inverse Compton emission is particularly intense, or the 21-cm line of H I does not trace the column density of interstellar gas. The analysis of Chen et al. cannot distinguish between these possibilities. Chen et al. speculated that an enhanced density of cosmic rays or intensity of the interstellar radiation field in this region might be associated with the massive star formation underway at lower latitudes in Cygnus.

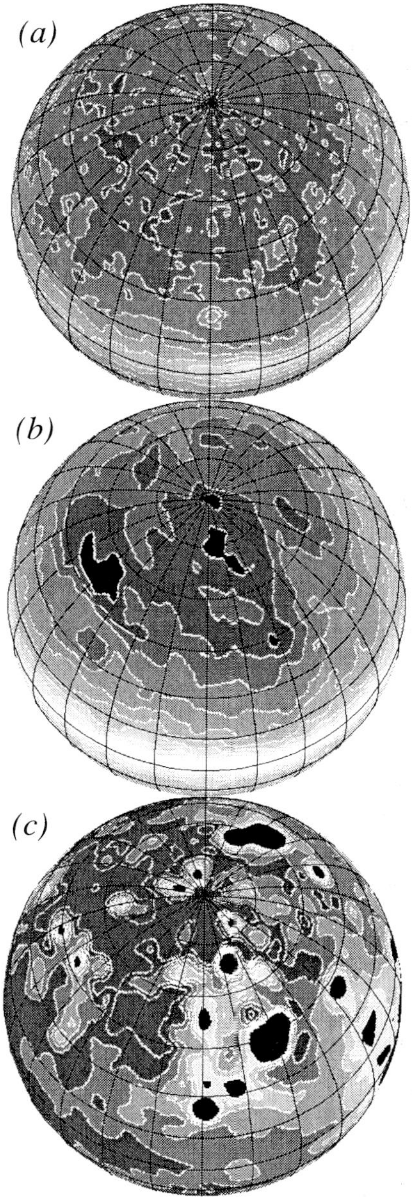

Figure 5 (*a*) Intensity (> 100 MeV) observed by EGRET in Phases 1–5. The sky map is projected onto a globe and rotated so that the center of the map is $(l, b) = (90°, +60°)$. Grid lines are spaced by 15°, and the intensity contours are logarithmic from 10^{-5} cm^{-2} s^{-1} sr^{-1} to 10^{-3} cm^{-2} s^{-1} sr^{-1}. (*b*) Column density of atomic hydrogen from the composite survey of Dickey & Lockman (1990), with the same projection and orientation as (*a*). The contours are logarithmic, from 10^{20} cm^{-2} to 10^{22} cm^{-2}. (*c*) Gamma-ray emissivity, derived as the ratio of (*a*) to (*b*) after subtraction of an isotropic intensity. The large area of enhanced emissivity near $(l, b) = (75°, +45°)$ was reported by Chen et al. (1995). The image is smoothed to approximately 6° resolution and the brightest peaks, which correspond to point sources, are blacked out. The contours are 1,2,. . . , 6 $\times 10^{-26}$ s^{-1} sr^{-1}, although emissivity values must be considered approximate owing to the incomplete accounting for inverse Compton emission in this simple ratio map.

4.7. HALO IN GAMMA RAYS

Dixon et al. (1998) reported the detection of an extended halo for the Milky Way in high-energy gamma rays. Essentially, they found a significant residual after subtracting the EGRET team's interstellar emission model (Hunter et al. 1997) from the observations and filtering the difference to enhance large-scale structure. The same residual was also reported from a study of the extragalactic isotropic emission (Sreekumar et al. 1998).

Interpretations in terms of diffuse emission from cosmic-ray collisions with cold baryonic dark matter have been offered (e.g., De Paolis et al. 1999). The speculation that dark matter in the Galaxy could be molecular gas that is too cold to detect in CO emission dates to the proposal of Pfenniger, Combes & Martinet (1994) that the outer disk may contain such gas. In principle, even if a molecular cloud is too cold to be detected in CO, it should still be a source of diffuse gamma-ray emission. However, Kalberla, Shchekinov, & Dettmar (1999) point out that the densities of the postulated cold, dense molecular clumps could be sufficient to stop cosmic rays from fully penetrating them, as well as absorb most of the gamma rays that are produced. Sciama (2000) found that stability against collapse and the effects of heating by cosmic rays could not rule out the existence of such cold clumps. The gamma-ray observations of the halo therefore cannot rule out the presence of large quantities of cold molecular gas, although they do not require it.

The apparent gamma-ray halo most likely is an artifact owing to the incomplete incorporation of inverse Compton emission above $|b| = 10°$ in the model of Hunter et al. (Hunter, priv. comm.). Moskalenko & Strong (2000) in fact show that the halo can be explained as inverse Compton emission, if the anisotropic nature of the interstellar radiation field, which increases the relative intensity of the high-latitude component, is taken into account.

4.8. SOURCE CONFUSION

Various lines of reasoning suggest that the contribution of unresolved point sources to the diffuse emission is small. Hunter et al. (1997) argue that even a 10% contribution to the flux would significantly alter the shape of the spectrum of the inner Galaxy measured with EGRET data. Pohl et al. (1997) and Zhang & Cheng (1998) both conclude that pulsars could contribute significantly to the > 1 GeV emission, but Pohl et al. point out that the latitude distribution of pulsars is significantly broader than that of the excess. Further progress in measuring the unresolved source fraction in diffuse emission observed by EGRET will likely have to wait for *GLAST* (Sec. 5).

5. ANTICIPATED PROGRESS WITH *GLAST*

GLAST, the Gamma-ray Large Area Space Telescope, described in the review by Gehrels elsewhere in this volume, will enable great advances in the study of interstellar emission. Table 3 compares the anticipated astronomical performance of *GLAST* relative to EGRET. The table compares a simulated one-year sky survey with *GLAST* to a simulated two-week pointed observation with EGRET, approximately equivalent observing times on consideration of the observing modes and relative fields of view of EGRET and *GLAST*. The background intensity was assumed to be flat with an intensity representative of the high-latitude sky (see Table 3), and the sources were assumed to have unbroken spectra with photon spectral indicies -2. For *GLAST*, preliminary versions of the point-spread function and effective area were assumed (Michelson et al. 1999). As the table shows, *GLAST* will have a great sensitivity advantage, and also will be much more capable of resolving structure in the interstellar emission and separating point sources from extended emission.

The advantages for study of interstellar emission are illustrated in the maps of Orion shown in Figure 6. With *GLAST*, even after just the initial year of observations, the giant molecular clouds in Orion will be resolved in detail, which will permit much more confident detection and removal of potentially-confusing background point sources. The improved sensitivity and resolution will also permit questions of the uniformity of the cosmic-ray density and of the X-ratio on the scale of an interstellar cloud to be investigated.

Table 3 Relative Astronomical Performance of EGRET and *GLAST*[a]

Figure of Merit	*Source Flux*	*EGRET*	GLAST
Flux Limit (> 100 MeV, 5σ)		8×10^{-8}	3.5×10^{-9}
Diam. 95% Confidence Region:	1×10^{-6}	0.17°	0.005°[b]
	5×10^{-7}	0.23°	0.005°[b]
	1×10^{-7}	0.68°	0.02°
Mininum Separation[c]	1×10^{-7}	1.25°	0.09°
Resolving Extended Sources:			
Min. diameter for resolving	5×10^{-7}	1.5°	0.08°
Max. diameter for detecting	5×10^{-7}	7.5°	> 10°

[a]For a two-week on-axis pointing of EGRET and a 1-year sky survey with *GLAST*. A two-week observation with EGRET approximates the exposure EGRET would obtain in a 1-year all-sky survey. Standard response functions for EGRET (Thompson et al. 1993) and preliminary *GLAST* response functions (Michelson et al. 1999) were assumed. Sources were assumed to have photon spectral indicies of -2 and the background intensity was assumed to have index -2.1 and integrated intensity 2×10^{-5} cm^{-2} s^{-1} sr^{-1} (> 100 MeV). All fluxes are integral for energies > 100 MeV in cm^{-2} s^{-1}.

[b]Limited by spacecraft pointing knowledge.

[c]Minimum separation for distinguishing two sources of the indicated flux at the 5σ level.

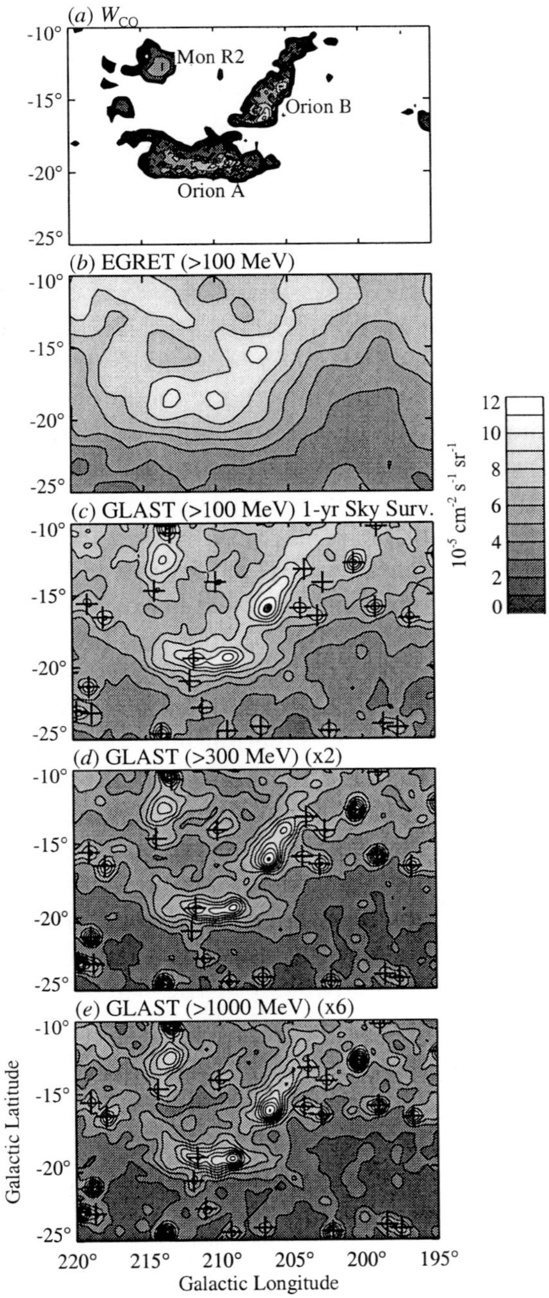

Figure 6 (*a*) Integrated intensity of the 2.6-mm CO line in Orion (Maddalena et al. 1986), showing the well-known Orion A and Orion B giant molecular clouds and the Mon R2 cloud. (*b*) Intensity of gamma rays with energies > 100 MeV observed by EGRET and analyzed by Digel et al. (1999). (*c–e*) Simulated intensity maps from the *GLAST* sky survey for the energy ranges indicated. To smooth statistical fluctuations, the EGRET data were convolved with a gaussian of FWHM 1.5°, and the *GLAST* intensities with FWHM 0.75°. Crosses mark the positions of background sources with fluxes greater than 10^{-8} cm^{-2} s^{-1} (> 100 MeV). The intensity scale refers to (*b–e*) with scale factors as noted for (*d–e*).

6. SUMMARY

Gamma-ray emission from cosmic-ray collisions with interstellar gas and photons makes the Milky Way a high-energy source that spans the entire sky. Study of interstellar gamma-ray emission has advanced greatly with EGRET, both because of superior sensitivity and background rejection relative to prior missions and because supporting data at other wavelengths are now available to aid interpretation of the gamma-ray observations. EGRET completed the first all-sky survey in high-energy gamma rays. It also detected the gamma-ray emission from π^0 decay in the spectrum of the inner Milky Way. Studies of the emission from individual, nearby interstellar clouds have been possible with EGRET and the results are starting to map the distribution of cosmic rays, and the variation of the mass-calibrating ratio $N(\mathrm{H}_2)/W_{\mathrm{CO}}$, within 1 kpc of the sun. Models of the interstellar emission of the whole Galaxy show promise for constraining models of cosmic-ray production and distribution. The EGRET team's model of the interstellar emission (Hunter et al. 1997) significantly underpredicts the emission above GeV. This 'GeV excess' appears to be correlated with the interstellar gas and may be interpreted in terms of spatial variations of cosmic-ray spectra, which are held fixed in the EGRET model. A gamma-ray halo for the Milky Way reported by Dixon et al. (1998) is apparently well-explained with models for inverse Compton emission that take into account the anisotropic nature of the interstellar radiation field (Moskalenko & Strong 2000). Variations of the gamma-ray emissivity at high latitudes are apparently real and may be correlated with cosmic-ray production near the plane. The modest angular resolution of EGRET, and the low fluxes of celestial gamma rays, make the question of point-source contamination important for each study of interstellar emission. Gamma-ray spectral evidence, together with the spatial distribution of Galactic sources suggest that unresolved point sources contribute significantly less than 10% to diffuse intensities. With greatly improved sensitivity and angular resolution, *GLAST* promises further large advances in the study of interstellar gamma rays.

References

Bertsch, D. L., et al. (1993) *ApJ*, **416**, 587.
Chen, A. W., Dwyer, J., Kaaret, P. (1995) *ApJ*, **445**, L109.
De Paolis, F., et al. (1999) *ApJ*, **510**, L103.
Dermer, C. D. (1986) *A&A*, **157**, 223.
Dickey, J. M., Lockman, F. J. (1990) *ARA&A*, **28**, 215.
Dickman, R. L. (1978) *ApJS*, **37**, 407.
Digel, S. W., et al. (1996) *ApJ*, **463**, 609.
Digel, S. W., et al. (1996) *ApJ*, **520**, 196.
Digel, S. W., et al. (2001) *ApJ*, in press

Dixon, D. D., et al. (1998) *New Astron.*, **3**, 539.

Elmegreen, B. G. (1989) *ApJ*, **338**, 178.

Hunter, S. D., et al.(1994) *ApJ*, **436**, 216.

Hunter, S. D., et al. (1997) *ApJ*, **481**, 205.

Kalberla, P. M. W., Shchekinov, Yu. A., Dettmar, R.-J. (1999) *A&A*, **350**, L9.

Koch, H. W., Motz, J. W. (1959) *Rev. Mod. Phys.*, **31**, 920.

Kraushaar, W. L., et al. (1972) *ApJ*, **177**, 341.

Maddalena, R. J., Morris, M., Moscowitz, J., & Thaddeus, P. (1986) *ApJ*, **303**, 375.

Michelson, P., et al. (1999), Proposal for *GLAST* Large Area Telescope Flight Investigation (http://glast.stanford.edu/pubfiles/proposals/bigprop/).

Mori, M. (1997) *ApJ*, **478**, 225.

Morrison, P. (1958) *Nuovo Cimento*, **8**, 858.

Moskalenko, I. V., Strong, A. W. (2000) *ApJ*, **528**, 357.

Parker, E. N. (1966) *ApJ*, **145**, 811.

Pfenniger, D., Combes, F., Martinet, L. (1994) *A&A*, **285**, 79.

Pohl, M., et al. (1997) *ApJ*, **491**, 159.

Pohl, M., Esposito, J. A. (1998) *ApJ*, **507**, 327.

Sciama, D. W. (2000), *MNRAS*, **312**, 33.

Sreekumar, P., et al. (1992) *ApJ*, **400**, L77.

Sreekumar, P., et al. (1993) *Phys. Rev. Lett.*, **70**, 127.

Sreekumar, P., et al. (1998) *ApJ*, **494**, 523.

Strong, A. W., Mattox, J. R. (1996) *A&A*, **308**, L21.

Strong, A. W., Moskalenko, I. V., & Reimer, O. (2000) *ApJ*, **537**, 763.

Thompson, D. J., et al. (1993) *ApJS*, **86**, 629.

van de Hulst, H. C., Muller, C. A., Oort, J. H. (1954) *B. A. N.*, **12**, 117.

Zhang, L., Cheng, K. S. (1998) *MNRAS*, **301**, 841.

NONLINEAR SHOCK ACCELERATION AND PHOTON PRODUCTION IN YOUNG SUPERNOVA REMNANTS

Donald C. Ellison

Department of Physics, Box 8202, North Carolina State University, Raleigh, NC 27695

don_ellison@ncsu.edu

Keywords: Supernova remnants, shock acceleration, cosmic rays

Abstract A number of young supernova remnants (SNRs) are now known to have nonthermal X-ray spectra. The steepness of the X-ray emission suggests that it is synchrotron from TeV electrons, and if this is the case, efficient shock acceleration is likely occurring in these objects. Here we use a model of nonlinear diffusive shock acceleration to fit the broad-band emission from SN1006, Tycho's, and Kepler's SNRs. Our fits confirm that all of these SNRs are producing TeV particles, but also show that the electron and ion spectra do not extend as a power law above a few TeV, well below the cosmic ray 'knee' at 10^{15} eV. Broad-band modeling also constrains emission in the EGRET energy range and these estimates are given by our models.

1. INTRODUCTION

The evidence that collisionless shocks accelerate particles with high efficiency is convincing. There are direct spacecraft observations confirming it (e.g., Ellison et al. 1990; Baring et al. 1997; Terasawa et al. 1999), plasma simulations show efficient acceleration consistent with spacecraft observations (e.g., Scholer et al. 1992; Giacalone et al. 1997), and all theoretical models allow it (e.g., Drury 1983; Jones & Ellison 1991). Other indirect evidence comes from the implied efficiency of galactic cosmic-ray production, radio emission from SNRs, equipartition arguments in AGNs and γ-ray bursts, etc. (e.g., Blandford & Eichler 1987). There is also clear evidence that shocks can produce strong self-generated turbulence. This has long been seen in heliospheric shocks (e.g., Lee 1982; Baring et al. 1997; Ellison et al. 1999) and there is strong evidence that it occurs at SNRs as well (i.e., Achterberg et al. 1994).

A. Carramiñana et al. (eds.),
The Nature of Unidentified Galactic High-energy Gamma-ray Sources, 213–226.

The recent discovery of nonthermal X-ray emission in shell-like SNRs like SN1006 (Koyama et al. Koyama et al. 1995), Cas A (e.g., Allen et al. 1999), G347.3-0.5 (Slane et al. 1999), Kepler's (Decourchelle & Ballet 1994), and RCW86 (Borkowski et al. 1999) has been interpreted as synchrotron emission from TeV electrons, giving more direct evidence that SNR shocks can accelerate electrons to cosmic ray energies.

Here, I outline a previously published algebraic model of nonlinear diffusive shock acceleration (Berezhko & Ellison 1999; Ellison et al. 2000) and use it to match the broad-band emission from three well-studied young remnants: SN1006, Tycho's, and Kepler's. While X-ray observations strongly imply that these SNRs are producing TeV electrons (e.g., Allen et al. 1999), the simultaneous fitting of X-ray and radio restricts (through a determination of the magnetic field strength) the maximum allowable electron and ion energies to a few TeV – well below the galactic cosmic ray 'knee' near 10^{15} eV (see also Reynolds & Keohane 1999; Slane et al. 2000). Since it is unlikely that these remnants will produce much higher energy particles in the future, this low maximum ion energy poses an important problem for cosmic ray origin.

Our broad-band spectra also allow estimates of the emission in the EGRET energy band and these are consistent within a factor of two with observed upper limits for SN1006.

2. NONLINEAR SHOCK MODEL

The 'simple' nonlinear shock acceleration model used here has been described in detail elsewhere (i.e., Berezhko & Ellison 1999; Ellison et al. 2000). Briefly, it synthesizes the essential features of two complementary approaches to nonlinear shock acceleration, namely the semi-analytical diffusion equation method of Berezhko et al. (e.g., Berezhko et al. 1996) and others (e.g., Kang & Jones 1991; Dorfi 2000) and the Monte Carlo technique of e.g., Ellison et al. (1996). The premise is that the essential physics of nonlinear (NL) shock acceleration can be understood in the framework of a plane-wave, steady-state assumption with simple approximations for the accelerated particle spectrum. The model includes the effects of Alfvén wave heating of the upstream plasma, particle escape, and shock modification when the backpressure from energetic ions modifies the shock structure and produces a precursor where the flow speed (in the shock frame and in the absence of instabilities) is monotonically decreasing. Since particle diffusion lengths are almost certainly increasing functions of momentum, high momentum particles sample a broader portion of the precursor, and hence experience larger compression ratios than low momentum particles. Consequently, higher momentum particles have a flatter power-law index than those at lower momenta, thereby dominating the pressure in a nonlinear fashion and producing a concave upward spectral curvature (until the particle spectra turn over at the highest energies).

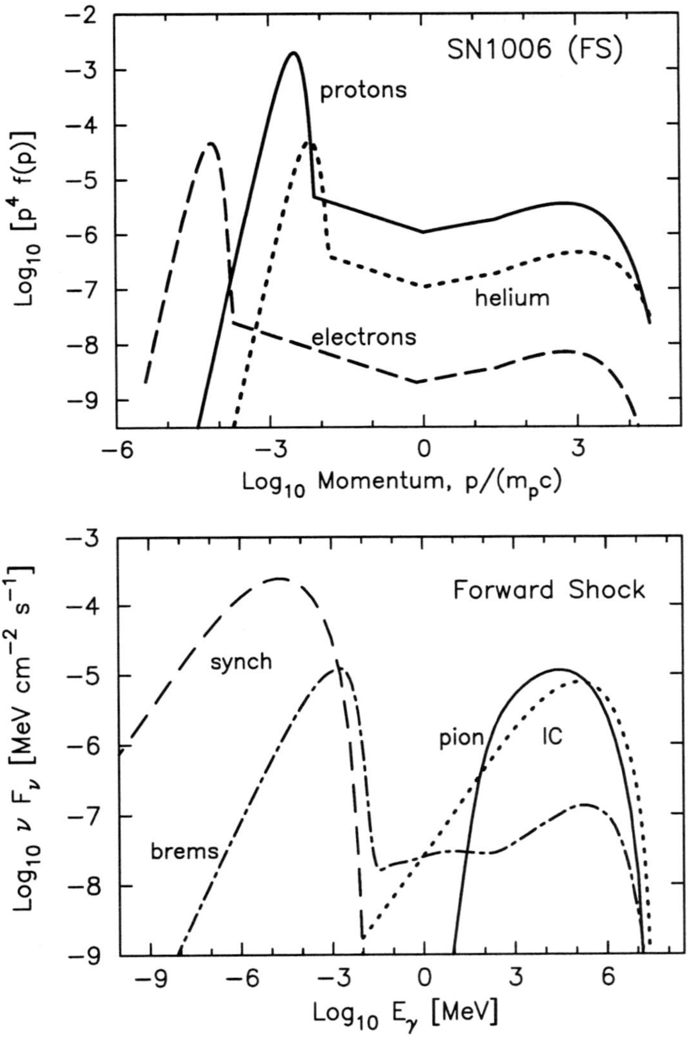

Figure 1 Top panel: Phase space distributions produced by the forward shock in SN1006. Bottom panel: Photon emission from the forward shock.

To model an evolving SNR, we use the model of SNR evolution given by Truelove & McKee (1999) which continuously maps between the free expansion and Sedov phases to give the forward and reverse shock parameters as a function of explosion energy, $E_{\rm sn}$, ejecta mass, $M_{\rm ej}$, and remnant age, $t_{\rm snr}$. We only consider uniform distributions for the ambient ISM mass, appropriate to Type Ia progenitors, but do allow for a power-law density profile for the ejecta, as in Chevalier (1982). We caution that we apply our model to Type II SNRs, neglecting any changes produced by modifications in the pre-SN density distribution.

In this model, the injection efficiency, together with the 'shock size,' set the shock structure and determine the overall acceleration efficiency. The injection efficiency, $\eta_{\rm inj,p}$ is the fraction of protons which end up as superthermal particles, and the shock size is determined from the shock radius and age, as described in Baring et al. (1999). An important parameter in this determination is $\eta_{\rm mfp}$, the ratio of scattering mean free path to gyroradius ($\eta_{\rm mfp} \sim 1$ is the so-called Bohm limit of strong scattering).

Typical results of the model are shown in Figure 1. The top panel shows the shocked particle momentum phase-space distributions, $f(p)$ (multiplied by p^4), and the bottom panel shows the photon spectra calculated from the $f(p)$'s. The parameters used for both panels are those for SN1006, given in Table 1, and only spectra from the forward shock are shown. The synchrotron, bremsstrahlung, inverse-Compton, and pion-decay emission are labeled and calculated using the techniques described in Baring et al. (1999). As mentioned above, nonlinear shocks produce spectra which differ somewhat from power laws and this spectral curvature is evident in the $f(p)$'s as a broken power law above and below $m_p c$.

3. THREE YOUNG SNRS

In Figure 2 we show the observations and models for SN1006, Tycho's, and Kepler's SNRs. The parameters used are given in Table 1. As mentioned above, the X-ray emission from these SNRs implies a nonthermal component and this is most likely synchrotron photons from TeV electrons. However, to match the X-ray observations, the synchrotron at X-ray energies must be falling well below a power law extrapolated from the radio (Reynolds 1998). In each of the three cases, we are able to find parameters which give reasonable fits to both the radio and X-ray continuum. However, the shocked magnetic field strengths, B_2, inferred for these fits (Table 1) are low enough that the electron synchrotron losses are not significant, giving the same maximum energy, $E_{\rm max} = p_{\rm max}\, c$, for electrons and protons.

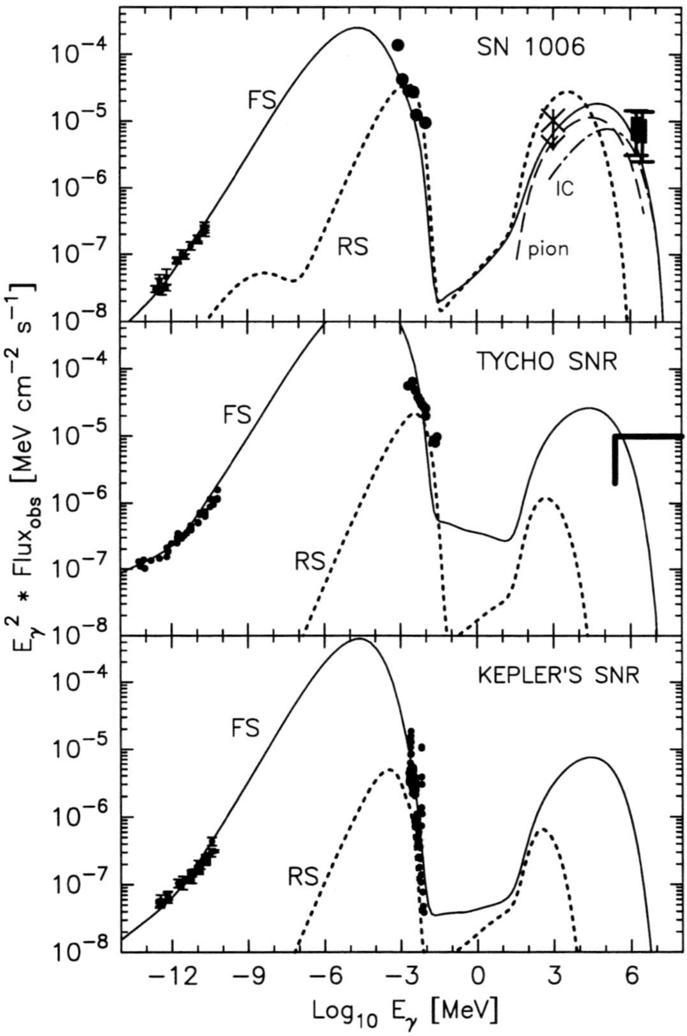

Figure 2 In all panels, the solid and dotted lines are the sum of synchrotron, bremsstrahlung, inverse-Compton, and pion-decay emission. The solid lines show the contribution from the forward shock and the dotted lines from the reverse shock. In the top panel, the dashed line is pion-decay and the dot-dashed is inverse-Compton, showing that the TeV points are primarily IC. For SN1006, the radio and X-ray data are adapted from Reynolds (1996), the EGRET upper limit (cross) is from Mastichiadis & de Jager (1996), and the CANGAROO TeV points (squares) are from Tanimori et al. (1998). For Tycho, the radio data is from Reynolds & Ellison (1992), the X-ray data is adapted from Hwang & Gotthelf (1997) and Allen et al. (1999), and the γ-ray limit is from Buckley (1994). For Kepler, the radio data is from Reynolds & Ellison (1992) and the X-ray data is from Ellison (2000).

As described in Ellison et al. (2000), particle spectra in the simple model are cut off at the highest energies such that

$$f(p) \propto \exp\left[-\frac{1}{\alpha}\left(\frac{p}{p_{\rm max}}\right)^{\alpha}\right], \qquad (15.1)$$

where α is a free parameter and is given in Table 1. In the simple model used here, α is simply chosen to give the best fit to the observations in the X-ray range and $0.5 < \alpha < 1$ is typically required. Reynolds (1998), on the other hand, has shown that turnovers consistent with this can result from a convolution of shock obliquities. For each example, $E_{\rm max} \sim 2$ TeV, more than a factor of 100 below 10^{15} eV. If the remnants evolve in a standard way and no significant magnetic field amplification occurs in the future, these SNRs will never produce protons or other cosmic ray ions with energies much above what they produce now.

Nonlinear shocks are complicated and the parameters we have chosen to fit these SNRs are not unique (e.g., see Ellison 2000 for a fit to Kepler using somewhat different parameters). However, we believe the conclusion that these remnants have a low $E_{\rm max}$ for both electrons and ions is quite insensitive to the particular set of parameters as long as the radio and X-ray emission is modeled.

For SN1006 and Tycho, γ-ray observations are also shown in Figure 2. The parameters we have chosen satisfy these constraints as well, except for the EGRET upper limit which is about a factor of two under the inverse-Compton emission from the reverse shock in SN1006. Considering the uncertainties in the modeling, this factor of two is probably not significant.

While different in details, the overall radio and X-ray observations are quite similar for these three remnants. In Figure 3 we compare Tycho's SNR with SN1006. We have adjusted the radio intensities to roughly match and it's clear that the radio to X-ray ratios are very similar. Even so, small changes in the model parameters can account for the factor of 3 higher TeV γ-ray intensity (relative to radio) believed to occur in SN1006.

For comparison, we have added the radio and X-ray observations of Cas A to Figure 3, adjusted to match the X-ray intensities of SN1006 and Tycho. The high radio to X-ray ratio in Cas A is a clear sign of a high magnetic field and values as high as 1000μG have been reported. The high B allows a fit to the X-rays by an electron spectrum undergoing synchrotron losses, while simultaneously allowing for the production of 10^{15} ions.

While the simple model only produces continuum emission, X-ray lines are also observed in these objects. In Figure 4 we show results

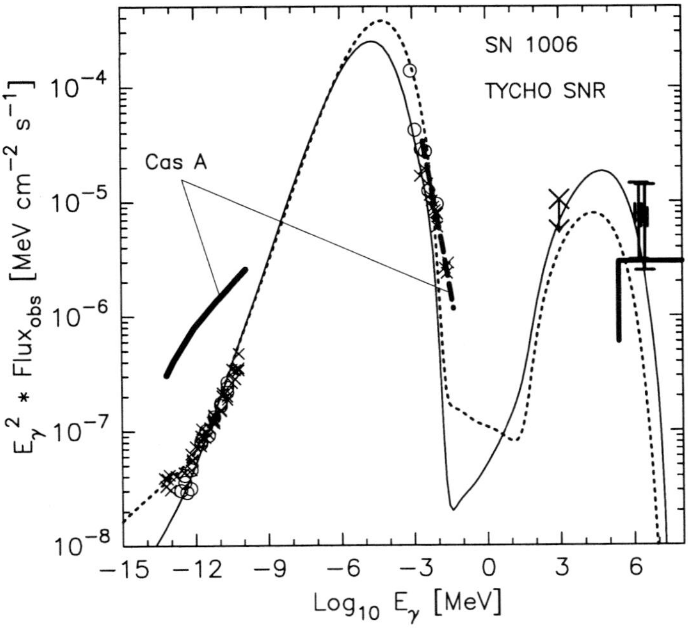

Figure 3 This figure compares Cas A, SN1006, and Tycho. The x's are Tycho observations while the open circles are SN1006 data. The solid curve is the forward shock model fit to SN1006 while the dotted curve is the forward shock fit to Tycho. The Tycho intensity has been adjusted to roughly match SN1006 in the radio band. The heavy lines are Cas A observations adjusted to match the X-ray intensities of SN1006 and Tycho. See Ellison et al. (1999) for references to Cas A observations.

from Decourchelle et al. (2000) who have developed a model which couples self-similar hydrodynamics to nonlinear diffusive shock acceleration, and X-ray emission including non-equilibrium ionization effects (Decourchelle & Ballet 1994). The preliminary results of Decourchelle et al. (2000) for Kepler's SNR demonstrate that shocks which accelerate particles efficiently (at least at the forward shock) can still produce X-ray lines in agreement with observations. This is significant because the inferred shock characteristics (e.g., the supernova explosion energy, shock speed, ambient density) can be very different from the test-particle characteristics which are commonly assumed.

When the acceleration is nonlinear, the shock compression ratio increases and the shocked temperature decreases compared to test-particle values. To get a temperature high enough to produce the iron K-line and the high energy continuum seen in Kepler, all nonlinear fits require large shock velocities and low ambient densities. These imply more kinetic energy in the ejecta and a larger inferred distance, $D_{\rm snr}$, than the test-particle case. To obtain a distance consistent with Kepler observations (4.8 ± 1.4 kpc, Reynoso & Goss 1999), a low efficiency at the reverse shock is required. This can be done in three ways: with $B_0 \gtrsim 25\mu$G at the reverse shock large enough to dampen acceleration, with $B_0 \gtrsim 0.1\ \mu$G and $\eta_{\text{inj,pRS}} \gtrsim 10^{-4}$ to allow high Mach number, unmodified solutions, or with no reverse shock acceleration ($\eta_{\text{inj,pRS}} \simeq 0$). A low magnetic field is expected from the expansion of the ejecta, but magnetic field amplification could increase it. The fit shown in Figure 4 is for efficient acceleration at the forward shock and test-particle conditions at the reverse shock.

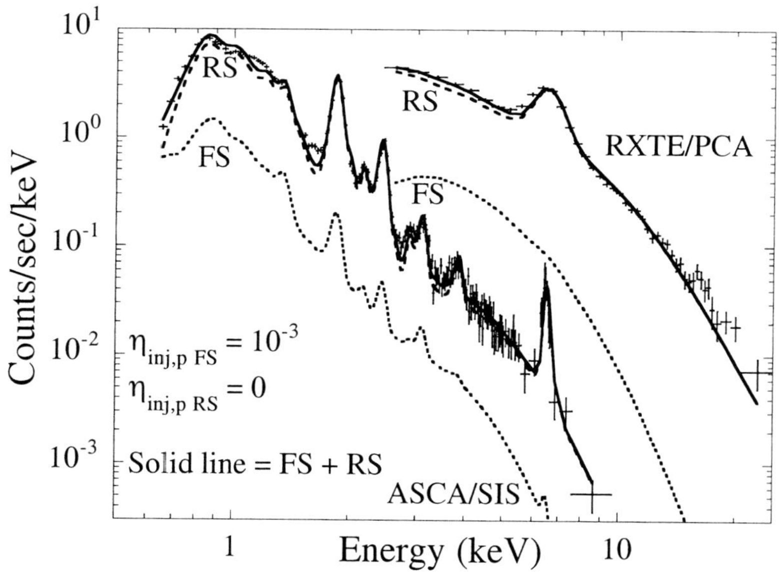

Figure 4 Fit to X-ray emission from Kepler's SNR. See Decourchelle et al. (2000) for details.

3.1. FITTING PARAMETERS

The terms in Table 1 are explained in Ellison et al. (2000). Briefly:

- The unshocked proton number density, n_{p0}, is arbitrarily chosen for the forward shock (FS) and determined for the reverse shock (RS) with eqs. (20), (28), and (30) in Truelove & McKee (1999).

- The unshocked magnetic field, B_0, is arbitrarily chosen and the shocked field (which produces the synchrotron emission) is taken to be $B_2 = r_{\text{tot}}\, B_0$. B_0 for the reverse shock is taken to be considerably smaller than for the FS.

Table 1 Forward shock parameters

Input	SN1006	Tycho	Kepler
Type	Ia	Ia	II
$t_{\rm snr}$ [yr]	994	428	396
n_{p0} [cm^{-3}]	0.1	0.5	0.5
B_0 [μG]	5	20	20
$E_{\rm sn}$ [10^{51} erg]	1	0.7	1
$M_{\rm ej}$ [$M_\odot$]	1.4	1.4	5
T_{p0} [K]	10^4	10^4	10^4
$T_{\rm e2}/T_{\rm p2}$	1	1	0.3
$\eta_{\rm inj,p}$	4×10^{-4}	2×10^{-3}	8×10^{-4}
$(e/p)_{\rm rel}$	2×10^{-3}	2×10^{-3}	3×10^{-3}
$\eta_{\rm mfp}$	20	50	60
n	7	7	9
α	0.67	0.67	1
Output values			
$V_{\rm sk}$ [km s^{-1}]	3700	4100	4100
$R_{\rm sk}$ [pc]	7	3	2.5
$M_{\rm S0}$	230	250	260
$M_{\rm A0}$	130	80	80
$r_{\rm tot}$	9.2	8.5	8.3
$r_{\rm sub}$	3.4	2.9	3.2
B_2 [μG]	50	170	165
$E_{\rm max,p}$ [TeV]	2.4	2	1.6
$T_{\rm p2}$ [K]	2.7×10^7	3.2×10^7	5.6×10^7
$T_{\rm tp}$ [K]	1.7×10^8	2.0×10^8	3.0×10^8
$\epsilon_{\rm rel}$	0.8	0.7	0.7
Flux parameters			
$D_{\rm snr}$ [kpc]	1.7	2.3	5
$V_{\rm emis}$ [pc^3]	150	20	20

- The supernova energy, $E_{\rm sn}$, and ejecta mass, $M_{\rm ej}$, are standard values.

- $T_{p0} = 10^4$ K is the unshocked proton temperature.

- The shocked electron to proton temperature ratio, T_{e2}/T_{p2}, is a free parameter in the NL shock model and can be adjusted to improve the X-ray fit from bremsstrahlung.

- The injection efficiency, $\eta_{inj,p}$, is the fraction of protons that become superthermal.

- The electron to proton ratio at relativistic energies, $(e/p)_{rel}$, is an arbitrary parameter in the simple model.

- The maximum energy cosmic rays obtain in the NL model depends on the scattering mean free path, λ, which is assumed to be,

$$\lambda = \eta_{mfp} \, r_{g,max} \, (r_g/r_{g,max})^{\alpha} \ , \tag{15.2}$$

  where $r_{g,max}$ is the gyroradius at E_{max}, η_{mfp} is taken to be independent of particle momentum, and α is given in the Table. Our fits require quite large values of η_{mfp}, giving maximum proton energies, $E_{max,p}$, well below that obtained in the Bohm limit. A similar effect was noticed first by Mastichiadis & de Jager (1996) for SN1006.

- The initial density profile in the ejecta has a power law density distribution, $\rho \propto r^{-n}$, and the unshocked ISM is assumed uniform.

The output values of the shock speed, V_{sk}, and radius, R_{sk}, are taken directly from the Truelove & McKee (1999) solution. Once these are obtained, the sonic and Alfvén Mach numbers (M_{S0} and M_{A0}) are determined, and then the model determines the total compression ratio, r_{tot}, the subshock ratio, r_{sub}, shocked temperatures, T_{p2}, and particle spectra. T_{tp} is the test-particle shocked temperature. Using the particle spectra, the continuum emission is calculated. All of the forward shocks in these examples are highly efficient and nonlinear. They yield total compression ratios > 4, shocked temperatures nearly 10 times lower than the corresponding test-particle shocks, and place the majority of the total energy flux, ϵ_{rel}, into relativistic particles (mainly protons).

Finally, our model includes a rough estimate of the emission volume

$$V_{emis} \approx (4\pi/3) \, R_{sk}^3 / r_{tot} \ , \tag{15.3}$$

as described in Ellison et al. (2000). This volume is considerably less than the total remnant volume. For SN1006, the value of V_{emis} given in the Table is the value given by eq.(15.3), for Tycho it is 1.5 times (15.3), and for Kepler it is 3 times (15.3).

4. CONCLUSIONS

While supernova remnants are generally believed to be the sources of galactic cosmic rays, little or no direct evidence for the production of relativistic ions in SNRs exists. A clear signature of ion acceleration would be the presence of a pion-decay bump at $\sim$ 100 MeV. As of this writing, no such feature has been unambiguously identified in any SNR. Relativistic electrons have long been directly associated with SNRs through radio emission, and the recent understanding that non-thermal X-ray emission is likely the result of synchrotron emission shows that TeV electrons can be generated by SNRs. However, the same X-ray fits, when combined with radio and constraints at γ-ray energies, also show that the SNRs considered here (SN1006, Tycho, and Kepler) *do not* accelerate electrons or ions to energies above a few TeV. Furthermore, unless something akin to magnetic field amplification occurs, these SNRs will not in the future produce particles with energies much above what they are producing now.

While high field objects like Cas A can produce 10^{15} eV particles, they are some subclass of all SNRs (there is no reason to believe, for instance, that Type Ia supernovae have such high fields). Important questions for the origin of cosmic rays are posed if, in fact, a significant fraction of SNRs do not produce cosmic rays up to the knee.

If nonlinear shock acceleration effects are important in SNRs, the entire particle distribution from thermal to the highest energies is interconnected. Furthermore, there is a direct linkage between protons and electrons. This makes emission in all energy ranges important for a complete understanding of the source, and there can be no doubt that SNRs produce emission in the EGRET energy range. Even without a clear detection, the EGRET upper limits provide important constraints. As shown by Ellison et al. (2000), the distinctive pion-decay bump will be most prominent for remnants in high n_{p0} and high B_0 environments; given the flux levels predicted, we anticipate positive detections of such spectral features (generally below EGRET sensitivities for $n \gtrsim 10\ \mathrm{cm}^{-3}$), by the GLAST experiment in the not too distant future.

Acknowledgments

I would like to thank the organizers and particularly Alberto Carramiñana for having a very enjoyable and productive workshop. I also thank G. Double for collecting some of the data used here.

References

Achterberg, A, Blandford, R. D. and Reynolds, S. P., 'Evidence for enhanced MHD turbulence outside sharp-rimmed supernova remnants,' 1994, A&A, 281, 220

Allen, G.E., Gotthelf, E.V., & Petre, R. 'Evidence of 10-100 TeV Electrons in Supernova Remnants,' 1999, astro-ph/9908209

Baring, M.G., Ellison, D.C., Reynolds, S.P., Grenier, I.A., & Goret, P., 'Radio to Gamma-Ray Emission from Shell-Type Supernova Remnants: Predictions from Nonlinear Shock Acceleration Models,' 1999, Ap.J., 513, 311

Baring, M.G., Ogilvie, K.W., Ellison, D.C., & Forsyth, R.J., 'Acceleration of Solar Wind Ions by Nearby Interplanetary Shocks: Comparison of Monte Carlo Simulations with Ulysses Observations,' 1997, Ap.J., 476, 889

Berezhko, E.G., Yelshin, V.K., & Ksenofontov, L.T., 'Cosmic Ray Acceleration in Supernova Remnants,' 1996, JETP, 82(1), 1

Berezhko, E. G. and Ellison, D. C., 'A Simple Model of Nonlinear Diffusive Shock Acceleration,' 1999, Ap.J., 526, 385

Blandford, R., & Eichler, D., 'Particle Acceleration at Astrophysical Shocks - a Theory of Cosmic-Ray Origin,' 1987, Phys. Repts., 154, 1

Borkowski, K. J., Rho, J., Dyer, K. K. and Reynolds, S. P., 'Thermal and Nonthermal X-ray Emission in SNR RCW 86,' 1999, AAS Meeting 195, presentation 43.13

Buckley, J.H., 'Observations of TeV Gamma Rays from Supernova Remnants,' 1994, AAS Meeting, 185, #120.07.

Chevalier, R.A. 'Self-similar solutions for the interaction of stellar ejecta with an external medium,' 1982, Ap.J., 258, 790

Decourchelle, A.,& Ballet, J. 'Modeling the X-ray spectrum of Kepler's supernova remnant,' 1994, A&A, 287, 206

Decourchelle, A., Ellison, D. C. and Ballet, J. 'Thermal X-ray Emission and Cosmic Ray Production in Young Supernova Remnants', 2000, *ApJL*, in press.

Dorfi, E.A. 'Cosmic Ray Production Rates in Supernova Remnants,' 2000, Astr. Sp. Sci., 272, 227

Drury, L.O'C. 'An Introduction to the Theory of Diffusive Shock Acceleration of Energetic Particles in Tenuous Plasmas,' 1983, Rep. Prog. Phys., 46, 973

Ellison, D.C., 'The Cosmic Ray – X-ray Connection: Effects of Nonlinear Shock Acceleration on Photon Production in SNRs,' 2000, in Acceleration and Transport of Energetic Particles Observed in the

Heliosphere, Eds. Richard A. Mewaldt et al. AIP Conf. Proceedings, p. 386.

Ellison, D.C., et al. 'Photon and Particle Production in Cassiopeia A: predictions from Nonlinear Diffusive Shock Acceleration,' 1999, Proc. 26th Int. Cosmic Ray Conf. (Salt Lake City), 3, 468.

Ellison, D.C., Baring, M.G., and Jones, F.C., 'Non-Linear Particle Acceleration in Oblique Shocks,' 1996, Ap.J., 473, 1029

Ellison, D.C., Berezhko, E.G., & Baring, M.G. 'Nonlinear shock acceleration and photon emission in supernova remnants,' 2000, Ap.J., 540, 292

Ellison, D.C., Jones, F.C., and Baring, M.G., 'Direct Acceleration of Pickup Ions at the Solar Wind Termination Shock: The Production of Anomalous Cosmic Rays,' 1999, Ap.J., 512, 403

Ellison, D.C., Möbius, E., & Paschmann, G., "Particle Injection and Acceleration at the Earth's Bow Shock: Comparison of Upstream and Downstream Events," 1990, Ap.J., 352, 376

Giacalone, J., Burgess, D., Schwartz, S.J., Ellison, D.C., & Bennett, L. 'Injection and Acceleration of Thermal Protons at Quasi-Parallel Shocks: A Hybrid Simulation Parameter Survey,' 1997, J.G.R., 102, 19,789

Hwang, U., and Gotthelf, E.V. 'X-Ray Emission-Line Imaging and Spectroscopy of Tycho's Supernova Remnant,' 1997, Ap.J., 475, 665

Jones, F.C., & Ellison, D.C. 'The Plasma Physics of Shock Acceleration,' 1991, Space Sci. Rev., 58, 259

Kang, H. & Jones, T. W. 'Numerical studies of diffusive particle acceleration in supernova remnants,' 1991, M.N.R.A.S., 249, 439

Koyama, K.,Petre, R., Gotthelf, E. V., Hwang, U., Matsura, M., Ozaki, M. and Holt, S. S., 'Evidence for Shock Acceleration of High-Energy Electrons in the Supernova Remnant SN:1006,' 1995, Nature, 378, 255

Lee, M.A., 'Coupled Hydromagnetic Wave Excitation and Ion Acceleration Upstream of the Earth's Bow Shock,' 1982, J.G.R., 87, 5063

Mastichiadis, A. & de Jager, O. C. 'TeV emission from SN 1006,' 1996, A&A, 311, L5

Reynolds, S.P. 'Synchrotron Models for X-Rays from the Supernova Remnant SN 1006,' 1996, Ap.J.(Letts), 459, L13

Reynolds, S. P., 'Models of Synchrotron X-Rays from Shell Supernova Remnants,' 1998, Ap.J., 493, 375

Reynolds, S. P. and Ellison, D. C., 'Electron acceleration in Tycho's and Kepler's supernova remnants - Spectral evidence of Fermi shock acceleration,' 1992, Ap.J.(Letts), 399, L75

Reynolds, S. P. and Keohane, J.W., 'Maximum energies of shock-accelerated electrons in young shell supernova remnants' 1999, Ap.J., 525, 368

'A New Determination of the Distance to Kepler's Supernova Remnant,' Reynoso, E. M., & Goss, W. M. 1999, A.J., 118, 926

Scholer, M., Trattner, K.J., & Kucharek, H. 'Ion Injection and Fermi Acceleration at Earth's Bow Shock: The 1984 September 12 Event Revisited,' 1992, Ap.J., 395, 675

Slane, P., Gaensler, B.M., Dame, T.M., Hughes, J.P., Plucinsky, P. P., and Green, A. 'Nonthermal X-Ray Emission from the Shell-Type Supernova Remnant G347.3-0.5,' 1999, Ap.J., 525, 357

Slane, P., Hughes, J.P., Edgar, R.J., Plucinsky, P.P., Miyata, E., Tsunemi, H., and Aschenbach, B. 'RXJ 0852.0-0462: Another Nonthermal Shell-Type SNR,' 2000, submitted to ApJ.

Tanimori, T. et al.., 'Discovery of TeV Gamma Rays from SN 1006: Further Evidence for the Supernova Remnant Origin of Cosmic Rays' 1998, Ap.J.(Letts), 497, L25

Terasawa, T., et al. 'Observation of a 'Cosmic-Ray-Modified' Interplanetary Shock,' 1999, Proc. 26th Int. Cosmic Ray Conf. (Salt Lake City), 6, 528.

Truelove, J.K., & McKee, C.F., 'Evolution of Nonradiative Supernova Remnants,' 1999, Ap.J.Suppl., 120, 299

GUIDING THE WAY TO GAMMA-RAY SOURCES: X-RAY STUDIES OF SUPERNOVA REMNANTS

Patrick Slane
Harvard-Smithsonian Center for Astrophysics
60 Garden Street
Cambridge, MA 02138
USA
slane@cfa.harvard.edu

Keywords: Supernova remnants, gamma-ray sources

Abstract Supernova remnants have long been suggested as a class of potential counterparts to unidentified γ-ray sources. The mechanisms by which such γ-rays can arise may include emission from a pulsar associated with a remnant, or a variety of processes associated with energetic particles accelerated by the SNR shock. Imaging and spectral observations in the X-ray band can be used to identify properties of the remnants that lead to γ-ray emission, including the presence of pulsar-driven nebulae, nonthermal X-ray emission from the SNR shells, and the interaction of SNRs with dense surrounding material.

1. INTRODUCTION

Supernova remnants and their associated pulsars represent an energetic class of objects which is intimately connected to sources of γ-ray emission in our Galaxy. The connection between SNRs and the energetic cosmic rays that pervade the Galaxy has long been assumed, for example; shock acceleration by the SNR blast wave provides ample energy for the production of multi-TeV particles, and the presence of nearby material in dense clouds from which the remnant progenitors collapsed forms a natural target for pion production with subsequent $\pi^0 \rightarrow 2\gamma$ decay. Nonthermal bremsstrahlung of electrons off ambient material, as well as inverse Compton scattering of electrons off ambient photons, can also lead to γ-ray production – and in many cases are the dominant mechanisms. SNRs are thus strong candidates for the emission of

A. Carramiñana et al. (eds.),
The Nature of Unidentified Galactic High-energy Gamma-ray Sources, 227–242.

γ-rays. Recent observations have revealed remnants whose emission is dominated by nonthermal X-rays from these energetic particles, providing direct evidence of TeV electrons, and leading to the detection of two remnants as VHE γ-ray sources.

Gamma-ray studies to date have clearly identified pulsars as strong candidates for γ-ray emission (see Thompson – these proceedings), and SNRs hold another key in this regard. Recent X-ray observations (e.g. Slane et al. 1997; Harrus, Hughes, & Slane 1998; Harrus & Slane 1999) have revealed numerous new synchrotron nebulae in SNRs which, to the best of our understanding, must house magnetospherically active neutron stars. Studies with *Chandra* and XMM-Newton can realistically be expected to expand much of the ground breaking work done with ASCA in this regard, thus improving our ability to identify promising candidates for energetic γ-ray production. Here I review recent and ongoing observational X-ray work on composite remnants, SNR interactions with molecular clouds, and nonthermal emission from shell-type SNRs, and discuss how these might act as guides to sources of γ-ray emission.

2. GAMMA-RAYS FROM SNRS

The large positional uncertainties for EGRET sources, coupled with the extended nature of SNRs, make the clear identification of direct associations between the two populations a difficult task (Figure 1, left). Moreover, statistical assessment of the likelihood of a given association is complicated by the fact that the distribution of potential source candidates (e.g. SNRs, pulsars, OB associations) are not unrelated (Kaaret & Cottam 1996; Romero et al. 1999; Grenier, 2000). These results render positional coincidences uncompelling as the only evidence for an association between a particular SNR and EGRET source. X-ray observations probe several characteristics of SNRs and their environments which are directly related to γ-ray production, and are thus of considerable importance in establishing such associations. In addition, such observations can provide candidate SNRs for further, more sensitive γ-ray observations.

There are three primary means by which γ-rays might be associated with an SNR. Though only indirectly connected, the most prevalent is emission directly from an associated pulsar and/or a pulsar-driven synchrotron nebula. This is likely to be a viable mechanism for only somewhat young pulsars, and it is thus possible that the associated SNRs will still be detectable. A more direct mechanism is π^0 decay through interaction of the SNR shell with dense material in a nearby molecular cloud. Finally, nonlinear shock acceleration at the SNR shell may result

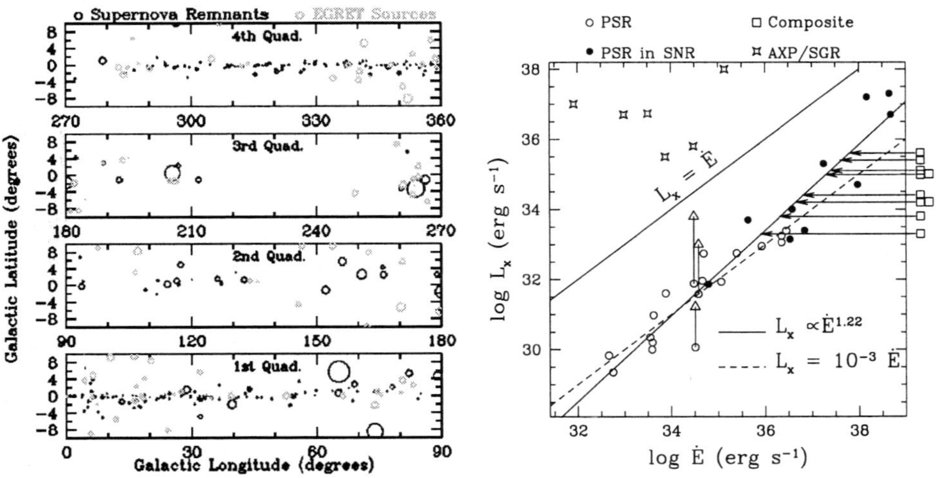

Figure 1 Left: Distribution of SNRs (black) and EGRET sources (grey) in the Galactic plane. Circles correspond approximately to SNR size or EGRET position error circles. Right: L_x vs. $\dot{E}$ for pulsars and plerions. Triangles correspond to pulsars with identified cooling components, and the associated circles have this component subtracted. Luminosities include emission from associated synchrotron nebula where known. For composite SNRs whose pulsars are not detected, L_x is plotted along the right axis. Arrows indicate where these objects would reside on the L_x vs. $\dot{E}$ diagram. Note the good agreement between these values and those for young pulsars in SNRs (filled circles).

in inverse Compton and/or nonthermal bremsstrahlung radiation from energetic electrons accelerated by the SNR. Each of these scenarios can be associated with particular characteristics of the SNR X-ray emission. I discuss these cases briefly, and present possible examples of each.

2.1. PULSARS AND SYNCHROTRON NEBULAE

Since the majority of SNe result from massive star collapse, and most such collapses are expected to form neutron stars (NSs), it is reasonable to expect that young NSs should be associated with most young SNRs. If all young NSs are rapidly rotating, magnetospherically active pulsars, the corollary would then suggest that these SNRs are likely to be sources of energetic γ-ray emission since pulsars are a well-established class of γ-ray sources. In practice, however, there are numerous examples of young SNRs for which no associated NS is observed. Moreover, recent observations have revealed compact X-ray sources, apparently associated with SNRs, whose properties differ considerably from those of Crab-like pulsars (see Caraveo, Bignami, & Trümper 1996; Kaspi 2000). These include the Anomalous X-ray Pulsars (AXPs) such as 1E 2259+586 in the remnant CTB 109, the radio-quiet neutron stars such as PKS 1209−51/52 in Puppis A (Zavlin, Truümper, & Pavlov 1999), and the point source in Cas A, whose properties are unlike any of the above (Pavlov et al. 2000). In cases where an X-ray emitting synchrotron nebula (or "plerion") is observed, however, we infer the presence of a young pulsar powering the nebula. Remnants displaying such a component are thus good candidates for γ-ray emission.

Empirically, it is found that there exists a relationship between $\dot{E}$, the pulsar energy loss rate, and L_x, the total X-ray luminosity from the pulsar (not including any cooling component) and its associated synchrotron nebula (Seward & Wang 1988; Becker & Trümper 1997). Thus, even in the absence of direct detection of a pulsar, its properties can be inferred from the nonthermal X-ray spectrum of the plerion. This is illustrated in Figure 1 (right) where I have plotted values for known pulsars. Synchrotron nebulae in composite SNRs for which no pulsar has yet been identified are shown as well. It is important to note, however, that variations in the L_x vs. $\dot{E}$ relation are large. This is due, at least in part, to the variety of conditions inherent to the systems used to establish the relation (bow shock nebulae, plerions interacting with supernova ejecta, etc.). This has been addressed in recent work by Chevalier (2000), who has developed a one-zone model for the X-ray properties of pulsar nebulae which accounts for differences in the relative efficiency with which L_x is produced, based on the cooling regime in which the electrons find themselves after passing through the wind termination shock.

2.1.1 G21.5−0.9. Although we infer the presence of an active, young NS with any X-ray synchrotron nebula, it is by no means the case that pulsars have been identified for each. G21.5−0.9, for example, is a well-defined plerion for which radio observations show no direct evidence of a compact central source. *Chandra* observations (Slane et al. 2000) reveal a compact source at the center of the nebula, but the high resolution images clearly demonstrate that the source is slightly extended. Using the radio spectrum to estimate the magnetic field strength of the nebula under the assumption of equipartition between the field and the particles, and requiring that this associated magnetic pressure act to confine a pulsar wind driven by the spin-down power, $\dot{E} = 3.5 \times 10^{37} d_5^{1.6}$ erg cm^{-2} s^{-1}, of a central pulsar (using the empirical $\dot{E}$ vs. L_x relationship, with d_5 as the distance in units of 5 kpc) indicates that the extended size of the central emission is consistent with

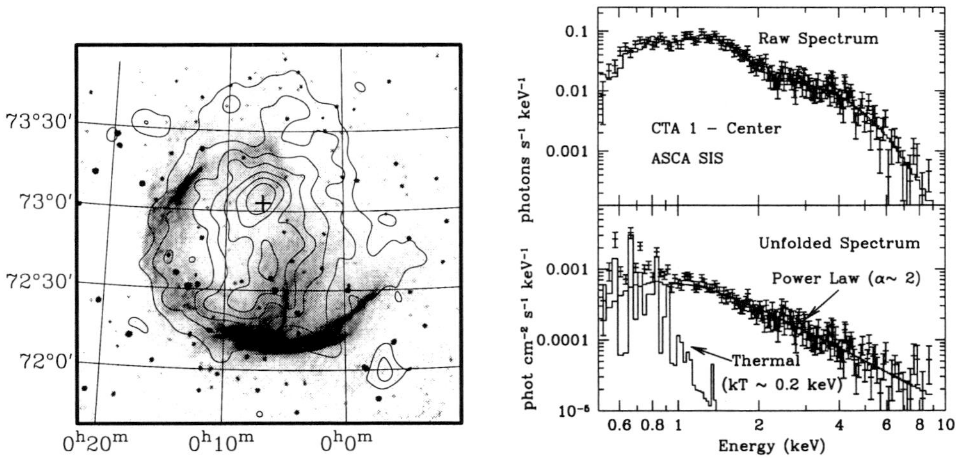

Figure 2 Left: Continuum image of CTA 1 at 1420 MHz with X-ray contours from the ROSAT PSPC. (Radio image kindly provided by T. Landecker.) Note the incomplete shell-like radio morphology contrasted with the centrally bright X-ray morphology. The position of the faint point source J000702+7302.9 is indicated with a cross. Right: ASCA data extracted from SIS chip centered on central region of CTA 1. Upper: Raw spectrum with best-fit power-law plus thermal model. Lower: Unfolded spectrum and model components.

the radius of the stand-off shock where the wind is confined. A smaller $\dot{E}$ is predicted by the model of Chevalier (2000), resulting in a smaller estimated shock radius, but one still in rough agreement with the observations. As the particle wind traverses the standoff shock, the nebular structure changes from a particle-dominated to a magnetic-dominated flow (Kennel & Coroniti 1984). The finite lifetime of the synchrotron emitting electrons results in a steepening of the spectrum with radius, an effect clearly observed in G21.5−0.9 (Slane et al. 2000). Thus, the X-ray observations indicate the presence of an energetic pulsar in G21.5−0.9. The inferred value of $\dot{E}/d^2$ (a rough indicator of the expected γ-ray flux) is larger than that for many of the known γ-ray pulsars (or for only several, using the value estimated by Chevalier) suggesting that sensitive γ-ray observations of this source may be of interest.

2.1.2 CTA 1 and 2EG J0008+7303. Another example of the pulsar-plerion connection of particular interest in the context of γ-ray sources in that of CTA 1. In the radio band, CTA 1 (G119.5+10.3) consists of an incomplete shell (Figure 2) with faint emission extending to the northwest in what may be a breakout region (Pineault et al. 1997). The X-ray morphology is centrally brightened with emission bounded by the radio shell in the south and west, and with faint emission extending into the breakout region. ROSAT PSPC observations indicate a two-component spectrum for which the hard component dominates in the central regions but becomes weaker in the outer regions of the remnant (Seward, Schmidt, & Slane 1995). A faint X-ray source (J000702+7302.9) is found at the peak of the central X-ray emission (Figure 2).

ASCA observations of CTA 1 (Slane et al. 1997) reveal a nonthermal central spectrum clearly indicating the composite nature of this remnant (Figure 2). This is bolstered by the presence of the EGRET source 2EG J0008+7303 whose error circle includes the position of the faint X-ray source (Brazier et al. 1997) and whose spectrum, when extrapolated to the soft X-ray band, is consistent with that measured for J000702+7302.9 (although the latter fact is not necessarily expected from emission models).

The nonthermal X-ray luminosity is $5.6 \times 10^{33} D_{1.4}^2$ erg s^{-1} leading to an inferred pulsar energy loss rate of $\dot{E} = 1.7 \times 10^{36} D_{1.4}^{1.4}$ erg s^{-1}. If correct, the value of $\dot{E}/d^2$ is exceeded by only 4 pulsars, all of which (along with 4-5 with smaller values) are known γ-ray sources. The X-ray measurements thus strongly indicate that 2EG J0008+7303 is associated with the X-ray source and synchrotron nebula in CTA 1. Modeling of outer gap γ-ray production (Zhang & Cheng 1998) indicates that a γ-ray pulsar scenario for this source is indeed feasible. Deeper X-ray observations of this source are currently underway with XMM-Newton.

2.1.3 Other Plerionic Candidates. In addition to the Crab and Vela pulsars, both associated with synchrotron nebulae, and both well-known γ-ray pulsars, there are several EGRET sources for which X-ray observations suggest a pulsar (or pulsar-driven nebula) counterpart. Halpern et al. (2001) use X-ray, radio, and optical data to show that 3EG J2227+6122 may be associated with a pulsar-powered source. An incomplete circular shell within the EGRET error circle is seen in VLA sky survey data, and ASCA data from this region reveals emission well-described by a power law. The radio image resembles a bow-shock nebula or wind-blown bubble, and the X-ray properties are consistent with such an interpretation.

ASCA observations of the region around 2EG J1418−6049 (Roberts and Romani 1998) also suggest a pulsar wind nebula interpretation for this source. The X-ray image reveals an extended nonthermal object whose properties are consistent with being a nebula energized by a pulsar with a characteristic age of $\sim 10^{4.5}$ yr.

Extended nonthermal X-ray emission is also observed within the error circle of 2EG J1811−2339 (Oka et al. 1999). These authors suggest that the source may be a synchrotron nebula associated with a currently unknown pulsar, and that the nebula may be interacting with a dense cloud of material seen in CO observations. In this case, the γ-ray emission may only be indirectly associated with the pulsar, which may act as a source of electrons that undergo shock acceleration at the cloud interaction site. Several other SNRs with nearby pulsars have been suggested as counterparts for EGRET sources (e.g. W44 and W28), but it appears that the γ-ray emission from these sources is also associated with the remnants and not the pulsars. Such interaction processes are discussed in more detail below.

2.2. PARTICLE ACCELERATION BY SNR SHOCKS

Beyond the early γ-ray emission fueled by the radioactive decay of unstable isotopes formed in supernovae, the mechanical energy released in the explosions can be transformed into γ-ray emission over a longer period of time. As the blast wave from the explosion expands, material from the surrounding circumstellar material (CSM) and ISM is swept up into a shell of hot gas. For an ideal $\gamma = 5/3$ gas, the shock jump conditions yield an increase in density by a factor of 4 and a postshock temperature

$$T = \frac{3\mu m}{16k} V_s^2$$

where V_s is the shock speed, m is the proton mass, and μ is the mean molecular weight of the gas ($\mu \approx 0.6$). This shock-heated gas yields the familiar X-ray emission, characterized by a thermal bremsstrahlung spectrum accompanied by strong emission lines.

As increasing amounts of material are swept up by the blast wave, the shock is decelerated and the ejecta from the explosion encounter the dense shell. A reverse-shock is generated, heating the ejecta. At early times, then, the X-ray spectrum is dominated by emission from the ejecta. As the amount of swept-up material increases, the spectrum becomes dominated by emission from material with ISM abundances.

In addition to thermal heating of the swept-up gas, some fraction of the shock energy density goes into nonthermal production of relativistic particles through diffuse shock acceleration. Particle acceleration by SNR shocks has, of course, been suggested for decades as a process by which cosmic rays are produced. Radio observations of SNRs provide ample evidence of electrons with GeV energies through synchrotron radiation in the compressed magnetic field at the remnant shell.

At higher particle energies, γ-ray production may result from nonthermal bremsstrahlung of electrons colliding with ambient gas, from inverse Compton scattering of ambient photons, and from the decay of neutral pions created by the collision of energetic protons. If the relativistic particle component of the energy density becomes comparable to that of the thermal component, the shock acceleration process can become highly nonlinear. The gas becomes more compressible, which results in a higher density and enhanced acceleration. This process has been considered in detail by Baring et al. (1999) who present a model for the radio to γ-ray emission. The relative contributions from the different γ-ray production mechanisms depend highly on ambient conditions, and X-ray studies of these SNRs reveal these conditions and can provide spectral measurements which highly constrain the models. Here I summarize the X-ray properties of several SNRs which may be associated with γ-ray emission, in an effort to illustrate these ideas.

2.3. PION DECAY AND MOLECULAR CLOUD INTERACTIONS

If the blast wave of an SNR is capable of accelerating protons (or other ions) to sufficiently high energies, then the interaction of these particles with ambient material can result in the production of neutral pions which subsequently decay into γ-rays. The γ-ray flux for such a process was calculated by Drury, Aharonian, & Völk (1994). For a power-law particle spectrum with spectral index $\alpha \sim 2.1 - 2.3$, the flux is

$$F_\gamma(\geq E_{\rm TeV}) \approx 5 \times 10^{-11} \epsilon E_{51} d_{\rm kpc}^{-2} n E_{\rm TeV}^{-\alpha+1} \ {\rm cm}^{-2}\ {\rm s}^{-1}$$

where ϵ is the fraction of the blast wave energy (E_{51}, in units of 10^{51} erg) converted into cosmic ray energy, $d_{\rm kpc}$ is the SNR distance in kpc, n is the number density of the ambient medium, and $E_{\rm TeV}$ is the γ-ray energy in TeV. The leading numerical factor actually depends on α; I have taken an average value which is accurate to within a factor of two or so over the stated range for α. This spectrum is plotted in Figure 3 (left) for two values of ϵ, the larger of which is probably an overestimate of what can typically be extracted from this process. Here I have assumed $n = 1\ {\rm cm}^{-3}$, and $E_{51} = d_{\rm kpc} = 1$. Also shown are approximate sensitivities for EGRET, GLAST, and VERITAS. Note that, for typical SNR parameters, the predicted flux is near or below the EGRET sensitivity, although above the sensitivity limits for GLAST or VERITAS.

For cases in which the SNR is encountering dense material (i.e. $n \gg 1$), such as in an interaction with a nearby molecular cloud, the γ-ray

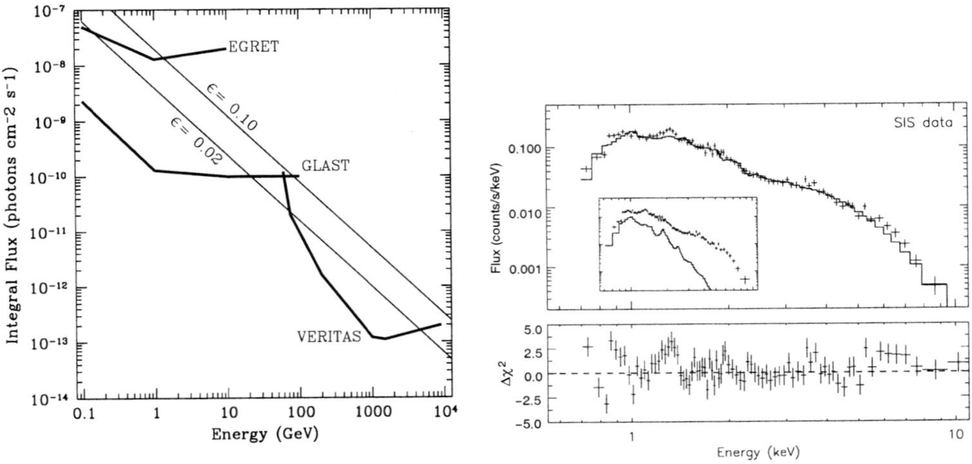

Figure 3 Left: Model pion-decay γ-ray flux from SNRs (see text). Also shown are detection sensitivities for EGRET, GLAST, and VERITAS. Right: X-ray spectrum from MSH 11-62 showing clear evidence of a synchrotron component.

flux can be enhanced considerably. In a model for the evolution of SNRs in molecular clouds, Chevalier (1999) argues that the γ-ray emission results from the radiative shell of the remnant where some fraction of the energy density in the postshock region goes into relativistic particles. Bremsstrahlung interactions in the dense shell yield the high energy emission, and synchrotron radiation from this population, in the compressed magnetic field, can simultaneously produce the observed radio emission. Because massive stars live their lives quickly, they expire nearby the molecular clouds from which they formed; the ensuing supernova explosion can thus lead to direct interaction with the molecular clouds. Such SNR/cloud interactions are most clearly identified through OH maser emission (e.g. Frail et al. 1996), although CO mapping also provides an important means of tracing the molecular gas around SNRs.

The X-ray properties of SNRs can also provide an indication of such interactions. SNR/cloud interactions can results in a significant brightening of the interaction region due to the increase in density. For example, molecular clouds may be associated with the so-called "mixed morphology" (MM) class of remnants (Rho & Petre 1998). These SNRs are characterized by a center-filled X-ray morphology, as might be expected for a shell-type remnant whose central region contains a pulsar-powered nebula, but in this case the emission is thermal in nature, and the radial temperature profile is relatively flat. Several models have been put forth to explain the brightness and temperature profiles for these SNRs. The cloud evaporation model derived by White and Long (1991) has been applied to the emission from W44 (Harrus et al. 1996), G272.2−3.2 (Harrus et al. 2001), MSH 11−6*1A* (Slane et al. 2001), and others. In this model the SNR evolves in a two-phase interstellar medium (ISM) consisting of cool, dense clouds and a low density intercloud medium. The SNR shock sweeps past the clouds quickly, leaving them to slowly evaporate in the hot SNR interior, thus contributing X-ray emitting material to the central regions. Harrus et al. (1996) also considered a model for W44 in which the remnant had entered the radiative phase, thus leaving the shell too cool to radiate significantly in X-rays, but allowing the emission from the hotter interior to persist. W44 is known to be interacting with an adjacent molecular cloud, which is consistent with deceleration of the shock and evolution into the radiative phase. Shelton et al. (1999) extended this picture significantly by including the effects of thermal conduction on smoothing out the temperature and density profile of the X-ray emission material. Slane et al. (2001) applied both models to emission from MSH 11−6*1A*.

Other SNRs which fit into this MM class include W28 (Long et al. 1991) and IC 443 (Petre et al. 1988, Asaoka & Aschenbach 1994), two remnants which (along with W44) have been suggested as counterparts to EGRET sources; each is known to be interacting with a molecular cloud. Romero et al. (1999) have compiled a list of SNRs with positional coincidences with EGRET sources, and at least one-third of these show some indication of the presence of molecular clouds. As noted above, these positional coincidences are far from convincing evidence that each of these SNRs are associated with γ-ray sources, but the presence of dense material provides some supporting evidence.

2.3.1 MSH 11−6 *1A* and 3EG J1101−6103. The unidentified EGRET source 3EG J1101−6103 provides an interesting case for which each of the topics above deserves consideration. Early catalogs associated this source with MSH 11−6*2*. X-ray observations (Harrus, Hughes, & Slane 1998) reveal the remnant to be composite in nature, with a two-component spectrum (Figure 3, right) consisting of a central nonthermal component and an extended thermal component. As noted above, the presence of a synchrotron nebula is indicative of a young pulsar, making the association a reasonable one. An improved position for the EGRET source revealed that MSH 11−6*2* falls outside the error circle, however (Figure 4, left).

Kaspi et al. (1997) identified a young pulsar PSR J1105−6107 which lies within the error circle, and suggested an association between the pulsar and the nearby remnant MSH 11−6*1A* (Figure 4). The value of $\dot{E}/d^2$

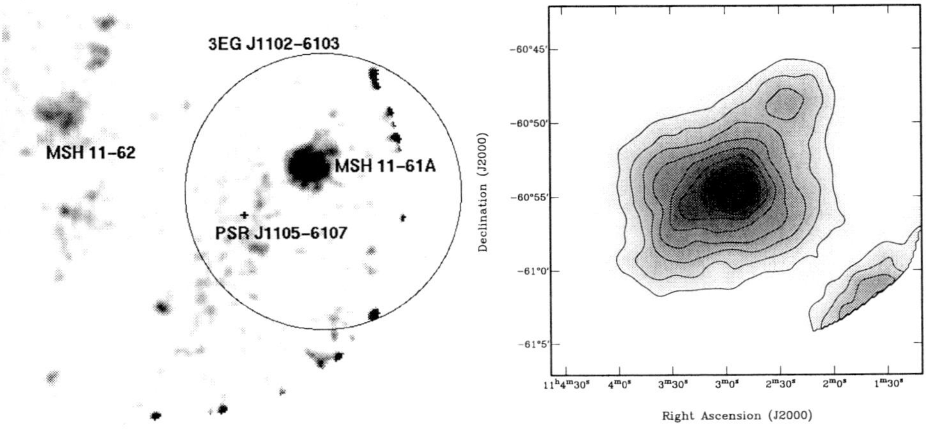

Figure 4 Left: X-ray emission from region surrounding 3EG J1102−6103. Right: ASCA GIS image of MSH 11–61*A*. Emission at the southwest edge of the field is associated with an off-axis source just outside the primary telescope field of view.

for this pulsar is quite low for a γ-ray source, however, though higher than that for PSR B1055−52 which is detected by EGRET. Moreover, X-ray studies indicate that an association between the pulsar and the SNR is unlikely (Slane et al. 2001).

It is of interest that MSH 11−61*A* lies along the line of sight to a dense molecular cloud whose distance ($\sim$ 7 kpc), from CO measurements appears consistent with that estimated for the remnant. Petruk (2000) has suggested that the mixed-morphology nature of this remnant (and others) is the result of enhanced emission from a cloud interaction directly along the line of sight. Further, this same picture is used to estimate the π^0-decay γ-ray luminosity. For a cloud density of $\sim$ 40 cm^{-2}, the predicted flux (assuming $\epsilon = 0.1$) matches that of 3EG J1101−6103. Efforts to model the observed X-ray temperature and brightness profiles under this scenario are underway. An important point, however, is that more complete modeling of the particle acceleration process in shock-cloud interactions indicates that emission from pion decay falls considerably below that from bremsstrahlung (Gaisser et al. 1998, Bykov et al. 2000).

2.4. NONTHERMAL X-RAY EMISSION FROM SNRS

While the shock-heated ejecta and CSM/ISM components of shell-type SNRs produce line-dominated X-ray spectra, as described above, a growing number of remnants also reveal evidence of hard, nonthermal X-ray emission components, apparently associated with high energy elec-

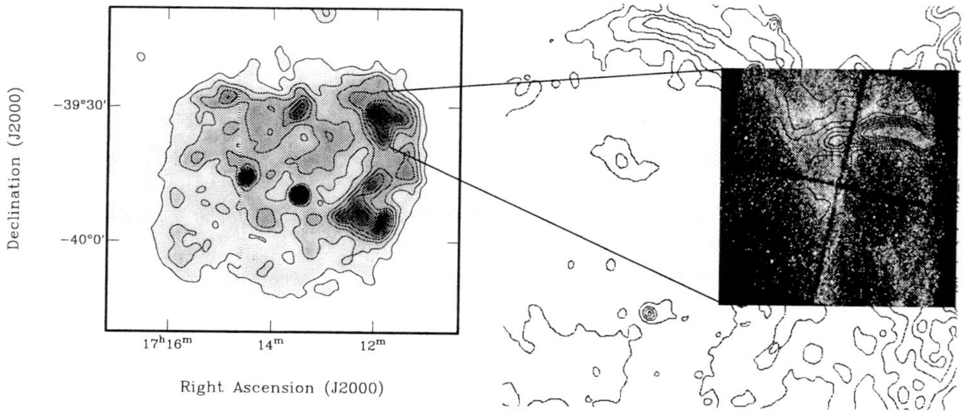

Figure 5 Left: ROSAT PSPC image of G347.3−0.5. Right: *Chandra* image of north-west rim, with radio contours from ATCA.

trons accelerated by the SNR shock. In three cases – SN 1006 (Koyama et al. 1995), G347.3−0.5 (Koyama et al. 1997; Slane et al. 1999), and G266.2−1.2 (Slane et al. 2001) – the nonthermal emission components completely dominate the thermal components, and the X-ray spectra from the shells are featureless.

Although the radio emission from SNRs provides direct evidence of GeV electrons, it is the recent detections of nonthermal X-rays from shell-type SNRs that provide direct evidence of particles accelerated to energies as high as 10 – 100 TeV, approaching the knee in the cosmic ray spectrum. Models for nonlinear shock acceleration (e.g. Ellison, Berezhko, & Baring 2000) predict a strong correlation between the radio, X-ray, and γ-ray flux that allow parameters of the acceleration process to be derived. It is thus of considerable interest to localize the regions of nonthermal X-ray emission, and to investigate the broad-band spectra (particularly in radio, X-rays, and γ-rays) from these regions.

In Figure 5, I present a preliminary *Chandra* image of the X-ray emission from the NW limb of G347.3−0.5, whose shell emission is nonthermal. Overlaid are radio contours from the Australia Telescope Compact Array (ATCA), showing good general correspondence between the components. This region of the SNR has also been detected as a source of ~TeV γ-rays (Muraishi et al. 2000). Preliminary modeling suggests a self-consistent picture in which the X-ray and radio flux is synchrotron radiation from the accelerated electrons, while the γ-ray flux represents inverse-Compton scattering of energetic electrons off of the microwave background (Ellison, Slane, & Gaensler – in preparation). The predicted GeV flux falls below the EGRET threshold, but should be de-

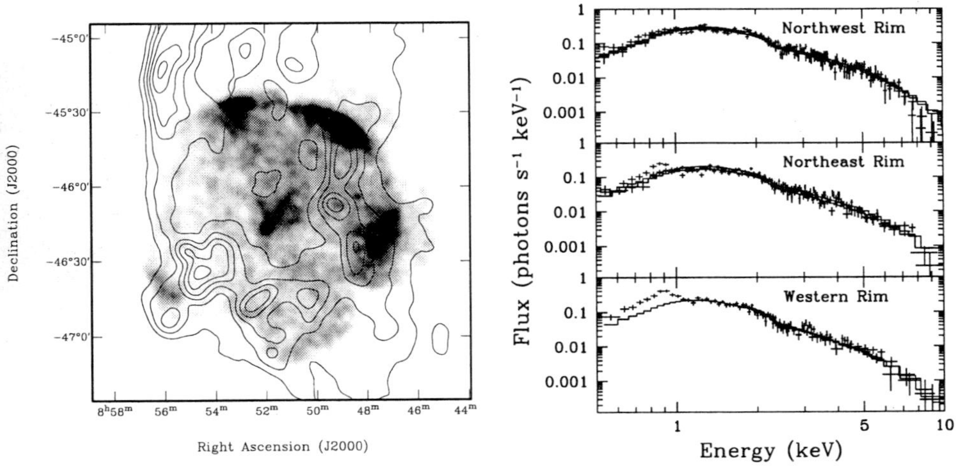

Figure 6 Left: ASCA GIS image of G266.2–1.2 ($E = 0.7 - 10$ keV). The image consists of a mosaic of 7 individual fields. Contours represent the outline of the Vela SNR as seen in ROSAT survey data with the PSPC. Right: ASCA spectra from both GIS detectors for regions of G266.2–1.2. The featureless spectrum is well described by a power law. Excess flux at low energies is presumably associated with soft thermal emission from the Vela SNR.

tectable with GLAST. Such measurements, along with more sensitive X-ray observations, will provide significant information for constraining the parameters of these models.

G266.2$-$1.2 was discovered by Aschenbach (1998) using data from the ROSAT All-Sky Survey. Situated along the line of sight to the Vela SNR, the X-ray emission stands out above the soft thermal emission from Vela only at energies above ~ 1 keV. ASCA observations show that the emission from G266.2$-$1.2 is nonthermal (Slane et al. 2001), providing another example of a shell-type SNR for which the X-ray emission is dominated by such processes. CO data (May et al. 1988) reveal a concentration of giant molecular clouds – the Vela Molecular Ridge – at a distance of $\sim 1-2$ kpc in the direction of Vela. The X-ray observations of G266.2$-$1.2 indicated a column density consistent with the remnant being located near, but probably not beyond, this molecular material. More sensitive X-ray measurements are required to investigate the shock structure and spectrum, for comparison with nonlinear shock models. Observations at TeV energies are of considerable interest, particularly given that SN 1006 and G347.3$-$0.5 have both been detected at such energies. Given the relatively small distance to this remnant, it would appear to be an excellent candidate for future observations with GLAST as well.

3. SUMMARY

Several mechanisms exist by which SNRs may be associated with the emission of energetic γ-rays. X-ray observations provide powerful clues by which the properties of these mechanisms can be revealed. Through the identification of pulsar-driven nebulae, nonthermal emission from particles accelerated at the SNR shell, or sites of interactions with dense clouds, X-ray studies of SNRs are helping to resolve at least some of the questions centered on the nature of unidentified EGRET sources. The prospects for continued success in this area are extremely encouraging. The capabilities of current X-ray observatories, coupled with upcoming γ-ray missions (both space and ground based), promise not only to help identify the counterparts to these sources of γ-rays, but also to provide new and powerful constraints on developing models for such emission.

Acknowledgments

The author wishes to thank Don Ellison and Glenn Allen for helpful discussions on nonthermal emission from SNRs. This work was supported in part by NASA through contract NAS8-39073 and grants NAG5-9106 and NAG5-9281.

References

Asaoka, I. & Aschenbach, B. 1994, *A&A*, **284**, 573

Aschenbach, B. 1998 *Nature*, **396**, 141

Baring, M. G., Ellison, D. C., Reynolds, S. P., Grenier, I. A., & Goret, P. 1999, *ApJ*, **513**, 311

Becker, W. & Trümper, J. 1997, *A&A*, **326**, 682

Brazier, K.T.S., Reimer, O., Kanbach, G., & Carraminana, A. 1997, *MNRAS*, **295**, 819

Bykov, A. M., Chevalier, R. A., Ellison, D. C., & Uvarov, Yu. A. 2000, *ApJ*, **538**, 203

Caraveo, P. A., Bignami, G. F., & Trümper, J. 1996, *A&A Rev.*, **7**, 209

Chevalier, R. A. 1999, *ApJ*, **511**, 798

Chevalier, R. A. 2000, *ApJ*, **539**, L45

Drury, L. O'C, Aharonian, F. A., & Völk, H. J. 1994, *A&A*, **297**, 959

Ellison, D. C., Berezhko, El G., & Baring, M. G. 2000, *ApJ*, **540**, 292

Frail, D. A., Goss, W. M., Reynoso, E. M., Giacani, E. B., Green, A. J., Otrupcek, R. 1996, *AJ*, **111**, 1651

Gaisser, T. K., Protheroe, R. J., & Stanev, T. 1998 *ApJ*, **492**, 21

Grenier, I. A. 2001, *A&A* **364**, L93

Halpern, J. P., Gotthelf, E. V., Leighly, K. M., & Helfand, D. J. 2001, *ApJ*, in press

Harrus, I. M., Hughes, J. P., & Helfand, D. J. 1996, *ApJ*, **464**, L161

Harrus, I. M., Hughes, J. P., & Slane, P. O. 1998, *ApJ*, **499**, 273
Harrus, I. M. & Slane, P. O. 1999, *ApJ*, **516**, 811
Harrus, I. M., Slane, P. O., Smith, R. K., & Hughes, J. P. 2001, *ApJ* – in press
Kaaret, P. & Cottam, J. 1996, *ApJ*, **462**, L35
Kaspi, V. M., Bailes, M., Manchester, R. N., Stappers, B. W., Sandhu, J. S., Navarro, J., & D'Amico, N. 1997, *ApJ*, **485**, 820
Kaspi, V. M. 2000, in ASP Conf. Ser. 202, Pulsar Astronomy 2000 and Beyond, ed. M. Kramer, N. Wex, & R. Wielebinski (San Francisco: ASP), 485
Kennel, C. F. & Coroniti, F. V. 1984, *ApJ*, **283**, 710
Koyama, K. et al. 1995 *Nature* **378**, 255
Koyama, K. et al. 1997, *PASJ*, **49**, L7
Long, K. S., Blair, William P., Matsui, Yutaka, White, Richard L. 1991, *ApJ*, **373**, 567
May, J., Murphy, D. C., & Thaddeus, P. 1988, *A&ASS* **73**, 51
Muraishi, H. et al. 2000, *A&A*, **354**, L57
Oka, T., Kawai, N., Naito, T., Horiuchi, T., Namiki, M., Saito, Y., Romani, R. W., & Kifune, T. 1999, *ApJ*, **526**, 764
Pavlov, G. G., Zavlin, V. E., Aschenbach, B., Trümper, J., & Sanwal, D. 2000, *ApJ*, **531**, L53
Petre, R., Szymkowiak, A. E., Seward, F. D., Willingale, R. 1988, *ApJ*, **335**, 215
Petruk, O, 2001, in Proceedings of NATO Advanced Study Institute "Astrophysical Sources of High Energy Particles & Radiation", (Erice, Italy, 11-21 November 2000, eds. J. P. Wefel, M. M. Shapiro, & T. Stanev.
Pineault, S., Landecker, T.L., Swerdlyk, C.M., & Reich, W. 1997, *A&A*, **324**, 1152
Rho, J.-H. & Petre, R. 1998, *ApJ*, **503**, L167
Roberts, M. S. E. & Romani, R. W. 1998, *ApJ*, **496**, 827
Romero, G. E., Benaglia, P., & Torres, D. F. 1999, *A&A*, **348**, 868
Seward, F.D., Schmidt, B., and Slane, P. 1995, *ApJ*, **453**, 284
Seward, F.D. & Wang, Z.-R. 1988, *ApJ*, **332**, 199
Shelton, R. L., Cox, Donald P., Maciejewski, Witold, Smith, Randall K., Plewa, Tomasz, Pawl, Andrew, & Rzyczka, Michal 1999, *ApJ*, **524**, 192
Slane, P., Seward, F.D., Bandiera, R., Torii, K., & Tsunemi, H. 1997, *ApJ*, **481**, 225
Slane, P. et al. 1999, *ApJ*, **525**, 357
Slane, P., Chen, Y., Schulz, N. S., Seward, F. D., Hughes, J. P., & Gaensler, B. M. 2000, *ApJ*, **533**, L29

Slane, P. et al. 2001, *ApJ* - accepted (see astro-ph/0010510)
Slane, P., Hughes, J. P., Smith, R. K., & Petre, R. 2001, *ApJ* - submitted
White, R. L. & Long, K. S. 1991, *ApJ*, **373**, 543
Zavlin, V. E., Trümper, J., & Pavlov, G. G. 1999, *ApJ*, **525**, 959
Zhang, L. & Cheng, K. S. 1998, *A&A*, **335**, 234

V

GALACTIC MICROQUASARS, BINARIES AND BLACK HOLES

MICROQUASARS IN THE GALAXY

Luis F. Rodríguez
Instituto de Astronomía, UNAM, Campus Morelia
Apdo. Postal 3-72, Morelia, Michoacán 58089 México
l.rodriguez@astrosmo.unam.mx

I. Félix Mirabel
CEA/DSM/DAPNIA/SAp, Centre d'Etudes de Saclay,
F-91191 Gif-sur-Yvette, France
mirabel@discovery.saclay.cea.fr

Keywords: Microquasars, Radio continuum stars, X-ray binaries

Abstract Microquasars are black-hole X-ray binaries with relativistic jets that mimic phenomena seen in quasars. We will present recent results on GRS 1915+105 as well as on other microquasars that show the close connection between instabilities in the accretion disk (seen in X-rays) and the ejection of relativistic plasma clouds (seen in radio). The possibility of a connection between these objects and gamma-ray sources is suggested by the possible association of a microquasar with an EGRET source.

1. INTRODUCTION

Microquasars are galactic X-ray binaries that mimic on a much smaller scale some of the phenomena seen in quasars. The microquasars combine two relevant aspects of relativistic astrophysics: accreting black holes (of stellar origin) which are a prediction of general relativity and are identified by the production of hard X-rays and gamma-rays from surrounding accretion disks, and relativistic jets of particles that are understood in terms of special relativity and are observed by means of their synchrotron emission.

There are about 250 X-ray binaries known in our Galaxy. Of these, about 25 (10% of the sample) have been detected as radio sources. Fi-

A. Carramiñana et al. (eds.),
The Nature of Unidentified Galactic High-energy Gamma-ray Sources, 245–254.

nally, synchrotron jets have been observed in about a dozen of these 25 sources; these are the objects classified as microquasars. In Table 1 we present a list of microquasars with the proposed identification for the compact object and, when observed, the velocity of the jets.

Table 1 Microquasars in our Galaxy[a]

Source	*Compact Object*	v_{ejecta}
GRS 1915+105	black hole	0.92c-0.98c
GRO J1655–40	black hole	0.92c
XTE J1748–288	black hole	
SS 433	neutron star ?	0.26c
Cygnus X-3	neutron star ?	~0.3c
CI Cam	neutron star ?	~0.15c
Scorpio X-1	neutron star	~0.5c
Circinus X-1	neutron star	≥0.1c
1E1740.7–2942	black hole	
GRS 1758–258	black hole	
XTE J1819–254	black hole	0.85c
LS 5039	?	≥0.15c

[a] As of 2000 January.

The field was greatly stimulated with the detection of superluminal motions in GRS 1915+105 (Mirabel & Rodríguez 1994). Recent, detailed reviews of the subject are given by Mirabel & Rodríguez (1998; 1999). Here we concentrate on results obtained over the last two years.

2. NEW MICROQUASARS

Two objects have been recently added to the list of microquasars: XTE J1819–254 (Hjellming et al. 2000) and LS 5039 (Paredes et al. 2000).

During 1999 September, XTE J1819–254 (=V4641 Sgr=SAX J1819.3–2525) underwent an event that was extremely rapid in its rise and decay across radio, optical, and X-ray wavelengths (Hjellming et al. 2000). The associated radio source was resolved even in the first Very Large Array (VLA) images. The inferred proper motions for the early extended emission and the radio morphology suggests that this is a relativistic jet source like GRS 1915+105 and GRO J1655–40. Note, however, that Chaty et al. (2001) have questioned the true nature of the proper motions. From optical spectroscopic observations Orosz et al. (2000) derive an optical mass function of $f(M) = 2.74 \pm 0.12$ M$_\odot$, and conclude that the mass of the compact object is in the range 8.73–11.70 M$_\odot$, well above the maximum mass of a stable neutron star and indicative of a black hole. These authors favor a distance of ~10 kpc for V4641 Sgr. The very short duration of the bright X-ray phase of V4641 Sgr suggests that many similar objects could be present in our Galaxy, most of which are not noticed when they are in outburst because of the short duration of these outbursts. A considerable number of the black holes present in our Galaxy might be contained in systems similar to V4641 Sgr.

LS 5039 is a very important object in the context of this meeting because it may provide a first link between the unidentified galactic gamma-ray sources and the microquasars. This source exhibits radio jets in VLBA observations and is the only X-ray source in the *ROSAT* catalog that coincides positionally with the source 3EG J1824–1514. A detailed review is given by Paredes et al. (2001) in these Proceedings. The study of this source should be pursued in detail since it may point to a solution for the nature of the unidentified sources.

3. THE SEARCH FOR OPTICAL AND INFRARED COUNTERPARTS TO MICROQUASARS

A better understanding of the nature of microquasars requires of multiwavelength observations. Unfortunately, many of them are very heavily obscured and in regions of the sky that are very crowded. In these Proceedings, Martí (2001) discusses VLT observations toward this goal. In the case of GRS 1758–258, two possible counterparts have been found (Martí et al. 1998). The infrared counterpart of GRS 1915+105 has been known for some years. However, the source has been proposed to be a low-mass X-ray binary (LMXRB) by Castro-Tirado et al. (1996) and Greiner et al. (2001) and a high-mass X-ray binary (HMXRB) by Chaty et al. (1996), Mirabel et al. (1997), and Martí et al. (2000a) who favor an Oe/Be spectral type. Finally, Martí et al. (2000b) have done a deep search for the near-infrared counterpart of 1E 1740.7–2942 (Mirabel et al. 1992) using the VLT and have found, for the first time, several counterpart candidates that may be consistent with the best radio and X-ray positions available for this source. However, the non-detection of variability between their two observing epochs and the positional uncertainty still remaining at the 1″ level prevent them from identifying an unambiguous counterpart. Alternatively, the VLT images can be used to set new upper limits significantly deeper than previously reported that constrain the binary companion to be later than B8 V or earlier than G5 III.

4. THE CONNECTION BETWEEN DISK AND JET

Collimated jets seem to be systematically associated with the presence of an accretion disk around a star or a collapsed object. In the case of microquasars, the X-rays probe the inner accretion disk region, while radio waves relate to the synchrotron emission from the relativistic jets. From the theoretical point of view the relation between disk and jet is generally understood in terms of the magnetohydrodynamical model of Blandford & Payne (1982). These authors proposed that the angular momentum of a magnetized accretion disk around the collapsed object is the responsible for the acceleration of the plasma. The magnetic field lines are taken to be frozen into the disk and the plasma is assumed to follow them like a "bead on a wire", at least close to the disk. If the field line forms an angle with the plane of the disk smaller than 60°, the displacements of the plasma from its equilibrium position become unstable. This happens because along these field lines the component of the centrifugal force will be larger than the component of the gravitational force and the plasma will be accelerated outwards. Then, in its origin, the outflow motion has an important "equatorial" component, while on larger scales the jets are observed to have a motion that is dominantly "poloidal". In other words, after the acceleration a collimating mechanism is required to change the wide-angle centrifugal outflow into a collimated jet.

This collimation is proposed to be achieved as follows. Inside an inner region, the magnetic field energy density is larger than the kinetic energy density of the flow but at some distance from the disk (the Alfvén surface), this situation reverses and the flow stops corotating with the disk. This causes that a loop of toroidal (azimuthal) field is added to the flow for each rotation of the footpoint of the field line. The tension of this wound-up toroidal field that is formed external to the Alfvén surface produces a force directed toward the axis (the "hoop stress") that eventually collimates the flow into a jet. Most models for the production of jets in the astrophysical context use elements of MHD acceleration and collimation.

However, from the observational point of view, there is relativey little work done to relate disk and jet. In an attempt to improve the state of things, Mirabel et al. (1998) did simultaneous observations in the X-ray, infrared, and radio wavelengths of the galactic superluminal source GRS

1915+105. They found that during episodes of rapid disappearance and follow-up replenishment of the inner accretion disk, evidenced by the X-ray oscillating flux, the ejection of relativistic plasma clouds in the form of synchrotron flares at infrared and radio wavelengths was observed.

In Figure 1 we show the radio, infrared, and X-ray light curves observed for GRS 1915+105 during the 1997 September 9 event. These results are very suggestive that the ejection of relativistic plasma takes place after the rapid disappearance and follow-up replenishment of the inner accretion disk seen, in the X-rays. Similar evidence for a temporary disappearance and subsequent restoring of the inner portion of the accretion disk during a radio flare is reported by Feroci et al. (1999).

Further evidence for a relation between disk and jet come from the periodic phenomena, with the same periodicity, in epochs when GRS 1915+105 shows X-ray and radio oscillations (see Figure 2).

The behavior of GRS 1915+105 is, however, extremely complex. Belloni et al. (2000) classify the *RXTE* observations in 12 separate classes. Eikenberry et al. (2000) present simultaneous infrared and X-ray observations of GRS 1915+105 during faint infrared flares using the Palomar 5m telescope and the *RXTE*. They conclude that these faint IR flares begin before the X-ray oscillations, implying an "outside-in" origin of the IR/X-ray cycle, opposite to the "inside-out" model of Mirabel et al. (1998).

5. VLBA OBSERVATIONS OF GRS 1915+105

Dhawan, Mirabel, and Rodríguez (2000) undertook a detailed VLBA (Very Long Baseline Array) study of GRS 1915+105 over a range of wavelengths (13, 3.6, 2.0, and 0.7 cm), in different states of the black hole binary and with milliarc second angular resolution. These observations always resolve the nucleus as a compact jet of length $\sim 10\ \lambda_{cm}$ AU (see Figure 3).

The brightness temperature of the core is $T_B \simeq 10^9$ K, and its properties are better fitted by a conically expanding synchrotron jet model than a thermal jet. The nuclear jet varies in 30 minutes during minor X-ray/radio outbursts and reestablishes within 18 hr of a major outburst, indicating the robustness of the X-ray/radio (or disk/jet) system to disruption.

At lower resolution (80-240 AU), Dhawan et al. (2000) observe more extended ejecta at 500 AU separation from the stationary core. Time-lapse images clearly detect the superluminal motion of the ejecta in a few hours. The measured velocity is (1.5 ± 0.1) c $(D/12$ kpc$)$ for the approaching component and is consistent with ballistic motion of

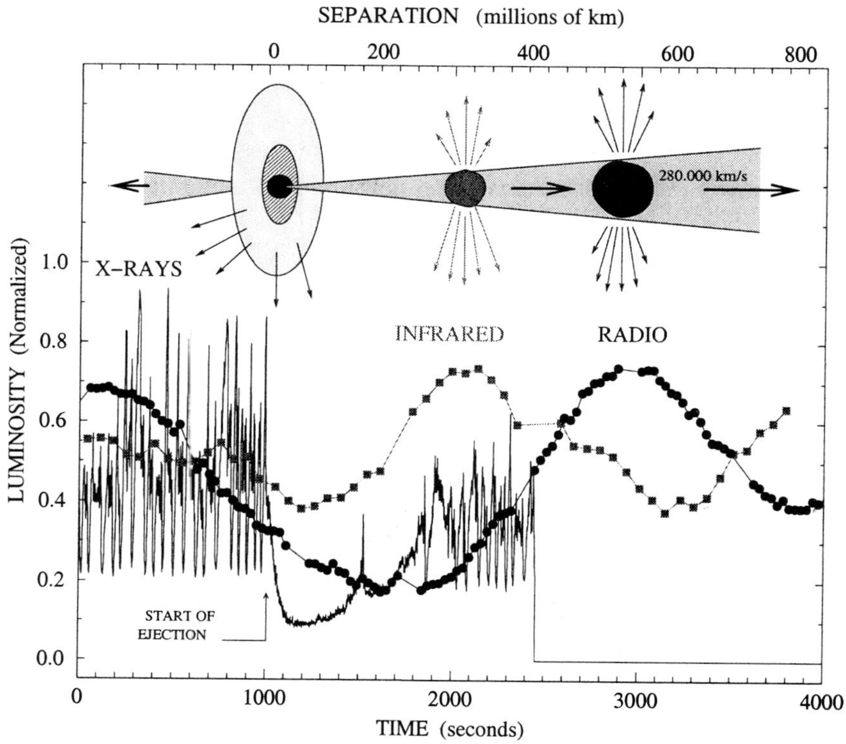

Figure 1 Blow up of the light curves at the time of the infrared flare detected in GRS 1915+105 on Sep 9, 1997. The infrared flare starts during the recovery from the X-ray dip, when an X-ray spike was observed. These observations show the connection between the rapid disappearance and follow up replenishment of the inner accretion disk seen in the X-rays, and the ejection of relativistic plasma clouds observed as synchrotron emission at infrared and radio wavelengths. A cartoon depicting the components responsible for the X-ray (disk), and infrared and radio (ejecta) components is shown at the top.

the ejecta from 500 AU outward, perhaps even since birth. The axis of the ejecta differs by $\leq 12°$ clockwise from the axis of the AU-scale jet, measured in the same observation. The reason for this misalignement is not known. Both axes are stable in time ($\pm$ 5°), the AU scale axis for 2 yr and the large scale axis for over 4 yr. Astrometry over 2 yr relative to an extragalactic reference locates the black hole to ± 1.5 mas, and its secular parallax due to Galactic rotation is $5.8\pm$ 1.5 mas yr^{-1}, consistent with a distance of 12 kpc. Finally, a limit of 100 km s^{-1} is placed on its proper motion with respect to its neighborhood.

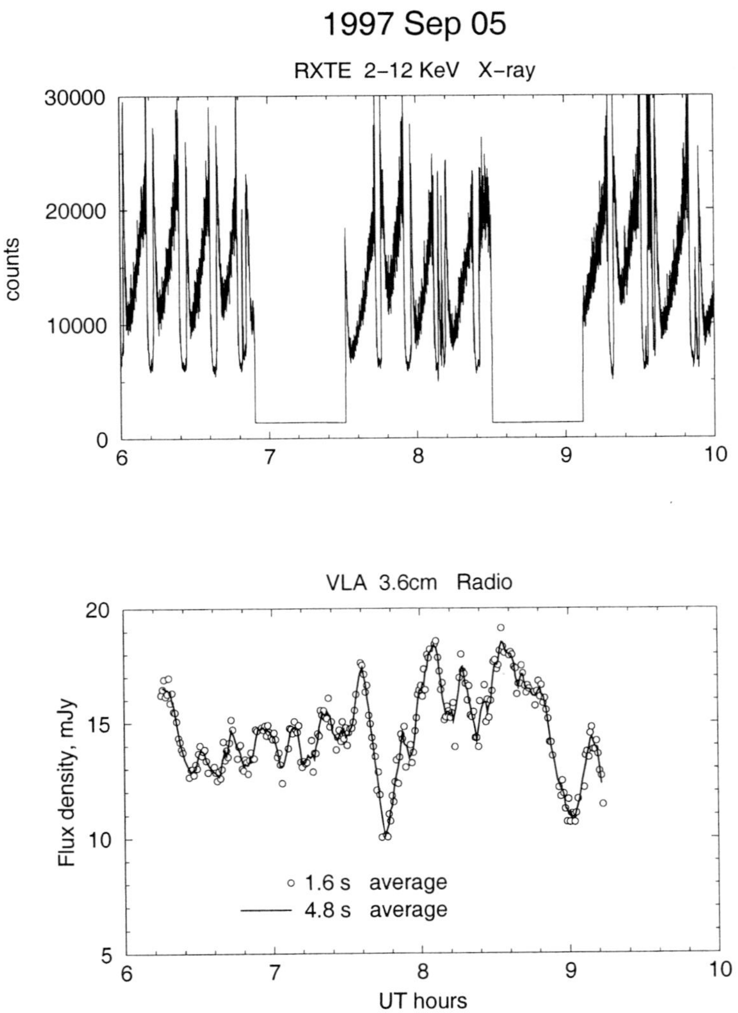

Figure 2 Top: *RXTE* 212 keV counts on 1997 September 5, 0610 UTC. Note the dips every 12 minutes. Bottom: VLA flux density at 3.6 cm, with the same periodicity (within errors) as the X-ray dips. This result supports the idea that the radio oscillations arise in a synchrotron jet, which is fed by plasma injected during the dips in the soft X-ray flux (Dhawan, Mirabel, and Rodríguez 2000).

6. PROSPECTS FOR THE FUTURE

The status of the field is presented in the proceedings of the Microquasar Workshop that took place in Granada, Spain and that will be published as a special volume of Astrophysics and Space Science Review. The study of microquasars will greatly benefit from future X-ray

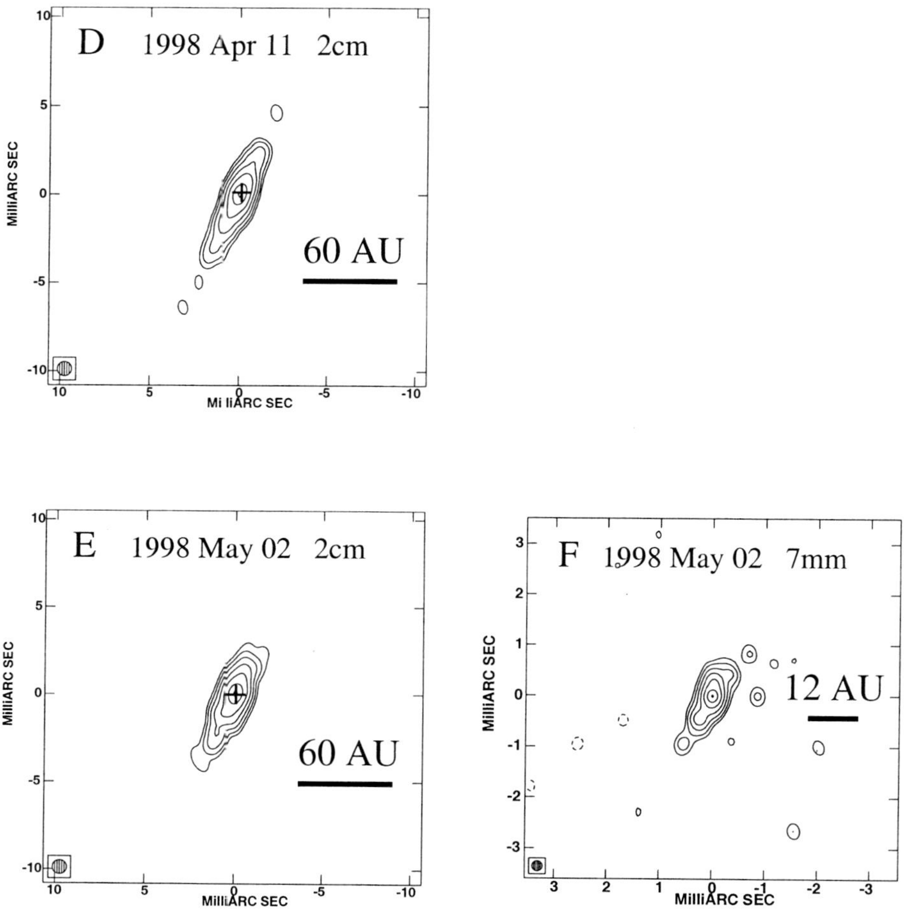

Figure 3 VLBA images during 1998 April-May, corresponding to the epochs shown in each panel. Contours are at −2%, 2%, 4%, 8%, 16%, 32%, 64%, and 96% of the peak. (D) 2 cm image at 7.5 AU resolution, showing the nuclear jet during a plateau state, 2 days before the flare on MJD 50,916; peak = 48 mJy beam-1. (E) 2 cm image at 7.5 AU resolution; peak = 64 mJy beam-1. (F) 0.7 cm image at 2.5 AU resolution; peak = 66.4 mJy beam-1. Epochs E and F are quasi-simultaneous data, centered at MJD 50,935.50. The position of E is identical to D within 1.0 mas. The black plus sign marks the astrometric position. For further details, see Dhawan, Mirabel, and Rodríguez (2000).

timing missions like *EXTRA* (Experiment for X-ray Timing and Relativistic Astrophysics), recently proposed to ESA and the related USA mission *RAE* (Relativistic Astrophysics Explorer). The study of the quasi-periodic events seen in microquasars may result in important advances in the study of general relativity phenomena in the vicinity of neutron stars and black holes. The Expanded Very Large Array (EVLA) will provide the capability of very sensitive, fast imaging of ejecta events. It has been speculated that most or all X-ray binaries have relativistic ejections and the EVLA will be able to test this conjecture.

The discovery and study of microquasars has produced major progress in our understanding of the nature of relativistic jets and of the connection between accretion onto compact objects and the formation of collimated ejecta. Many unanswered questions remain (see Mirabel 2001 for a detailed discussion of open questions), and we look forward to their answers.

Acknowledgments

LFR acknowledges the support of DGAPA, UNAM and CONACyT, México.

References

Belloni, T., Klein-Wolt, M., Méndez, M., van der Klis, M., and van Paradijs, J. 2000, *Astronomy and Astrophysics,* 355, 271-290

Blandford, R. D. and Payne, D. G. 1982, *Monthly Notices of the Royal Astronomical Society,* 199, 883-903

Castro-Tirado, A. J., Geballe, T. R., and Lund, N. 1996, During the September 1994 Outburst," *The Astrophysical Journal,* 461, L99-101

Chaty, S., Mirabel, I. F., Duc, P. A., Wink, J. E., and Rodríguez, L. F. 1996, *Astronomy and Astrophysics,* 310, 825-830

Chaty, S., Mirabel, I. F., Martí, J., and Rodríguez, L. F. 2001, to appear in *Proceedings of the First Integral Workshop*

Dhawan, V., Mirabel, I. F., and Rodríguez, L. F. 2000, *The Astrophysical Journal,* 543, 373-385

Eikenberry, S. S., Matthews, K., Muno, M., Blanco, P. R., Morgan, E. H., and Remillard, R. A. 2000, *The Astrophysical Journal,* 532, L33-36

Feroci, M., Matt, G., Pooley, G., Costa, E., Tavani, M., and Belloni, T. 1999, *Astronomy and Astrophysics,* 351, 985-992

Greiner, J., McCaughrean, M. J., Cuby, J. G., Castro-Tirado, A. J., and Menninchent, R. E., 2001, in *Microquasars, eds. A. J. Castro-Tirado, J. Greiner, and J. M. Paredes, Kluwer, ApSpSceRev,* in press

Hjellming, R. M. et al. 2000, *The Astrophysical Journal,* 544, 977-992

Martí, J., Mereghetti, S., Chaty, S., Mirabel, I. F., Goldoni, P., and Rodríguez, L. F. 1998, *Astronomy and Astrophysics,* 338, L95-99

Martí, J., Mirabel, I. F., Chaty, S., and Rodríguez, L. F. 2000a, *Astronomy and Astrophysics,* 356, 943-948

Martí, J., Mirabel, I. F., Chaty, S., and Rodríguez, L. F. 2000b, *Astronomy and Astrophysics,* 363, 184-187

Martí, J., Mirabel, I. F., Chaty, S., and Rodríguez, L. F. 2001, *these Proceedings*

Mirabel, I. F., Rodríguez, L. F., Cordier B., Paul J., and Lebrun F., 1992, *Nature,* 358, 215-217

Mirabel, I. F. and Rodríguez, L. F. 1994, *Nature,* 371, 46-48

Mirabel, I. F., Bandyopadhyay, R., Charles, P. A., Shahbaz, T., and Rodriguez, L. F. 1997, *The Astrophysical Journal,* 477, L45-48

Mirabel, I. F. and Rodríguez, L. F. 1998, *Nature,* 392, 673-676

Mirabel, I. F., Dhawan, V., Chaty, S., Rodríguez, L. F., Martí, J., Robinson, C. R., Swank, J., and Geballe, T. 1998, *Astronomy and Astrophysics,* 330, L9-12

Mirabel, I. F. and Rodríguez, L. F. 1999, *Annual Reviews of Astronomy and Astrophysics,* 37, 409-443

Mirabel, I. F. 2001, in Microquasars, eds. A.J. Castro-Tirado, J. Greiner, J.M. Paredes, Kluwer, ApSpSciRev (in press)

Orosz, J. A. et al. 2000, *American Astronomical Society Meeting 197,* #83.20

Paredes, J. M., Martí, J., Ribo, M., and Massi, M. 2000, *Science,* 288, 2340-2342

Paredes, J. M., Martí, J., Ribo, M., and Massi, M. 2001, *these Proceedings*

VLT OBSERVATIONS OF GALACTIC MICROQUASARS

Josep Martí
Departamento de Física, Escuela Politécnica Superior, Universidad de Jaén
Virgen de la Cabeza, 2, E-23071 Jaén, Spain
jmarti@ujaen.es

I. Félix Mirabel
CEA/DSM/DAPNIA/Service d'Astrophysique, Centre d'Études de Saclay
F-91191 Gif-Sur-Yvette, France
mirabel@discovery.saclay.cea.fr

Sylvain Chaty
Physics Department, The Open University
Walton Hall, Milton Keynes, MK7 6AA, United Kingdom
S.Chaty@open.ac.uk

Luis F. Rodríguez
Instituto de Astronomía, UNAM
Campus Morelia, Apdo. Postal 3-72
58089 Morelia, Michoacán, México
l.rodriguez@astrosmo.unam.mx

Keywords: Microquasars, infrared, X-rays, gamma-rays

Abstract We report the results of VLT observations at 2.2 μm with the ISAAC instrument of the galactic microquasars GRS 1915+105 and 1E 1740.7−2942. For GRS1915+105, we report the detection of a new Na I emission line and discuss the different scenarii consistent with the observed spectrum. For 1E 1740.7−2942, we present a deep VLT search for its infrared counterpart at the subarcsec accurate position determined by both the Very Large Array and the *Chandra* satellite at radio and X-ray wavelengths, respectively.

A. Carramiñana et al. (eds.),
The Nature of Unidentified Galactic High-energy Gamma-ray Sources, 255–262.

1. INTRODUCTION

The class of microquasar X-ray binaries has been recognized as such during the last decade. Its currently known population amounts to about a dozen confirmed cases, with some of them being really exotic systems (e.g. superluminal sources such as GRS 1915+105). In general, these galactic objects appear to be sources of both soft and hard X-rays as well as non-thermal radiation in the radio and infrared bands. Their most remarkable feature is probably their ability to produce both persistent and/or episodic ejections of relativistic plasma clouds into collimated jets. For a recent review on this field, the reader is referred to Mirabel & Rodríguez (1999), hereafter MR99.

In MR99 it is pointed out the possibility that microquasars may be related to sources of γ-rays. In particular, it has been proposed that Gamma Ray Bursts (GRBs) could be associated to extreme cases of microquasar behavior (the so-called 'microblazars'). This very tentative idea is based on recent studies of GRB afterglows suggesting that they are distant super-relativistic jets. On the other hand, the microquasar associated with the luminous star LS 5039 appears to be consistent with one of the unidentified sources in the 3rd EGRET catalog of high energy ($>$ 100 MeV) gamma-ray sources (Paredes et al. 2000; 2001). All these facts raise the possibility that microquasars are likely to have gamma-ray counterparts. Many questions come at this point: are all microquasars gamma-ray emitters? which are the physical mechanisms involved? do the gamma-ray photons have energies in the EGRET window? etc.

In this context, it is clear that microquasars deserve a further in depth study in the gamma-ray domain. Probably, the origin of gamma-rays is somehow connected with the nature of the binary system. For instance, a black hole compact companion may be required to produce them. Alternatively, if inverse Compton processes are involved, an early type star has to be present for efficient scattering of ultraviolet photons by the same relativistic electrons responsible for non-thermal radio emission. Unfortunately, the types of the binary system associated to galactic microquasars are still poorly known even in some prototypical cases, including those of the Galactic Center microquasar 1E 1740.7−2942 and the superluminal source GRS 1915+105.

In an attempt to improve this situation, we have conducted a series of infrared observations of these two microquasars. This type of research will certainly be helpful when trying to figure out possible scenarii for the origin of gamma rays. The use of infrared wavelengths is strongly recommended here in order to overcome the huge amount of interstellar absorption and to try to determine the stellar type of the companion.

The results presented here have been reported in more detail by Martí et al. (2000) and Martí et al. (2001).

2. OBSERVATIONS

The observations were conducted with the first 8.2 m unit (Antu) of the Very Large Telescope (VLT)[1]. The instrument used with the VLT was the ISAAC spectro-imager, working in the K-band (2.2 μm). The data were reduced using the IRAF package, including bias and dark current subtraction, flat field division and frame combination. For spectroscopic observations, the telluric absorptions due to the Earth atmosphere were removed by appropriately rationing the uncorrected spectrum with that of a solar type star.

3. GRS 1915+105

Low resolution spectra of the superluminal microquasar GRS 1915+105 were taken with the VLT+ISAAC on two different epochs. The first epoch was on 1999 May 29 (JD 2451327.84) under good weather conditions. The source was in a quiescent state according to the radio monitoring by the Green Bank Interferometer (GBI). The second observation took place on 1999 July 04 (JD 2451363.77) under less favorable weather conditions. Contrary to the first case, the second observation coincided with the decline of a radio flare as detected by the GBI. In both dates, we carried out low resolution ($R \sim 450$) spectroscopy with an integration time of 30 minutes.

The individual spectra are presented in Fig. 1. The most evident features in the data correspond to the He I (2.0587 μm) and Br γ (2.1661 μm) emission lines, both without significant Doppler shifts, as previously detected by other authors (Castro-Tirado et al. (1996); Mirabel et al. (1997); Eikenberry et al. (1998). A weak He II emission line seems to be present in the first epoch, thus giving support to the marginal detections by Castro-Tirado et al. (1996) and Eikenberry et al. (1998). Since we do not clearly see He II in the second epoch, this line is likely to be variable. On the other hand, a new emission feature is reliably detected and we identify it with the Na I doublet at 2.206-9 μm.

Interestingly, the He I line displays a dramatic change from a P Cygni profile to an emission blue wing from the first to the second epoch. This line is located in a region where the atmospheric absorption is very strong and difficult to correct. Therefore, additional confirmation of this behaviour should be advisable. Accepting its reality, these changes imply the existence of an expanding envelope around the source dur-

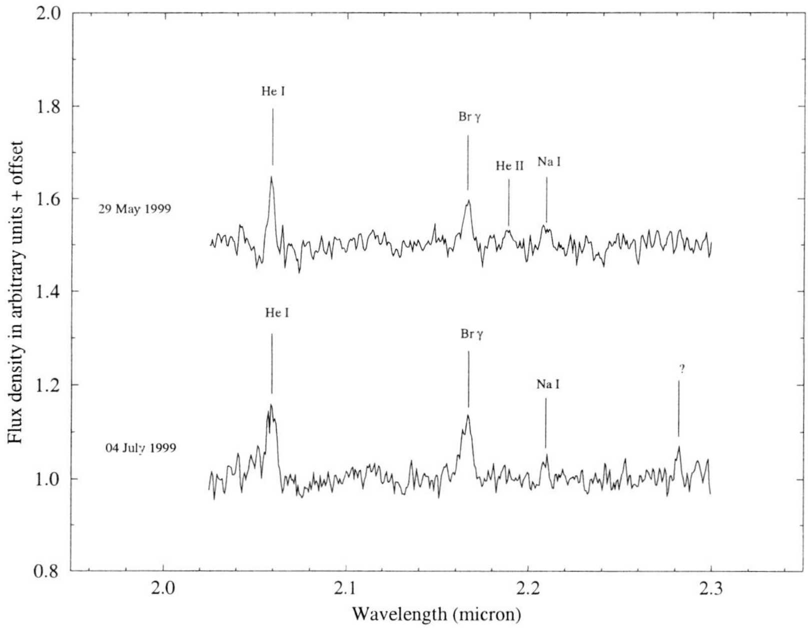

Figure 1 VLT spectra of GRS 1915+105 in two different source states. The first spectrum was taken with the source displaying very low activity in the radio, while the second one coincided with the decay of a radio flare reaching a ~ 0.25 Jy peak at cm wavelengths.

ing quiescence that is partially blown out during the X-ray and radio outbursts.

The combined spectrum of both epochs is presented in Figure 2. Such a combination clearly smears the variability effects discussed above, but has been performed to increase the Signal-to-Noise Ratio for the detection of weak features. Na I is clearly visible here. Figure 2 also includes several comparison stars in an attempt to provide a tentative spectral classification for the GRS 1915+105 companion. By comparing with infrared spectral atlases, our preferred classification is that of a Of/WN spectral type. A star of this kind easily accounts for the Na I line, whose low ionization potential requires some shielding from direct stellar radiation. This shielding could be provided by the circumstellar expanding envelope inferred from the P Cygni profile, that is also naturally understood in the context of a massive and luminous companion. GRS 1915+105 would be then a high mass X-ray binary (HMXRB) system. A Of/WN star also produces photons energetic enough to ionize the helium atoms responsible for the He II line. However, this does not

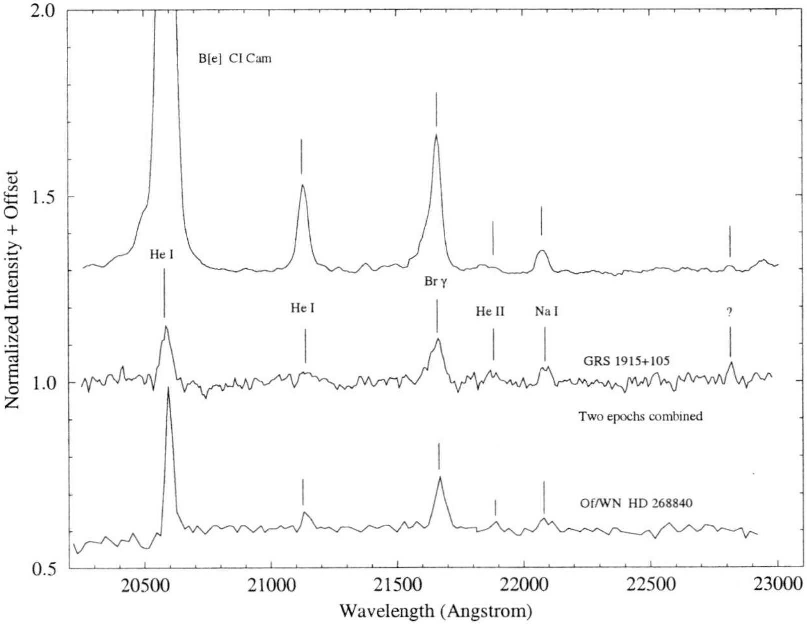

Figure 2 Comparison of the combined VLT spectrum of GRS 19115+105 with those of the B[e] system CI Cam (top) and the Ofpe/WN9 star HD 268840 (bottom) adapted from Clark et al. (1999) and Morris et al. (1996), respectively.

exclude the possibility that the He II emission from the accretion disk is also contributing to this feature and possibly accounting for most of its variability.

Finally, we point out that the recent detection of CO absorption bands by Greiner et al. (2001) is at variance with our interpretation and favors a low mass companion. However, the CO bands could also be associated with cold material surrounding GRS 1915+105. For instance, Mirabel et al. (1996) already found evidence of a dust component around this source. More recently, the prominent neutral K edges in the *Chandra* X-ray spectra (due to iron, silicon and magnesium) have been also interpreted in terms of cold material in the environment of GRS 1915+105 (Lee et al. 2001).

4. 1E 1740.7−2942

This Galactic Center microquasar was observed with the VLT+ISAAC in imaging mode and on two different epochs. The Ks filter centered at the 2.16 μm wavelength was used. In Figure 3, we show an expanded view of the VLT image covering the accurate radio and X-ray positions

Table 1 Photometry of the candidate counterparts

Object	Ks (mag)	Remarks
4	18.0 ± 0.1	Possibly extended
6	19.2 ± 0.3	
7	18.7 ± 0.3	
8	19 ± 1	Uncertain background

currently available. It corresponds to our second epoch image taken on 1999 May 02 (JD 2451300.873), the one with best seeing. The limiting magnitude of this image, at the 3σ level, is estimated to be $Ks = 19.5$. No significantly variable sources were detected in the field when comparing the two epochs separated by nearly one month.

The radio coordinates adopted by us are $\alpha_{\rm J2000} = 17^h43^m54\overset{s}{.}83$ and $\delta_{\rm J2000} = -29^\circ 44' 42\overset{''}{.}60$ with $\pm 0\overset{''}{.}1$ accuracy in each coordinate. This position is slightly different from that reported in Mirabel et al. (1992) since it has been derived from the expanded Very Large Array (VLA) data set. The X-ray position has been taken from the *Chandra* observations by Cui et al. (2001) and it corresponds to $\alpha_{\rm J2000} = 17^h43^m54\overset{s}{.}876 \pm 0\overset{s}{.}004$ and $\delta_{\rm J2000} = -29^\circ 44' 42\overset{''}{.}48 \pm 0\overset{''}{.}04$, where the errors represent roughly formal 90% confidence limits. There is clearly an offset of $0\overset{''}{.}6$ between both positions that will require some position revision in the future. Both positions are plotted on Figure 3 as star symbols. The uncertainty when plotting the symbols against the infrared star background is dominated by the error in the astrometric fit ($\sim 0\overset{''}{.}60$ at 3σ).

In any case, the VLT image clearly shows the presence of four infrared sources that could be consistent with both the VLA and *Chandra* position (allowing for the astrometric error quoted above). They are labeled as 4, 6, 7 and 8 following the notation of Martí et al. (2001) where further details are given. Their magnitudes are listed in Table 1. None of these candidates displayed evidences of variability between the two epochs. Therefore, we cannot select one of them on the basis of variability. In this situation, any possible association with 1E 1740.7−2942 remains to be confirmed. However, this is the first time that counterpart candidates are proposed for this microquasar on the basis of positional agreement.

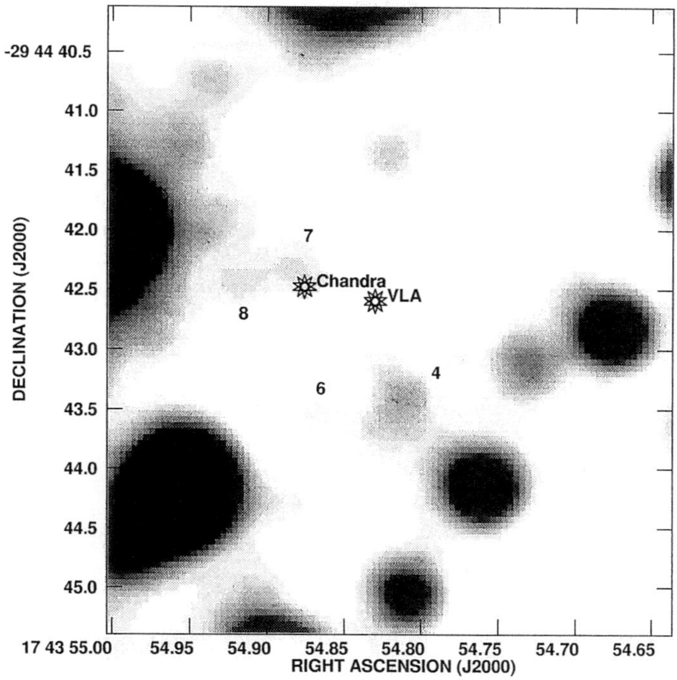

Figure 3 The field of 1E 1740.7−2942 observed with the VLT+ISAAC at the 2.2 μm wavelength. The star symbols represent the VLA radio and *Chandra* X-ray position of this microquasar. Taking into account the astrometric error, several infrared sources are detected that may be consistent with the microquasar. They are labeled as 4, 6, 7 and 8 according to Martí et al. (2001).

It is relevant to remember here the strong similarity in X-rays between 1E 1740.7−2942 and the classical black hole candidate Cygnus X-1. How absorbed should be Cygnus X-1 to look like one of the 1E 1740.7−2942 candidates at the Galactic Center distance? Let us adopt here K = 18.5-19.0 as a representative magnitude of sources in Table 1. The absolute magnitude of Cygnus X-1 is $K = -5.8 \pm 0.2$ (Beall et al. 1984). From standard extinction laws, the necessary absorption amounts then to $A_K = 9.7 - 10.2$, equivalent to a hydrogen column density of $N_H = (1.5 - 1.6) \times 10^{23}$ cm^{-2}. This value is in full agreement with those derived from X-ray spectral fits, i.e., $N_H > 8 \times 10^{22}$ cm^{-2} (Churazov et al. 1996; Sheth et al. 1996).

Finally, if none of the candidates is related to the microquasar, we can put some constraints on its stellar type based on the limiting magnitude achieved. In this case, the 1E 1740.7−2942 infrared counterpart should have an absolute magnitude $K > -0.1$, thus allowing only main sequence stars later than about B8 V or giants earlier than G5 III.

Acknowledgments

JM acknowledges partial support by DGICYT (PB97-0903) and Junta de Andalucía (Spain). IFM acknowledges financial support from CONICET/Argentina. SC acknowledges financial support from grant F/00-180/A from the Leverhulme Trust. LFR acknowledges the support from DGAPA, UNAM and CONACyT, México.

Notes

1. Based on observations collected at the European Southern Observatory, Chile (ESO No 63.H-0261).

References

Beall, J.H., Knight, F.K., Smith, H.A., Wood, K. S., Lebofsky, M., Rieke, G. 1984, ApJ, 284, 745

Clark, J.S., Steele, I.A., Fender, R.P., Coe, M.J. 1999, A&A, 348, 888

Castro-Tirado, A.J., Geballe, T.R., Lund, N. 1996, ApJ, 461, L99

Churazov E., Gilfanov M., Sunyaev R.A. 1996, ApJ, 464, L71

Cui, W., Schulz, N.S., Baganoff, F.K., Bautz, M.W., Dotyi, J.P., et al. 2001, ApJ, 548, 394

Eikenberry, S. S., Matthews, K., Murphy., T. W., Jr., Nelson, R. W., Morgan, E. H., Remillard, R. A., Muno, M. 1998, ApJ, 506, L31

Greiner, J., McCaughrean, M.J., Cuby, J.G., Castro-Tirado, A.J., Menninckent, R.E. 2001, in Microquasars, eds. A.J. Castro-Tirado, J. Greiner, J.M. Paredes, Kluwer, ApSpSciRev (in press)

Lee, J.C., Schulz, N.s., Reynolds, C.S., Fabian, A.C., Blackman, E.G. 2001, ASP Conference Series (in press, available at `astro-ph/0012111`).

Mirabel, I.F., Rodríguez, L.F., Cordier, B., Paul J., Lebrun, F. 1992, Nat, 358, 215

Martí, J., Mirabel, I. F., Chaty, S., Rodríguez, L. F. 2000, A&A, 356, 943

Martí, J., Mirabel, I. F., Chaty, S., Rodríguez, L. F. 2001, A&A (in press)

Mirabel, I.F., Rodríguez, L.F., Chaty, S., Sauvage, M., Gerard, E., Duc, P.-A., Castro-Tirado, A.J., Callanan, P. 1996) ApJ, 472, L111

Mirabel, I. F., Bandyopadhyay, R., Charles, P. A., Shahbaz, T., Rodriguez, L. F. 1997) ApJ, 477, L45

Mirabel, I. F., & Rodríguez, L. F. 1999, Annu. Rev. Astron. Astrophys., 37, 409

Morris P.W., Eenens P.R.J., Hanson M.M., et al. 1996, ApJ, 470, 597

Paredes, J. M., Martí, J., Ribó, M., & Massi, M. 2000, Science, 288, 2340

Paredes, J. M., Martí, J., Ribó, M., & Massi, M. 2001, this volume

Sheth, S., Liang, E., Luo, C., Murakami, T. 1996, ApJ, 468, 755

MICROQUASARS AND UNIDENTIFIED EGRET SOURCES: THE CASE OF LS 5039

Josep M. Paredes
Departament d'Astronomia i Meteorologia, Universitat de Barcelona
Av. Diagonal 647, E-08028 Barcelona, Spain
josep@am.ub.es

Josep Martí
Departamento de Física, Universidad de Jaén
Calle Virgen de la Cabeza 2, E-23071 Jaén, Spain
jmarti@ujaen.es

Marc Ribó
Departament d'Astronomia i Meteorologia, Universitat de Barcelona
Av. Diagonal 647, E-08028 Barcelona, Spain
mribo@am.ub.es

Maria Massi
Max Planck Institut für Radioastronomie
Auf dem Hügel 69, D-53121 Bonn, Germany
mmassi@mpifr-bonn.mpg.de

Keywords: Microquasars, X-ray binaries, LS 5039, gamma-rays, 3EG J1824−1514

Abstract LS 5039 is a massive X-ray binary with persistent radio emission at a distance of 3 kpc. Recent high resolution radio observations with the VLBA show that LS 5039 is resolved into bipolar jets emanating from a central core, revealing its microquasar nature. This system also appears to be associated with the EGRET source 3EG J1824−1514. The association with the γ-ray source is further supported by LS 5039 being the only X-ray and radio source within the EGRET 95% confidence contour.

A. Carramiñana et al. (eds.),
The Nature of Unidentified Galactic High-energy Gamma-ray Sources, 263–270.

1. INTRODUCTION

Only a small fraction (∼10%) of the catalogued X-ray binaries are radio emitters. Some of these radio emitting X-ray binaries (REXRBs) behave as scaled down versions of quasars and active galactic nuclei and are called microquasars. All the microquasars have relativistic jets and in a few of them superluminal motions have been observed. Among the dozen of known microquasars, the best representative examples are SS 433, with a precessing jet with a period of 163 days, GRS 1915+105, the first superluminal source found in our Galaxy, and Cygnus X-3, a source that experiences very strong outbursts (Mirabel & Rodríguez, 1999; Rodríguez, 2001).

During the last years we have been interested in the search of new REXRBs, specially those with radio jet morphology. The method used was based on the cross-correlation of radio, optical and X-ray catalogues under strict conditions, such as compact radio structure, high X-ray hardness ratio, etc. In a first step, the cross-correlation process was restricted to galactic latitudes less than 5 degrees. The catalogues used were: in the radio, the NRAO VLA Sky Survey (NVSS) (Condon et al. 1998); in the X-rays, the *ROSAT* All Sky Survey Bright Source Catalogue (Voges et al. 1999) and, in the optical, the Guide Star Catalogue (Lasker et al. 1990). At the end of the process, a list of nine candidates emerged. One of them, LS 5039, focused our attention immediately because its bright optical counterpart.

LS 5039 has been identified as a high mass X-ray binary candidate by Motch et al. (1997). The optical counterpart of the system has a visual magnitude of 11.2, a O7V(f) spectral type and an estimated distance of 3 kpc. In the radio domain, it is a relatively bright source with non-thermal synchrotron spectrum. At the VLA resolution, LS 5039 appears as an unresolved point source ($\leq 0\farcs1$) (Martí, Paredes, & Ribó, 1998). The unabsorbed X-ray luminosity (in the energy range 0.1 – 2.4 keV) is about 8.1×10^{33} erg s^{-1} and the hardness of the source is consistent with a neutron star or a black hole (Motch et al. 1997).

In this paper, we present the radio jet structure of LS 5039 obtained from high resolution interferometric observations and its proposed identification with the EGRET source 3EG J1824−1514. A more detailed description can be found in Paredes et al. (2000).

2. A NEW MICROQUASAR: LS 5039

The level of radio emission of LS 5039 and its persistent nature, allowed us to study this source with higher angular resolution than that of the VLA. We observed LS 5039 with the Very Long Baseline Array

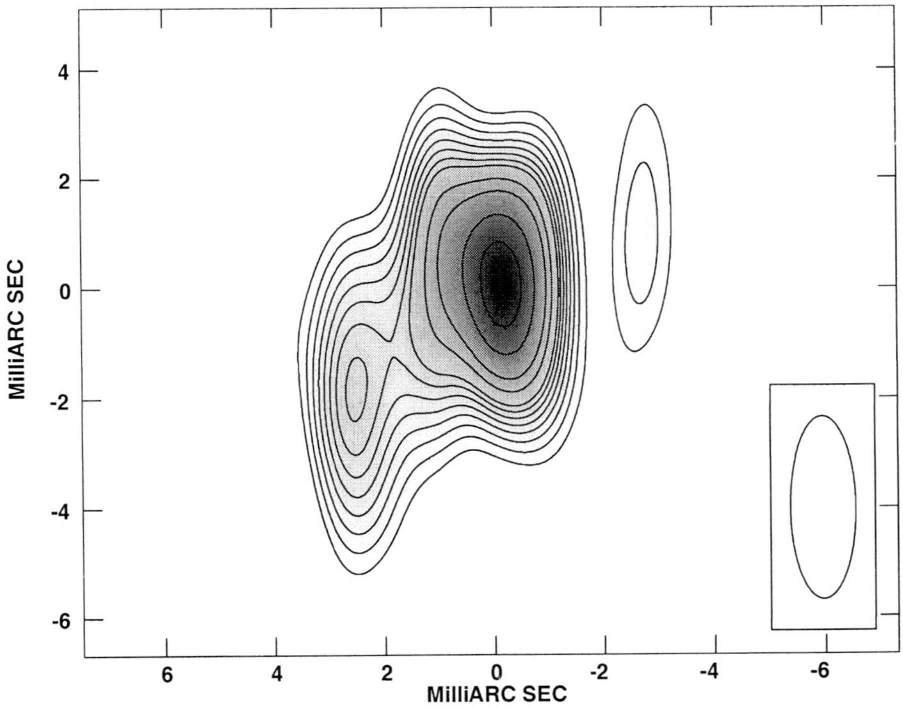

Figure 1 Radio image of LS 5039 obtained with the VLBA+VLA at 6 cm wavelength (Paredes et al. 2000). The displayed contour levels are 6, 8, 10, 12, 14, 16, 18, 20, 25, 30, 40 and 50 times 0.085 mJy/beam, the rms noise. The half power beam width, 3.3×1.2 mas^2, is shown in the lower right corner. North is at the top and east is at the left.

(VLBA) and the VLA of the National Radio Astronomy Observatory (NRAO)[1], in phased array mode, on 1999 May 8th at 6 cm wavelength. The data were edited and reduced using the AIPS and Difmap software packages. The resulting uniform weighted map is presented in Figure 1. This map clearly shows the presence of an elongated structure, with position angle (PA) of 125° with respect to the North, and with a full angular size of about 6×2 mas^2. This implies an overall projected length of 18 AU at 3 kpc distance. The total flux density is 16 mJy, being the flux density of the Southeast component 2.15 mJy and 1.05 mJy for the Northwest component. The most likely interpretation of Figure 1 is that we are seeing radio jets emanating from a bright central core in LS 5039. The brightness asymmetry between the jet components is likely to arise due to Doppler boosting effects, as in other microquasar systems (see e.g., Mirabel & Rodríguez (1999) for a review). This fact allows us to

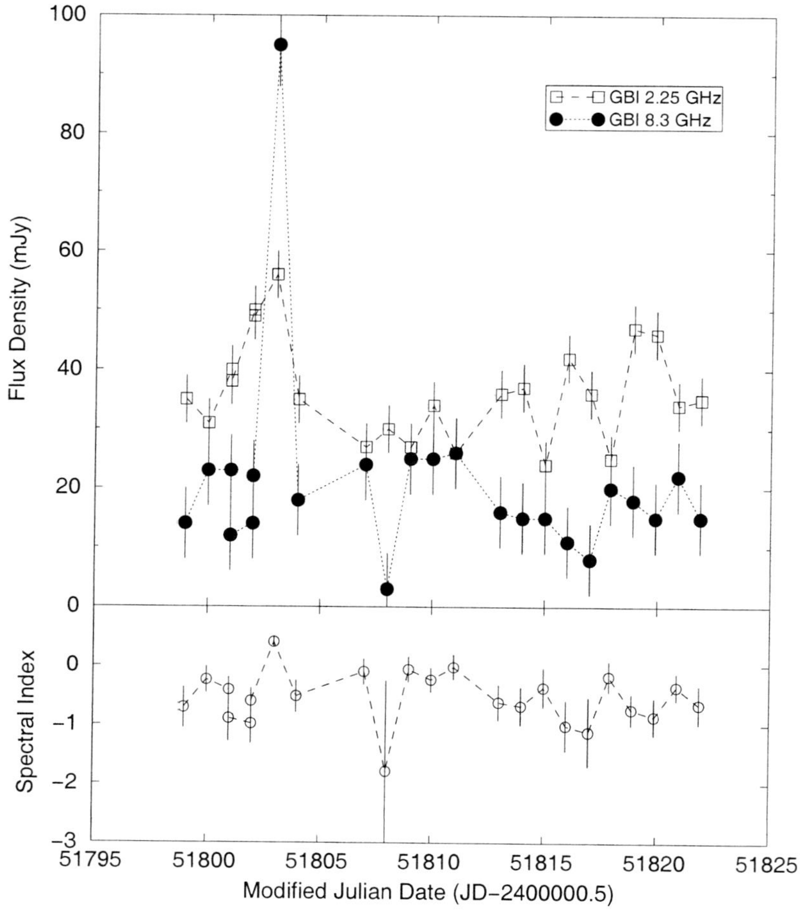

Figure 2 *Top:* GBI radio light curves of LS 5039 at the frequencies of 2.25 and 8.3 GHz (13.1 and 3.6 cm wavelength respectively). A strong burst, with a rising time of less than 12 hours at 8.3 GHz, took place on MJD 51803. *Bottom:* The corresponding spectral index α, $S_\nu \propto \nu^\alpha$. Representative $\pm$ 1 σ error bars have been plotted in both figures.

derive a lower limit of $v \geq (0.15 \pm 0.04)c$ for the jet velocity and an upper limit for the ejection angle with the line of sight of $\theta \leq 81^\circ \pm 2^\circ$.

At the time of our observations, and during the weeks before, the source was in its typical persistent level of radio emission with moderate variability, according to the Green Bank Interferometer (GBI)[2] monitoring (Ribó et al. 1999). The presence of radio jets in a state of normal persistent radio emission strongly suggests that they are a permanent feature of the system. However, as can be seen in Figure 2, a strong flare took place on September 16, 2000. This was the first time that a significant flare had been observed in LS 5039. The short rising time, less than 12 hours at 8.3 GHz, agrees with the galactic nature of this source. It is possible that this strong radio event created a pair of jet components moving at velocities greater than $0.15c$.

The overall size of the source, as seen in our VLBA+VLA map, and the total flux density implies a brightness temperature of $\sim 9.4 \times 10^7$ K, i.e., consistent with non-thermal synchrotron radiation. From equipartition arguments between the radio emitting relativistic electrons and the magnetic field, we derive an energy content of $\sim 4.8 \times 10^{39}$ erg with an equipartition magnetic field of ~ 0.2 G. These numbers must be considered as very rough estimates since we still do not have enough information for appropriate calculations in the rest frame of the ejecta.

3. ASSOCIATION BETWEEN 3EG J1824−1514 AND LS 5039

The third EGRET catalog of high-energy ($E_\gamma > 100$ MeV) γ-ray sources (Hartman et al. 1999) contains over one hundred unidentified emitters at low galactic latitudes. 3EG J1824−1514 is one of them and we show its location map in Figure 3. In the same figure, we also have plotted the *ROSAT* bright and faint sources as well as the NVSS sources with radio emission greater than 20 mJy. It is clear that LS 5039, at galactic coordinates $l = 16.88°$ and $b = -1.29°$, is well inside the 95% confidence contour of 3EG J1824−1514 and is the only one simultaneous X-ray and radio emitter. These arguments suggest a likely association between the microquasar LS 5039 and the EGRET source 3EG J1824−1514.

The EGRET observations of 3EG J1824−1514 reveal a permanent γ-ray emission for ~ 5 years. This can be seen in Figure 4, where we also have plotted the radio data from LS 5039. The average γ-ray flux for all EGRET viewing periods in this figure is $\Phi_\gamma = (35.2 \pm 6.5) \times 10^{-8}$ photon cm^{-2} s^{-1} with photon spectral index $p = 2.19 \pm 0.18$, where $\Phi_\gamma \propto E_\gamma^{-p}$. This photon index is steeper than the $p \lesssim 2$ values usually found for pulsars (Merck et al. 1996). Assuming a distance of 3 kpc, the corresponding integrated luminosity amounts to $L_\gamma(> 100 \text{ MeV}) \sim 3.8 \times 10^{35}$ erg s^{-1}.

The high energy γ-rays are likely to result from inverse Compton scattering of UV photons, emitted by the hot O-type companion, by the same radio emitting electrons (Figure 5). The details of the calculation are worked out in Paredes et al. (2000). They find that this scenario is conceivable provided that the jet flow is faster than $v > 0.05c$, so that

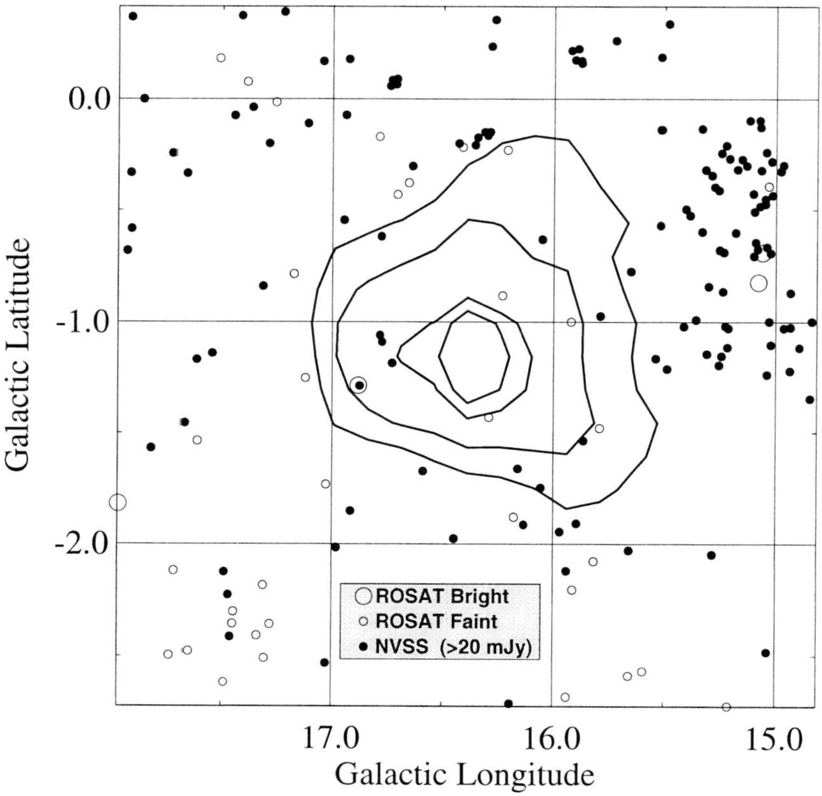

Figure 3 Location map of 3EG J1824−1514. The contours represent 50%, 68%, 95%, and 99% statistical probability that a γ-ray source lies within the given contour. The open big circles are *ROSAT* bright sources, the small open circles are the *ROSAT* faint sources and the the filled circles are radio sources from the NVSS survey. The only one source with X-ray and radio emission (filled circle inside an open big circle, $l = 16.88^\circ$ and $b = -1.29^\circ$), and well inside the 95% contour, is LS 5039.

the relativistic electrons are able to avoid strong inverse Compton losses and still be able to power the radio jets. In this scenario, both the radio and γ-ray emissions are expected to be persistent in time, as long as the microquasar is able to accelerate relativistic electrons with Lorentz factors of $\gamma \sim 10^3$. The long term flux history of LS 5039 at radio and γ-ray wavelengths (see Figure 4) gives support to this persistent activity.

4. SUMMARY

The milliarcsecond radio jets detected in the massive X-ray binary system LS 5039 revealed its microquasar nature. The procedure used to select LS 5039 suggests the possibility that several other microquasar

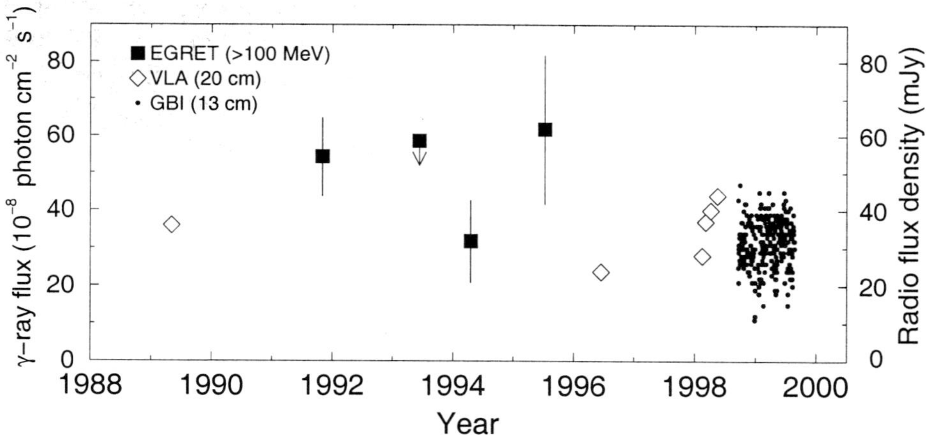

Figure 4 The persistent light curves of 3EG J1824−1514 and LS 5039 at γ and radio wavelengths. Both sources are proposed to be the same object.

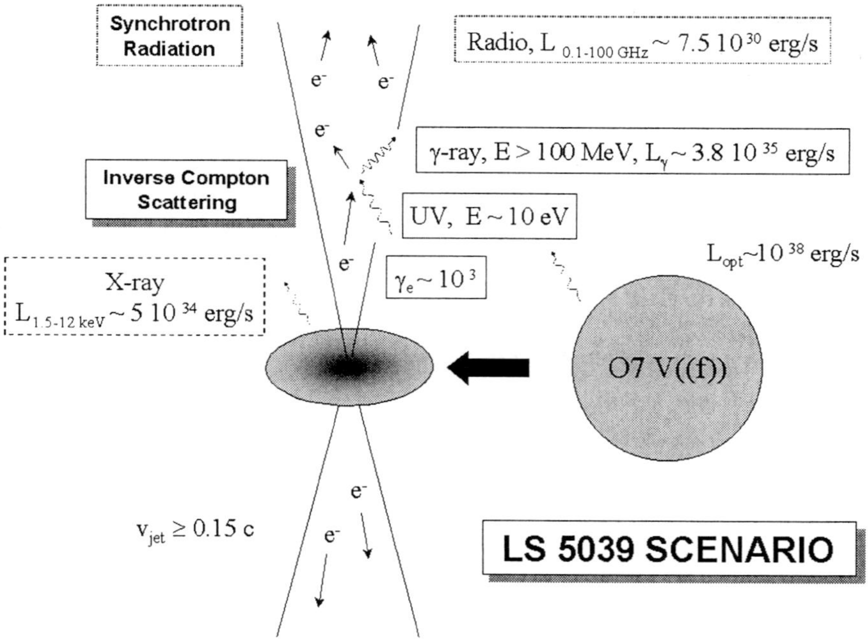

Figure 5 Scenario where the multiwavelength emission originates.

systems can be found in our Galaxy in a similar way. Moreover, LS 5039 is in positional agreement with the EGRET source 3EG J1824−1514. The physical association between the γ-ray source and the microquasar is further supported by LS 5039 being the only X-ray and radio source within the EGRET 95% confidence contour. This facts open the possibility of a relationship between galactic microquasars and some of the unidentified EGRET sources at low galactic latitudes.

Acknowledgments

This project has been partly supported by DGESIC (PB97-0903) in Spain. MR acknowledges receipt of a fellowship from Generalitat de Catalunya (ref. 1999 FI 00199). MM and JM acknowledge support received from the European Commission's TMR program and Junta de Andalucía, respectively.

Notes

1. NRAO is operated by Associated Universities, Inc., under cooperative agreement with the National Science Foundation.
2. The GBI is a facility of the NSF operated by NRAO with support from the NASA High Energy Astrophysics program.

References

Condon, J.J., Cotton, W.D., Greisen, E.W., et al. 1998, AJ, 115, 1693
Hartman, R.C., Bertsch, D.L., Boom, S.D., et al. 1999, ApJS, 123, 79
Lasker, B.M., Sturch, C.R., McLean, B.J., et al. 1990, AJ, 99, 2019.
Martí, J., Paredes, J.M. & Ribó, M. 1998, A&A, 338, L71.
Merck, M., Bertsch, D.L., Dingus, B.L., et al. 1996, A&ASS, 120, 465.
Mirabel, I.F., & Rodríguez, L.F. 1999, ARA&A, 37, 409
Motch, C., Haberl, F., Dennerl, K., Pakull, M., Janot-Pacheco, E. 1997, A&A, 323, 853.
Paredes, J.M., Martí, J., Ribó, M., & Massi, M. 2000, Science, 288, 2340.
Ribó, M., Reig, P., Martí, J., & Paredes, J.M. 1999, A&A, 347, 518.
Rodríguez, L.F. 2001, These Proceedings
Voges, W., Aschenbach, B., Boller, Th., et al. 1999, A&A, 349:389–405.

NP BLACK HOLES AS HIGH ENERGY GAMMA-RAY SOURCES

Brian Punsly
4014 Emerald Street no.116
Torrance
California
USA 90503
brian.punsly@gte.net

Keywords: Black holes, galactic gamma ray sources

Abstract It is been made clear in these proceedings that EGRET has detected at least one new class of celestial object that was previously inconspicuous in other observing bands. Many of these high energy gamma ray sources are highly variable which is suggestive of relativistic jet emission. This talk describes a possible central engine for these gamma ray jets, rapidly rotating, magnetized black holes; NP black holes.

1. INTRODUCTION

Non-pulsating (NP) black holes are very rapidly rotating, charged black holes (Kerr-Newman) black holes and their magnetospheres. They represent a new class of compact objects that are potential candidates for the unidentified EGRET population (Punsly 1998). These collapsed objects are electrodynamically similar to pulsars. Given the ubiquitous lack of radio pulsations from the unidentified EGRET source population (Nice and Sawyer 1997), the NP black holes offer an interesting alternative to pulsars as candidates, since NP black holes never pulse (hence the name NP black holes). The Kerr-Newman electromagnetic field is always aligned with the symmetry (rotation) axis of the hole, thus there is no "lighthouse effect" and therefore no pulsations. Furthermore, most of the galactic population of unidentified EGRET sources have high energy γ-ray variability and spectral indices that are statistically different from the current EGRET pulsar identifications (Romero 2001, in these proceedings).

A. Carramiñana et al. (eds.),
The Nature of Unidentified Galactic High-energy Gamma-ray Sources, 271–282.

The NP black hole attains its ability to power EGRET sources through a gravitohydromagnetic (GHM) coupling of the gravitational field of the black hole to a plasma-filled magnetosphere. It is the "frame dragging" of the rotating spacetime that provides this coupling in the rapidly rotating system, as all plasma must rotate in the same sense as the black hole as it approaches the event horizon (in the region known as the ergosphere, between the event horizon and the stationary limit). In fact, corotation is complete, asymptotically at the horizon. The GHM coupling can extract the enormous rotational energy of the black hole and drive bipolar jets (for a detailed treatment of this phenomenon, the reader should see the new book Punsly 2001). The bipolar jets are of interest because they are relativistic, like the γ-ray emitting jets in AGN. The synchrotron component provides weak radio and optical emission, but is partly inverse Compton scattered to gamma ray energies by the hot jet pair plasma (this is known as the synchrotron self Compton process, or SSC process employed in the AGN theory of EGRET sources).

There are a number of relevant issues that are associated with the postulated existence of NP black holes that are delineated below.

1.1. WHERE DO THEY COME FROM?

An interesting outstanding problem in astrophysics is the understanding the nature of the final state of the catastrophic collapse of a very massive ($M > 35M_{\odot}$) magnetized star. Such a star is probably too massive to create a neutron star and the final state should be a black hole. In this instance, what happens to the magnetic flux that is "frozen-into" the plasma during collapse? The NP black hole end point of the collapse solves this problem. Every other compact object has a magnetized subpopulation (i.e., neutron stars and white dwarfs) so why not black holes? Also, the NP black hole can be the ultimate consequence of the cannibalism of a magnetized white dwarf or neutron star by a black hole in a binary system.

1.2. WHY DOESN'T THE NP BLACK HOLE ELECTRICALLY DISCHARGE?

The existence of a magnetic field emanating from a black hole requires the existence of a net charge on the black hole. In the early days of black hole physics this was construed as evidence that electrostatic discharge was imminent. However, since that time we have learned to solve Maxwell's equations in the Kerr spacetime (the spacetime of a rotating uncharged black hole). These solutions indicate that another force, rotationally induced EMFs from the aforementioned frame dragging creates charge separation as charges flow in an attempt to cancel this electromotive force by creating an electrostatic force (somewhat analogous to moving laboratory conductors). Thus, the electrostatic force is not the only force present and a simple discharge scenario is not expected. In fact, I will show that a net charge on the central object is a universal consequence of rotating magnetized dynamical systems. This equates to the modification of the condition, $\vec{E} = 0$, in electrostatics, to the condition, $\vec{E} + \vec{v}/c \times \vec{B} = 0$, in relativity. The example of the neutron star in the axisymmetric pulsar model will be used to illustrate this circumstance in the next section.

It is also of interest to understand the time scale for electrical discharge in the NP black hole. For example, low angular momentum systems, or systems disrupted by strong accretion should discharge in less than one second. This type of rapid electrical discharge is the type of phenomenon that is needed to explain gamma ray bursts.

1.3. WHAT IS THE EMISSIVITY OF THE JET?

At the end of this presentation, I will show that a NP black hole with pulsar-like parameters can create SSC jets that reproduce typical EGRET luminosities, spectral indices and variability. These jets are also capable of creating modest or even weak emission in the other, more conventional, observing bands as you would expect for an unidentified source. Furthermore, a NP black hole in isolation can still power a jet (since the rotational energy of the black hole is the power source) which is consistent with the lack of interesting physical structures in many EGRET error boxes.

2. THE CHARGE ON A MAGNETIZED BLACK HOLE

It is very important to demystify the existence of a charge on a magnetized, rotating black hole. This is a general relativistic effect, as will be shown in Section 2.2, associated with "the dragging of inertial frames." However, we can make a very strong analogy to more conventional physics by first considering the charge on a rotating neutron star in the standard model of a pulsar as discussed in Ruderman and Sutherland (1975).

2.1. THE CHARGE OF A NEUTRON STAR IN A PULSAR

Consider the standard axisymmetric pulsar model. The interior of a neutron star can be assumed to possess a uniform magnetization for simplicity, then using the frozen-in condition, $\vec{E}+\frac{\vec{v}}{c}\times\vec{B}=0$, and Gauss's law, we find an interior charge density, ρ_e, and net charge, Q,

$$\rho_e = -\frac{\Omega B_0}{2\pi c}\,, \quad Q = -\frac{2}{3}\,\frac{\Omega R^3 B_0}{c}\,, \tag{20.1}$$

where B_0 is the uniform magnetic field strength, Ω is the angular velocity of the neutron star and R is the stellar radius. Similarly, the frozen-in condition yields the radial electric field at the stellar surface,

$$E_r^{(int)} = -\frac{B_0 \Omega R}{c}\sin^2\theta\,. \tag{20.2}$$

One can also construct a frozen-in magnetosphere filled with a tenuous or even a charge separated plasma (Goldreich and Julian 1969). The zeroth order magnetic field is considered to be dipolar. The magnetospheric charge density necessary to short out the proper electric field (i.e., that which is required to set $\vec{E}+\frac{\vec{v}}{c}\times\vec{B}=0$) in the pulsar magnetosphere is known as the Goldreich-Julian charge density. There is an additional source of the poloidal magnetic field that arises from the azimuthal current associated with the corotation of the Goldreich-Julian charge (or equivalently, the magnetosphere) with the neutron star. However, at the stellar surface, $\Omega R \ll c$, so the contribution from the bulk motion of Goldreich-Julian charge is small and we still have essentially a dipolar magnetic field near the star (of course this is not true near the light cylinder). The frozen-in condition can then be used to find the radial magnetic field in the magnetosphere at the stellar surface,

$$E_r^{(mag)} = \frac{B_0}{2}\frac{\Omega}{c}R\sin^2\theta\,. \tag{20.3}$$

Notice that (20.2) and (20.3) do not agree, so there is a surface charge density, σ_e, and a net surface charge, Q_{surf}, on the stellar surface,

$$\sigma_e = \frac{3B_0\Omega R}{8\pi c}\left(1-\cos^2\theta\right)\,, \quad Q_{surf} = \frac{\Omega R^3 B_0}{c}\,. \tag{20.4}$$

Combining the interior charge and surface charge in (20.1) and (20.4), the neutron star has a net charge,

$$Q_n = \frac{1}{3}\frac{\Omega R^3 B_0}{c}\,. \tag{20.5}$$

Presumably, the opposite sign of charge to (20.5) was ejected into the magnetosphere and wind zones by rotationally induced EMFs during the collapse phase of the pulsar. It is the rotationally induced EMFs that create the charge on the neutron star. The condition, $\vec{E} + \frac{\vec{v}}{c} \times \vec{B} = 0$, in relativistic conductive media tells us that a rotating magnetic system always has a charge separation. A neutron star cannot maintain (stay frozen-into) its magnetic flux unless it is charged as indicated by (20.5).

2.2. EMFS FROM BLACK HOLE ROTATION

The fact that frame dragging is always making "spacetime rotate" around the black hole creates unbalanced rotationally induced EMFs that drive a charge separation (analogous to the rotating neutron star interior and Goldreich-Julian magnetosphere of the pulsar). This is illustrated by the simple example of $\ell = 1, m = 0$ solution of Laplace's equations in the Kerr spacetime (see Figure 1). At asymptotic infinity, the magnetic field is uniform and the electric field goes to zero. However, near the black hole, the field is not pure magnetic as a significant electric field is produced by the the dragging of inertial frames.

In the late 1970s and early 1980s some people thought about introducing a tenuous or charge separated plasma in such a magnetosphere. A tenuous plasma is frozen onto the magnetic field lines in its gyro-orbits. It can flow along the magnetic field lines, but not across the magnetic field. Such charges can be accelerated by the semi-vacuum component of the electric field along the magnetic field lines, $E_{||}$, on the polar flux tubes into the black hole. Thus, some suggested the possibility of a global current system driven by the vacuum field. However, in a charge separated plasma, $\vec{J}$ and $\vec{E}$ are always in the same direction. As we move to lower latitudes on the event horizon to find a reversal of $\vec{E}$ that can drive the return current, a magnetic flux tube is found; however, $E_{||}$ changes sign along the flux tube which is inconsistent with a global current flow in one direction along the flux tube as noted above. This magnetic flux tube is open circuited as in the "Goldreich-Julian conundrum" of the axisymmetric charge starved pulsar models. Similarly, charges cannot cross the strong magnetic field lines in the equatorial plane to close the circuit since they are only free to flow along the field lines. Thus, there is no global current flow, only accretion of charges into the polar regions of the black hole which charges up the black hole in an effort to minimize the global electric potential induced by frame dragging.

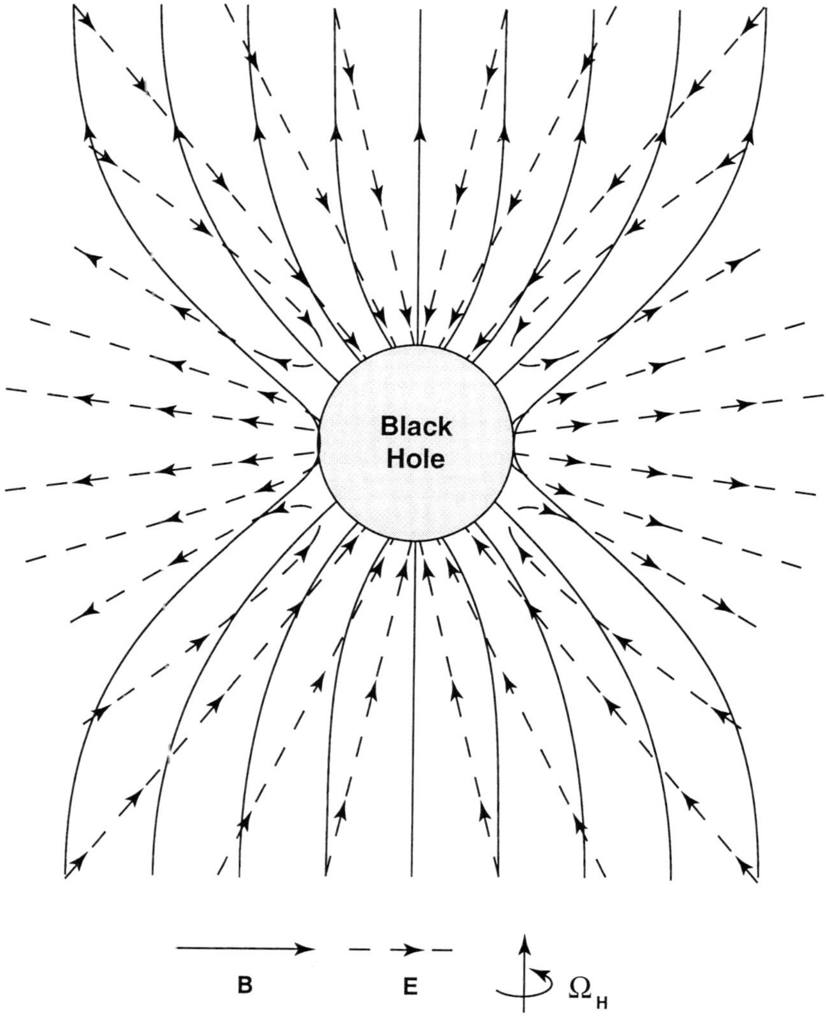

Figure 1 The electromagnetic field of the $\ell = 1$, $m = 0$ moment of Laplace's equations on the Kerr spacetime. The figure is from Punsly (2001).

3. THE STRUCTURE OF THE NP BLACK HOLE MAGNETOSPHERE

The accretion of charge as suggested in the last section creates the $\ell = 0$, $m = 0$ moment of the electromagnetic field, the Kerr-Newman electromagnetic field (see Figure 2).

This field exists as a test field on the Kerr metric which can be used in place of the more complicated Kerr-Newman metric, even for black hole

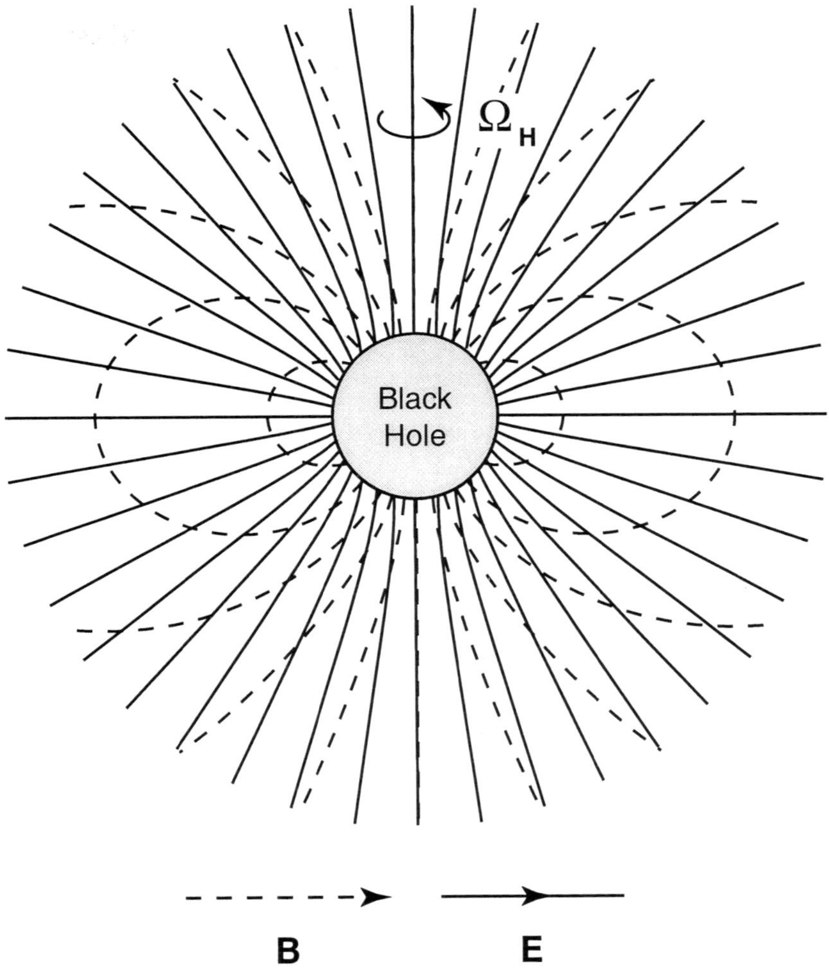

Figure 2 The Kerr-Newman electromagnetic field. The figure is from Punsly (2001).

charges that are astrophysically large (i.e., on the order of the charge on the neutron star in the Crab pulsar). For example, if r_+ is the radius of the event horizon, M is the mass of the black hole, and a is the angular momentum per unit mass of the black hole in geometrized units of a very rapidly rotating black hole, we have the following relative magnitudes,

$$r_+ \gtrsim M \gtrsim a \approx 10^6\text{cm} \,, M = 7M_\odot \,. \tag{20.6}$$

If the magnetic field at the pole of the horizon is 10^{10} G then $Q \approx 2 \times 10^{22}$ esu $\approx 5 \times 10^{-3}$ cm. Thus, $Q^2 \ll a^2 \lesssim M^2$ and the geometry is essentially Kerr spacetime. This is a tremendous simplification that facilitates calculations.

3.1. THE KERR-NEWMAN FIELD

Note that the moderate angular momentum system depicted in Figure 2 has an asymptotic radial electric field (Gauss' law). Furthermore, since every charge that approaches the event horizon must corotate with the horizon, a J^{ϕ} current source develops as the hole accretes charge and one necessarily has a dipolar magnetic field accompanying the electric field. The existence of both magnetic and electric components of the field is the result of frame dragging.

3.2. MAGNETOSPHERIC DYNAMICS

Figure 2 indicates that $\vec{E}\cdot\vec{B} \neq 0$ in the Kerr-Newman magnetosphere, which was the starting point for our understanding of pulsar magnetospheres. Furthermore, the electromagnetic forces are $\sim 10^{13}$ times the gravitational forces in the magnetosphere, everywhere except near the event horizon. This is significant for two reasons. Firstly, the magnetospheric dynamics should be very similar to those of an axisymmetric pulsar throughout the magnetosphere, everywhere except very close to the horizon. Secondly, the transition near the event horizon is associated with the extraction of the rotational energy of the black hole. The horizon boundary condition is that all plasma is inertially dominated. Gravitational forces are stronger than any other imposed forces; this is why it is a black hole as opposed to another type of compact object. There is no electromagnetic force that can overcome the plasma motion, spiraling inward approaching the speed of light in corotation with the horizon, as dictated by gravity. The frame dragging force is very large. One would need a rocket capable of overpowering the rotational inertia of the black hole $\sim 0.3Mc^2$ in order to alter this result! Thus, the plasma flow changes from being incredibly magnetically dominated to being completely inertially dominated in a very small region. This transition, which represents very intense physics, creates a GHM dynamo that can power outgoing bipolar winds. The fascinating details are addressed in the new book (Punsly 2001).

Other than the dynamo near the horizon, the magnetosphere is very similar to that found in axisymmetric pulsar models. Thus, we expect similar vacuum gap pair creation processes in the charge starved limit as in pulsars (see Punsly 1998 for details). The global structure of the NP black hole is depicted in Figure 3.

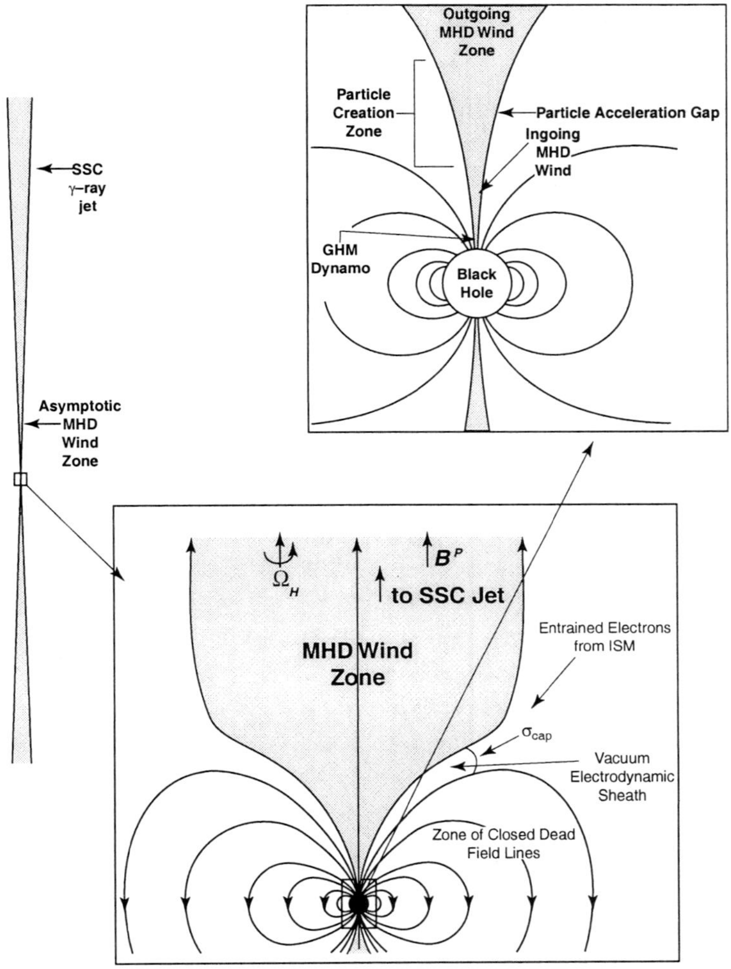

Figure 3 The Kerr-Newman black hole model of galactic hard γ-ray sources. On the left is a wide angle view of the twin SSC (synchrotron self-Compton) γ-ray emitting jets driven by a compact object. At the bottom, center, is an exploded view of the compact object. A black hole (indicated by the solid disk) supports a magnetic field that is primarily dipolar, except at the poles where an MHD wind emanates from the magnetosphere. The hole rotates with an angular velocity, Ω_H, about an axis that is aligned with both the dipole moment and the MHD wind bulk flow velocity. At the top, right, is an expanded view of the black hole magnetosphere. A GHM dynamo that powers the outgoing MHD wind exists just above the event horizon. Paired (ingoing and outgoing) MHD winds emerge from a "particle creation zone." The flow divides within the particle creation zone in a gap containing a strong radial electric field. The figure is from Punsly (1998)

4. JET LUMINOSITY

About 80% of the jet luminosity (intrinsic power) ends up in gamma rays. The apparent luminosity of γ-rays from the jets is estimated in Punsly (1998) to be

$$L_{apparent} \approx \delta^3(\theta) 3.5 \times 10^{32} \left[\frac{B_S}{10^{10}\mathrm{G}}\right]^2 \left[\frac{M}{7M_\odot}\right]^2 \mathrm{ergs/sec}\ , \tag{20.7}$$

where B_S is the magnetic field strength at the poles of the event horizon. The quantity $\delta(\theta)$ is the Doppler factor associated with the motion of the bulk jet flow relative to earth, which is a function of the line of sight to earth, θ,

$$\delta(\theta) = \frac{1}{\Gamma\left(1 - \frac{v}{c}\cos\theta\right)}\ . \tag{20.8}$$

Typical bulk Lorentz factors are $\Gamma \sim 5$ in the jet.

Figure 4 is an SSC jet model from Punsly (1999) of the compact companion in the binary system, LSI +61° 303. This is an interesting system since mass function estimates yield a constraint of $1.1M_\odot < M < 3.3M_\odot$ for the compact companion which is large for a neutron star. Furthermore, no pulsations in any frequency band have ever been found (see Punsly 1999 for details).

Figure 5 is another SSC jet model from Punsly et al. (2000) with a dense jet that produces a strong annihilation component. This soft gamma ray excess can create steep spectral indices.

5. SUMMARY

Models of SSC jets ejected from NP black holes can create EGRET spectra with large variability due to the $\delta^3(\theta)$ dependence in (20.8) as the jet wobbles as a consequence to plasma instabilities. The NP black hole can radiate as an isolated system (which appears to be the case for many mid-latitude galactic sources) or in a weakly accreting binary system.

We note the following important observational consequences of the NP black hole model of the unidentified EGRET sources:

1 The stronger sources can have 10 – 50 mJy of radio emission at 5 GHz. Many of these sources should show jets with the VLBA.

2 The weaker EGRET sources would tend to have very weak radio emission, order of 0.5 mJy – 5 mJy at 5 GHz.

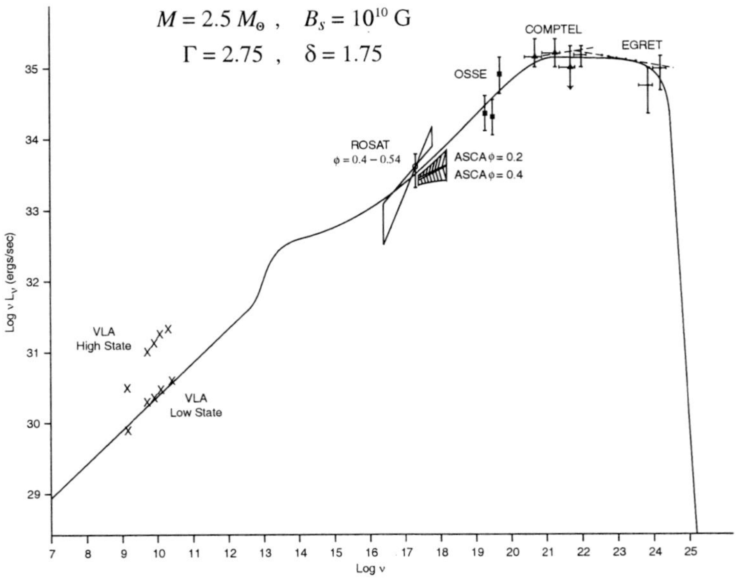

Figure 4 The SED of the compact object LSI + 61° 303 is fit by a NP black hole model. The figure is adapted from Punsly (1999)

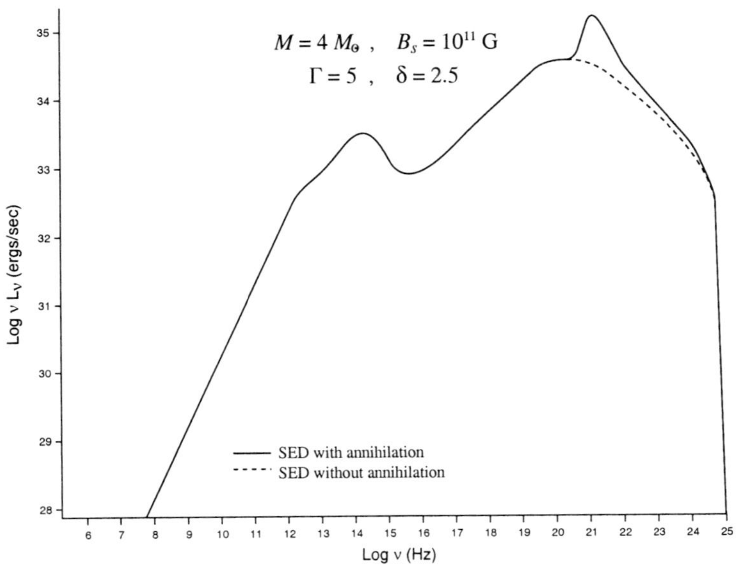

Figure 5 The SED of a NP black hole with a strong annihilation component. The figure is adapted from Punsly et al. (2000).

3 Optically, they should appear as an 18th to 23rd magnitude blue stars. However, many sources are in the galactic plane, so obscuration can be significant.

4 In isolated (non-binary) systems, the gamma ray luminosity should exceed the X-ray luminosity by about two orders of magnitude.

5 The steep EGRET spectrum (gamma ray photon number index larger than 2.5) sources should have a significant annihilation broad line.

References

Goldreich,P. and Julian, W.H. (1969), ApJ. **157**, 869

Ruderman, M.A. and Sutherland, P.G. (1975), ApJ. **196**, 51

Nice, D. and Sawyer, T. (1997), ApJ. **476**, 261

Punsly, B. (1998), ApJ.**506**, 790 (1998)

Punsly, B. (1999), ApJ. **527**, 624 (1999)

Punsly, B. (2001) *Black Hole Gravitohydromagnetics*, (Springer-Verlag, Heidelberg)

Punsly, B., Romero, G.E., Torres, D.F. and Combi, J.A. (2000) Astron. Astrophys. in press

VI

IDENTIFYING THE UNIDENTIFIED

TEV OBSERVATIONS OF SNRS AND UNIDENTIFIED SOURCES

Stephen Fegan
Fred Lawerence Whipple Observatory
Harvard-Smithsonian CfA
P.O. Box 97, Amado, AZ 85645-0097, U.S.A.

Department of Physics
University of Arizona
1118 E. 4th Street, Tucson, AZ 85721, U.S.A.
sfegan@egret.sao.arizona.edu

Keywords: Very high energy gamma rays, supernova remnants, unidentified sources

Abstract

A review of very high energy γ-ray astronomy is presented. Particular attention is paid to the atmospheric Cherenkov imaging technique whose employment has resulted in detections of both galactic and extra-galactic objects at energies above 300GeV. Next generation ground-based telescopes promise to increase our knowledge of extreme astrophysical objects as they begin to operate over the next few years.

1. INTRODUCTION

Very High Energy (VHE) γ-ray astronomy, or TeV γ-ray astronomy as it is sometimes known, traditionally describes observations in the energy range from 300GeV to 100TeV. Doing astronomy in this energy range was made possible by the development of the atmospheric Cherenkov imaging technique, which led to the detection of the first VHE source, the Crab Nebula, more than ten years ago (Weekes et al. 1989). Ground-based instruments operating in this energy region typically have large collecting areas, high angular resolution and good energy resolution (Ong 1998).

In the years since the detection of the Crab Nebula many other galactic and extra-galactic VHE γ-ray sources have been detected. There are observatories operating in the northern and southern hemispheres which have been able to independently confirm a number of the claimed de-

A. Carramiñana et al. (eds.),
The Nature of Unidentified Galactic High-energy Gamma-ray Sources, 285–297.

tections. Ground-based telescopes have participated in multi-wavelength observations with satellite-based detectors, yielding many interesting results. To date only 1% of the VHE sky has been observed at these energies.

The coming years will bring a new generation of ground-based VHE γ-ray observatories that will build on the techniques developed over the past three decades. A number of large, next generation, Cherenkov telescopes are either starting to operate or being built around the world. The first decade of the new millennium promises many new exciting results in this field.

2. GROUND-BASED TELESCOPES

At very high energies, γ-rays and cosmic-rays interact with the atmosphere to produce Extensive Air Showers (EAS), showers of charged secondary particles which propagate through the atmosphere, reaching the ground if the energy of the primary particle is sufficiently high. Detecting these showers and inferring the composition, energy and arrival direction of the primary particle is the challenge of ground-based VHE astronomy. Fortunately, the mechanism by which the shower is produced is very well known and can be modeled for different primaries at any energy.

Directly detecting the secondary particles is possible for instruments placed at high altitude and for sufficiently energetic primaries. For lower energy primaries, the cascade of charged particles does not reach the ground but can be indirectly detected from the Cherenkov radiation emitted as the relativistic charged particles traverse the atmosphere at speeds in excess of the speed of light in air. The Cherenkov photons are strongly beamed in the arrival direction of the incident photon and form a narrow cone of light as they travel down through the atmosphere. A telescope with fast light detectors can sample the $\sim$ 5ns pulse of Cherenkov light. The total number of photons recorded is roughly proportional to the energy of the primary, unless the telescope samples at the edge of the shower, where the number of photons falls off quickly.

Historically, the most successful ground-based detection method has been the Imaging Atmospheric Cherenkov Technique originally proposed in 1977 (Weekes & Turver 1977). A large telescope collects Cherenkov light from showers and focuses it onto a plane of photomultiplier tubes. Imaging the shower in such a fashion produces a "picture" of its development through the atmosphere. Typically, these images are characterized by a number of straight forward parameters (Hillas 1985) and it is on the basis of subtle differences in these parameters that hadronically produced showers can be differentiated from purely electromagnetic showers and rejection of cosmic-ray particles can be made. Although the isotropic background consists of $\sim$ 1000 times more cosmic-rays than γ-rays, a single imaging atmospheric Cherenkov telescope (IACT) can reject 99.7% of the events caused by these background showers, leaving a statistically detectable γ-ray signal, if one is present and sufficiently strong. An array of such telescopes working in coincidence can reject substantially more (Daum et al. 1997).

Since the detection of the Crab Nebula in 1989, VHE gamma-ray astronomy has advanced in two major ways. First, the development of high resolution cameras has allowed better imaging of the showers and more refined rejection of hadronic background events. A number of observatories such as the Whipple 10m telescope (Fig. 1), CAT, a French telescope in the Pyrenées (Barrau et al. 1998) and CANGAROO, a Japanese-Australian telescope in Woomera, Australia (Hara et al. 1993) employ this approach. The second major development is the combination of a number of IACTs to form an array that can view the shower development from a number of different points on the ground. Such stereoscopic viewing increases the suppression of hadronic background and allows for better reconstruction of the arrival direction and energy of the primary photon. To date the most significant example of such an array is HEGRA, five small imaging telescopes on La Palma run by an Armenian-German-Spanish collaboration (Daum et al. 1997).

Figure 1 Left, the Whipple observatory 10m imaging atmospheric Cherenkov telescope. Right, at the focus, a 490 pixel high resolution camera operating since 1999.

3. SUPERNOVA REMNANTS

3.1. PLERIONS

The Crab Nebula has been detected in VHE γ-rays by many independent Cherenkov observatories in the northern hemisphere and by one in the southern. Routine observations of the Crab Nebula by IACTs continue and it has come to be regarded by the VHE community as the "standard candle" and is often used to provide a check on the calibration of new instruments. Modern IACTs can detect the Crab Nebula at the 5-6σ level after one hour of observations, a testament to how far the field has advanced since the original detection in 1989, which was a 9σ excess based on 60 hours of observations. Recently the STACEE and CELESTE collaborations, have published significant detections of the Crab Nebula in the energy region $E > 190 \pm 60$GeV (Oser et al. 2000) and $E > 50$GeV (De Naurois et al,. 2000) respectively, an energy range that extends lower than other ground-based instruments. At the other end of the VHE spectrum, the Crab was the first γ-ray source source detected by an air shower array, when it was seen by the Tibet Air Shower Array at $E > 3$TeV (Amenomori 1999).

The energy spectrum between 300GeV and 50TeV has been well established by a number of groups (Hillas et al. 1998, Aharonian et al. 2000). To date no pulsation has been seen in the VHE γ-ray signal from the Crab (Gillanders et al. 1997; Burdett et al. 1999; Aharonian et al. 1999). The VHE flux is thought to arise, in the most part, from inverse Compton scattering of synchrotron photons by relativistic electrons, so called synchrotron self-Compton or SSC emission (De Jager & Harding 1992; Hillas et al. 1998). By fitting the SSC model to the energy spectrum that results from combining the X-ray, EGRET and VHE data, a magnetic field of $160\mu G$ is derived (Hillas et al. 1998).

No other plerions have been detected by IACTs in the northern hemisphere; however the situation is different in the southern hemisphere (Table 1) where the CANGAROO group have detected two. In 1993 they reported the detection of PSR 1706-44 based on 60 hours of observations (Kifune et al. 1995). Confirmation by the Durham group based on 10 hours of observation was subsequently made (Chadwick et al. 1997). The VHE source is associated with a pulsar of period 102ms and appears to be associated with a supernova remnant. No pulsations have been detected in the VHE data. The CANGAROO group has also reported the detection of a VHE signal in the neighborhood of the Vela pulsar (Fig. 2). The detection is at the 6σ level based on $\sim$ 120 hours of observation and the measured flux above 2.5TeV is $0.25 \times 10^{-11}\mathrm{cm}^{-2}\mathrm{s}^{-1}$. The VHE signal, which is offset from the location of the pulsar by 0.14°,

Table 1 TeV Observations of Plerions

Object Name	**Exposure time (hours)**	**Flux/Upper Limit x $10^{-11} cm^{-2} s^{-1}$**
EVERYONE		
Crab Nebula	$\rightarrow \infty$	7.0 (> 400GeV)
CANGAROO		
PSR 1706-44	60	0.15 (>1TeV)
Vela Pulsar	116	0.26 (E/2 TeV)$^{-2.4}$ TeV^{-1}
Durham		
PSR 1706-44	10	1.2 (>300GeV)
Vela Pulsar	8.75	<5.0 (>300GeV)

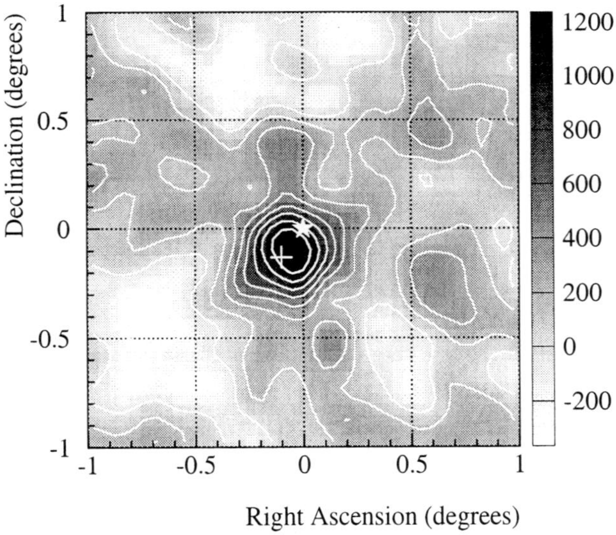

Figure 2 VHE emission in the neighborhood of the Vela Pulsar as detected by the CANGAROO experiment. The contours show the number of excess events per deg^2. The present location of the pulsar is marked with a star.

is thought to originate from a synchrotron nebula, powered by a population of relativistic electrons which were created in the supernova explosion and which have survived since then due to the low magnetic field in the nebula. The nebula is assumed to be centered on the birthplace of the pulsar, which was ejected at the time of the supernova explosion.

3.2. SHELL-TYPE

It has long been hoped that VHE γ-ray astronomy could provide a probe of the origin of cosmic-rays with E< 10^{14}eV. Supernovae are regarded as the most likely producers of these cosmic-rays as they are the only galactic objects capable of supplying the power required to account for the observed cosmic-ray energy spectrum. Furthermore, diffusive shock acceleration (Blanford & Ostriker 1978, Bell 1978) provides a natural mechanism to convert the kinetic energy of the SNR shock-front into a spectrum of accelerated charged particles with dN/dE=$E^{-2.1}$. This source spectrum, after correcting for diffusion in the galaxy (Swordy et al. 1990), fits the locally observed spectrum of $E^{-2.7}$. If this mechanism is correct, then interactions between the relativistic charged particles and the interstellar medium surrounding the SNRs should produce γ-rays through the decay of secondary π^0 particles (Drury, Aharonian & Völk 1994 – DAV). These models predict that fluxes of γ-rays should be high enough to be visible to the current generation of satellite-based and ground-based detectors. Detection of the signature π^0 bump at MeV energies and a spectrum extending to tens of TeV would be a clear indication that cosmic-ray acceleration does take place in SNRs. However, the experimental situation is complicated by the presence of a population of γ-rays produced by the inverse-Compton reaction of relativistic electrons and the cosmic microwave background. Separating these two components requires that the spectrum be measured continuously from 10 MeV to 10 TeV.

For most of its history, VHE γ-ray astronomy provided no detections of shell-type supernova remnants (Fig. 3). Observations of SNRs that are considered to be good candidates for neutral pion decay have been undertaken by numerous groups. In particular, observations of W44, W51, γ-Cygni, W63 and Tycho's SNR, selected due to their possible association with molecular clouds which should provide an enhanced target for π^0 decay, their possible associations with EGRET sources and their small angular extent (Buckley et al. 1998), have failed to produce detections in VHE γ-rays (Table 2). Buckley et al. 1998 suggest that, assuming the EGRET emission is from these shell-type SNRs and that π^0 decay dominates, the spectra of cosmic-rays produced would have to be softer than suggested by DAV. The required differential source spectrum is $E^{-2.5}$ for γ-Cygni and $E^{-2.4}$ for IC443. Even if no reference is made to the EGRET data, the VHE upper-limits for some of the objects (IC443) push the allowable parameter space of DAV. Gaisser et al. (1998) performed fits of the EGRET and Whipple results concluding that the EGRET data must be dominated at lower energies by electron bremsstrahlung but the source spectrum must still be steep to account for the Whipple upper limits.

Table 2 TeV Observations of Shell-type SNR

Object Name	Exposure time (hours)	Flux/Upper Limit x $10^{-11} cm^{-2} s^{-1}$
CANGAROO		
RXJ 1713.7-3946	66	0.53 ($\geq$1.8 TeV)
SN1006	34	0.46 ($\geq$1.7 TeV)
W28	58	<0.88 (> 5 TeV)[a]
HEGRA		
Cas A	232	0.058 (> 1 TeV) [b]
γ-Cygni	47	<1.1 (>500GeV)[c]
Durham		
SN1006	41	<1.7 (>300GeV)
Whipple		
Monoceros	13.1	<4.8 (>500GeV)
Cas A	6.9	<0.66 (>500GeV)
W44	6	<3.0 (>300GeV)
W51	7.8	<3.6 (>300GeV)
γ-Cygni	9.3	<2.2 (>300GeV)
W63	2.3	<6.4 (>300GeV)
Tycho	14.5	<0.8 (>300GeV)
CAT		
CasA	24.4	<0.74 (>400GeV)

[a] A different definition of Energy Threshold is used
[b] Evidence for emission at the 4.9σ level (Pühlhofer et al. 2001)
[c] Limits converted from Crab units using flux of Hillas et al. 1998

Recent observations of shell-type SNRs by the CANGAROO group have resulted in the detections of SN1006 (Tanimori et al. 1998) and RXJ1713.7-3946 (Muraishi et al. 2000). Observations of SN1006 in 1996 and 1997 show a significant excess from the NW rim of the SNR. The excess is consistent with the location of non-thermal X-rays detected by ASCA (Koyama et al. 1995). Similarly, RXJ1713.7-3946 has recently had a ROSAT and ASCA X-ray source associated with it (Pfeffermann & Aschenbach 1996). The TeV γ-ray signal from these objects, if confirmed, would not necessarily imply that they are accelerators of hadronic cosmic-rays. It is considered likely that the γ-rays are linked through inverse-Compton scattering with a population of rel-

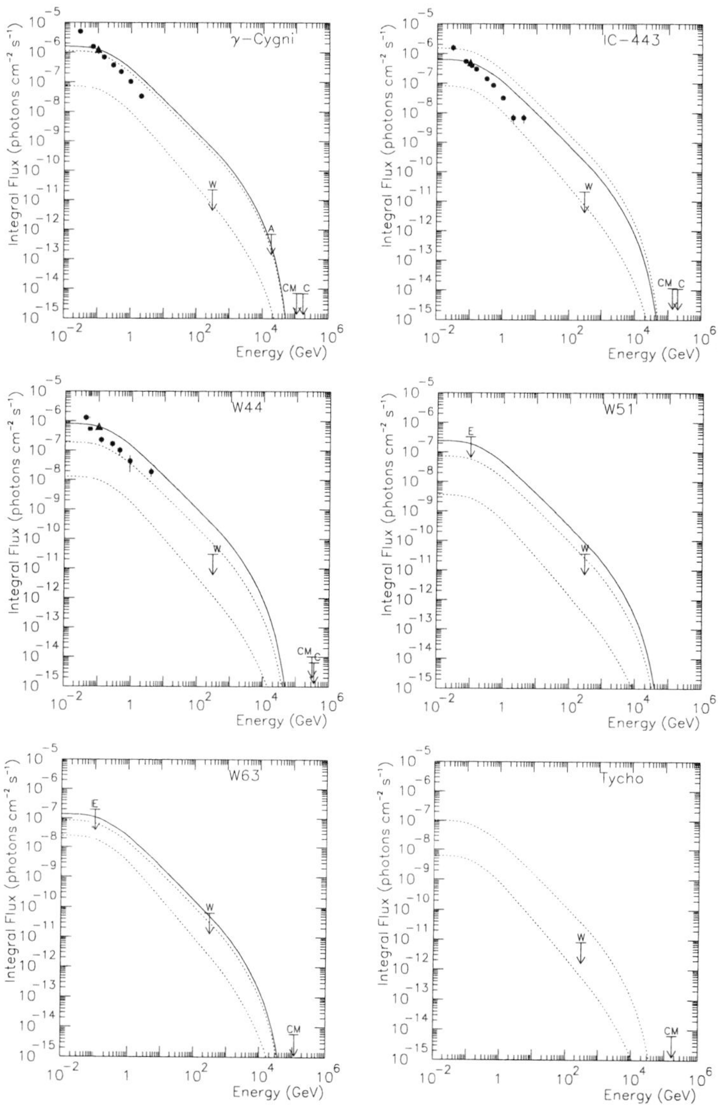

Figure 3 MeV - TeV observations of shell-type SNR. Whipple upper limits marked as (W), associated EGRET fluxes as points or EGRET upper-limits (E). Also shown are CASA-MIA (CM), Cygnus (C) and AIROBICC upper-limits. The solid curve shows the extrapolation of the DAV model from the EGRET integral data points at 100 MeV, shown as a triangle. The dashed curve show the reasonable flux ranges of the DAV model without making any assumptions about the EGRET data.

ativistic electrons which are also responsible for producing the X-rays. In any case, further observations and the determination of the energy spectrum with more sensitive instruments will be required either to confirm or to completely rule out the presence of a hadronic component to the TeV signal.

The latest shell-type SNR to be detected at TeV energies is Cassiopeia A, recently announced as a source by the HEGRA collaboration (Pühlhofer et al. 2001). Based on 232 hours of observations they report an excess at the 4.9σ level, and calculate a flux of $F = 5.8 \pm 1.2_{stat} \pm 2_{syst} \times 10^{-13} \mathrm{cm}^{-2}\mathrm{s}^{-1}$ at (E> 1 TeV). Whether the γ-rays from Cas A arise from inverse-Compton interactions or through neutral pion decay has not yet been determined. There are reasons to think that both components may be present at some level. First, Cas A is associated with a bright source of hard X-rays which indicates a population of non-thermal electrons with energies up to 100 TeV (Allen et al. 1999). Second, Cas A is situated in a region of high ambient matter density which has been associated with the wind system left over from the progenitor. Measurement of the energy spectrum will be required before any determination between the two components can be made.

4. UNIDENTIFIED SOURCES

IACTs are most sensitive when observing on-axis point sources, i.e. when pointing directly towards an object of small angular extent. Observations of extended sources or of sources where the location is not well known are possible, however, as IACTs do have good sensitivity across most of their field of view. It is therefore possible to create a two dimensional VHE map of the sky within, typically, $0.5 - 1.0°$ of the center of the field of view. The angular resolution of a single IACT is $0.15°$ (Lessard et al. 2000), and better for an array of IACTs, providing the ability to resolve the location of a source within the field of view to high accuracy.

VHE observations of unidentified EGRET sources are usually made on the basis of their luminosity, spectrum and size of their error box or if they have a good candidate association, such as an SNR or pulsar which allows sensitive, point source observations to be made.

Results of previous observations of unidentified EGRET sources with the Whipple 10m telescope (Buckley et al. 1997) and new results from ongoing observations are given in Table 3. No unidentified sources have been detected in VHE γ-rays but for many of them, upper limits have been placed on VHE emission from within the EGRET error circle.

Table 3 TeV Observations of Unidentified Sources

Object Name	**Exposure time (mins)**	**Flux Upper Limit x $10^{-11} cm^{-2} s^{-1}$**
J0241+6119	972	1.02[a]
J0433+2907	499	4.4[a]
J0545+3943	108	6.72[a]
J0618+2234	1188	0.911[a]
J0635+0521	108	5.59[a]
J0749+17	486	0.813[a]
J1746-2852	270	0.45[b]
J1825-1307	702	1.55[a]
J1857+0118	351	2.79[a]
J2016+3657	287	5.8[a]
J2020+4026	513	0.990[a]
J2227+6122	274	6.0[a]

[a]Integral Flux Above 400 GeV.
[b]Integral Flux Above 2.0 TeV.

5. THE NEXT GENERATION

The future of ground-based γ-ray astronomy lies with the next generation of observatories that will build on the advances made to date. Significant discoveries will come from extending the energy range observable from the ground and by improving the flux sensitivity.

Extending the observable energy range below 200GeV will be achieved by building instruments with larger mirror area to gather more light from the shower. Two approaches have been suggested. The first uses existing solar furnace facilities which have fields of large heliostats that lie unused at night. The arrival of the Cherenkov wavefront at groups of heliostats is precisely measured and this information is used to differentiate γ-rays from cosmic-rays (Ong 1998). STACEE (Chantell et al. 1998), CELESTE (Quebert et al. 1995) and Solar-2 (Tümer et al. 1999), each of which use this technique, have started operating in the last few years and hope to observe at ~20-30 GeV. MAGIC (Barrio et al. 1998) takes a different approach; by extending the size of a traditional IACT to 17m they plan to achieve a significantly lower energy threshold than is possible with current generation telescopes.

At the other end of the VHE spectrum, air shower arrays on very high mountains, such as the Tibet Air Shower Array (Amenomori 1999) and Cherenkov telescopes operating at large zenith angles have been

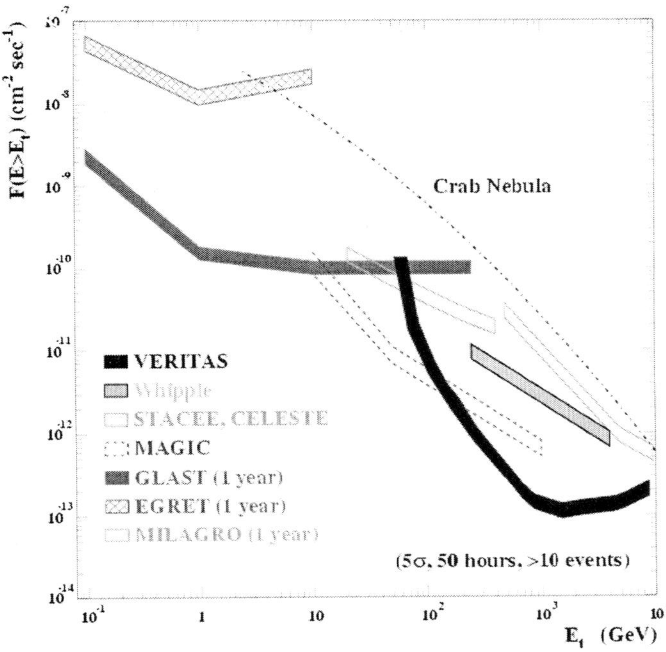

Figure 4 Point source sensitivity of some of the current and next generation of γ-ray instruments; Whipple (Weekes et al. 1989), MAGIC (Barrio et al. 1998), STACEE & CELESTE (Chantell et al. 1998 & Quebert et al. 1995), HEGRA (Daum et al. 1997), *GLAST* (Gehrels & Michelson 1999), EGRET (Thompson et al. 1993) and MILAGRO (Sinnis et al. 1995).

joined by MILAGRO (Sinnis et al. 1995), a water Cherenkov detector in New Mexico that views all of the visible sky with 45° of the zenith and operates 24 hours a day and hopes to achieve a better energy overlap with traditional ground-based techniques.

To significantly increase flux sensitivity and angular resolution while achieving a broad operating energy range of 100GeV to 10TeV a number of groups have undertaken to build arrays of large ($\sim 10 - 12$m) IACTs operating in coincidence. VERITAS (Bradbury et al. 1999), a system of seven 10m telescopes in Arizona will build on the experience gained by the Whipple group. HESS (Hoffmann 1997), an array of four (and possibly sixteen when completed) 12m telescopes in Namibia will extend the ideas pioneered by HEGRA. The Japanese-Australian CANGAROO group are building an array of four 10m telescopes at the site of the present CANGAROO experiment, one telescope has been completed. Collectively, these three arrays and the MAGIC telescope are often referred to as the Next Generation Gamma-Ray Telescopes (NGGRTs).

Figure 4 shows how all of these upcoming experiments will overlap with the EGRET and Whipple sensitivities and also with the sensitivity of the upcoming GLAST mission. It is evident that the NGGRTs will provide some overlap with GLAST and the solar arrays, and at the high energy side with the Tibet Air Shower array and MILAGRO.

The future of ground-based γ-ray astronomy will be very interesting. Continued cooperation with space-based missions will ensure that the energy range from 1MeV to 10TeV will be well covered in both the northern and southern hemispheres. Better flux sensitivity will mean that new sources will be discovered; excellent angular resolution will ensure that accurate source locations and associations will be found. Detailed energy spectra in this range will allow for better models of the emission regions.

6. ACKNOWLEDGEMENTS

Research in VHE γ-ray astronomy at the Whipple Observatory is supported by a grant from the U.S. Department of Energy. My thanks to Deirdre Horan and Trevor Weekes for their helpful comments.

References

Aharonian, F.A., et al. (1999), *A&A*, **346**, 913

Aharonian, F.A., et al. (2000), *ApJ*, **539**, 317

Allen, G.E., Gotthelf, E.V., Petre, R. (1999), *Proc. 26th Internat. Cosmic Ray Conf. (Salt Lake City)*, **3**, 480

Amenomori, M., et al. (1999), *ApJ*, **525**, L93

Barrau, A. et al. (1998), *Nucl. Instrum. Methods A*, **416**, 278

Barrio, J.A., et al. (1998), *"The MAGIC Telescope", design study*, **MPI-PhE/98-5**

Bell, A.R. (1978) *MNRAS*, **182**, 147

Blanford, R.D. & Ostriker, J.P. (1978), *ApJ*, **221**, L29

Bradbury, S.M., et al. (1999), *Proc. 26th Internat. Cosmic Ray Conf. (Salt Lake City)*, **5**, 280

Buckley, J.H., et al. (1997), *Proc. 25th Int. Cosmic Ray Conf. (Durban)*, **3**, 237

Buckley, J.H., et al. (1998), *A&A*, **329**, 639

Burdett, A., et al. (1999), *Proc. 26th Internat. Cosmic Ray Conf. (Salt Lake City)*, **3**, 448

Chadwick, P.M. et al. (1997), *Proc. 25th Int. Cosmic Ray Conf. (Durban)*, **3**, 189

Chantell, M.C., et al. (1998), *Nucl. Instrum. Methods A*, **408**, 468
Daum, A., et al. (1997), *Astropart. Phys.*, **8**, 1
De Jager, O.C. & Harding, A.K. (1992), *ApJ*, **396**, 161
De Naurois, M., et al. (2000), *Proc. Int. Symp. on High Energy Gamma-Ray Astro. (Heidelberg)*, AIP Conf. Proc. 558, ed. F.A. Aharonian, H.J. Volk, p. 540.
Drury, L.O'C., Aharonian, F.A., Völk, H.J. (1994), *A&A*, **287**, 959
Gaisser, T.K., Protheroe, R.J., Stanev, T. (1998), *ApJ*, **492**, 219
Gehrels, N. & Michelson, P. (1999), *TeV Astrophysics of Extragalactic Sources; Astropart. Phys.*, **11**, 277
Gillanders, G., et al. (1997), *Proc. 25th Int. Cosmic Ray Conf. (Durban)*, **3**, 185
Hara, T. et al. (1993), *Nucl. Instrum. Methods A*, **332**, 300
Hillas, A.M., et al. (1985), *Proc. 19th Int. Cosmic Ray Conf. (La Jolla)*, **3**, 445
Hillas, A.M., et al. (1998), *ApJ*, **503**, 774
Hoffmann, W., (1997), *Proc. Workshop on TeV γ-ray Astrophys. (Kruger Park)*, 405
Kifune, T., et al. (1995), *ApJ*, **438**, L91
Koyama, M., et al. (1995), *Nature*, **378**, 255
Lessard, R.W., et al. (2001), *Astropart. Phys.*, **15**, 1
Muraishi, H., et al. (2000), *A&A*, **354**, L57
Ong, R.A. (1998) *Phys Rep*, **305**, 93
Oser, S., et al. (2000), *ApJ*, **547**, 949
Pfeffermann, E. & Aschenbach, B., (1996), *Röntgenstrahlung from the Universe, In. Conf. on X-ray Astron. and Astrophys., MPE Report*, **263**, P267
Pühlhofer, G. et al. (2001), *Proc. Int. Symp. on High Energy Gamma-Ray Astro. (Heidelberg)*, AIP Conf. Proc. 558, ed. F.A. Aharonian, H.J. Volk, p. 749.
Quebert, J., et al. (1995), *Towards a Major Atmospheric Cherenkov Detector - IV (Padova)*, 428
Sinnis, G., et al. (1995), *Nucl. Phys. B (Proc. Suppl.)* **43**, 141
Swordy, S.P., et al. (1990), *ApJ*, **349**, 625
Tanimori, T., et al. (1998), *ApJ*, **497**, L25
Thompson, D.J., et al. (1993), *ApJS*, **86**, 629
Tümer, T., et al. (1999), *TeV Astrophysics of Extragalactic Sources; Astropart. Phys.*, **11**, 271
Weekes, T.C., et al. (1989) *ApJ*, **342**, 379
Weekes, T.C. & Turver, K.E. (1977), *Proc. 12th ESLAB Symp. (Frascati)* ESA SP-124, 279

A FIRST EGRET-UNID-RELATED AGENDA FOR THE NEXT-GENERATION CHERENKOV TELESCOPES

Dirk Petry
Dept. of Physics and Astronomy
Iowa State University
Ames, IA 50011
petry@iastate.edu

Keywords: Cherenkov Telescopes, Unidentified Gamma-Ray Sources, Very-High-Energy Gamma-Rays

Abstract The next generation of Imaging Atmospheric Cherenkov Telescopes (IACTs) will open the regime between $\approx$ 30 GeV and 200 GeV to ground-based gamma observations with unprecedented point source sensitivity and source location accuracy. I examine the prospects of observing the unidentified objects (UNIDs) of the Third EGRET Catalog using the IACT observatories currently under construction by the CANGAROO, HESS, MAGIC and VERITAS collaborations. Assuming a modest spectral steepening similar to that observed in the inverse Compton component of the Crab Nebula spectrum and taking into account the sensitivity of the instruments and its zenith angle dependence, a detailed list of 78 observable objects is derived which is then further constrained to 38 prime candidates. The characteristics of this agenda are discussed.

1. INTRODUCTION

The EGRET experiment (1991-1999) has given us the first detailed view of the entire high-energy gamma-ray sky. Its successful history is described by other authors (e.g. D.J. Thompson, these proceedings). In parallel with EGRET's observations, observers also began to explore the gamma-ray sky at even higher energies. The first successful observation of a gamma-ray source above 500 GeV (the Crab Nebula, Weekes et al. 1989) using a Cherenkov telescope was made only few years before EGRET's launch and one year after it followed the detection of the second source, Mkn 421 (Punch et al. 1992). Both sources were also

A. Carramiñana et al. (eds.),
The Nature of Unidentified Galactic High-energy Gamma-ray Sources, 299–319.

detected by EGRET and it has become common practice to show the EGRET flux measurement as a reference in plots of Cherenkov telescope flux measurements.

However, it soon became clear that an EGRET detection did not imply that the source would also be detected at very high energies. The class of EGRET Blazars has been thoroughly observed by Cherenkov telescopes, but none has been detected except Mkn 421 although extrapolations of the EGRET spectra are in many cases comfortably above the point source sensitivity of the present instruments. The second extragalactic source ever to be detected at very high energies, Mkn 501, was initially not an EGRET source. Only by raising the energy threshold and looking for flares, Kataoka et al. (1999) were able to report a marginal detection. Today, the situation remains essentially the same. About 11 very high energy (VHE) gamma-ray sources have been detected: 4 Blazars (Mkn 421, Mkn 501, 1ES2344+514, PKS2155-304) and 7 SNRs/Plerions (Crab Nebula, PSR1706-44, SN1006, Cas A, Vela, PSR B1509-58, SNR 347.3-00.5) (see Figure 1). About half of these are EGRET sources.

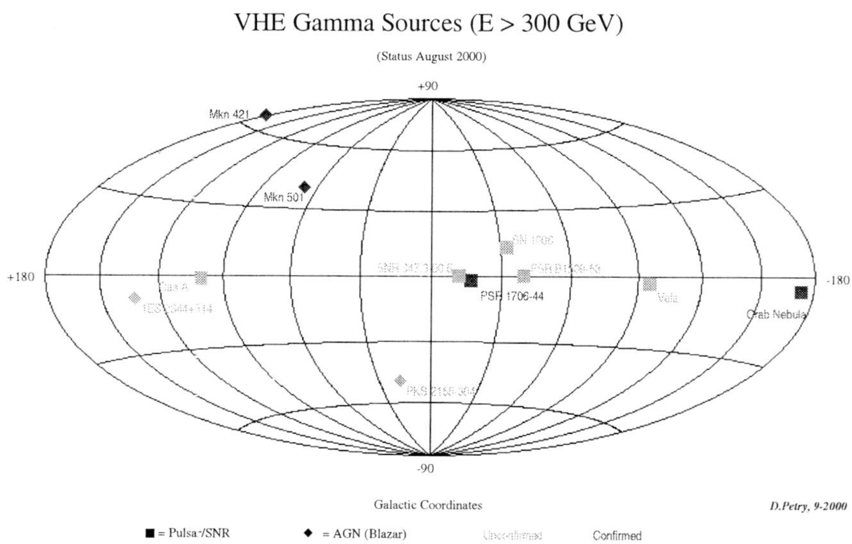

Figure 1 A sky map of the presently known VHE gamma-ray sources.

We are therefore looking at three classes of very high energy sources: (a) steeply cutting-off sources which are detected by EGRET but become unobservable above a few 100 GeV, (b) flat spectrum sources like Mkn 501 which only become observable at higher energies, and (c) intermediate cases like Mkn 421 or the Crab Nebula. Most known sources must belong to type (a) given the way observation technology has developed. Whether this is really a selection effect is unclear. It seems that the universe becomes abruptly very much darker above a few 100 GeV. The number of detectable sources decreases rapidly with rising energy threshold even though the point source sensitivity of present Cherenkov telescopes is sufficiently many orders of magnitude better than that of EGRET to allow for a spectral index of 2.1 in differential photon flux. In other words, the cut-off of most high-energy source spectra seems to take place in the range 1 - 200 GeV. The first decade of energy in this range has been covered by EGRET and 57 sources have been detected (Lamb & Macomb 1997). The remaining 10-200 GeV have never been explored until today and it is this "gap" that forms the major incentive behind the projects for future, more sensitive Cherenkov telescopes.

According to the Third EGRET Catalog (Hartman et al. 1999), 197 out of the 271 objects in the catalog could so far not be identified with an optical, radio or X-ray counterpart with certainty. These sources are the Unidentified EGRET objects (UNIDs), 38 of which have already a tentative identification. The main reason for the non-identification of so many sources is the fact that their position is only poorly known because of EGRET's wide point spread function and difficulties with modeling the background gamma radiation in the galactic plane (see again Thompson, these proceedings).

For EGRET, the point spread function is described by the angle θ_{68}, the half-opening angle of a cone which contains 68 % of all spark chamber events caused by photons from one point-source. This angle is weakly energy dependent. It is 5.85° at 100 MeV and 1.7° at 1 GeV. The expected accuracy of the source location is estimated by $\theta_{68}/\sqrt{N}$ where N is the number of detected photons. However, due to systematic errors of the background model also stronger sources can have location errors of more than 1°.

Imaging Atmospheric Cherenkov Telescopes (IACTs) are ground-based detectors using the atmosphere as a tracker and calorimeter. They typically have point spread functions with $\theta_{68} < 0.16°$. The number of detected primary gamma-photons is typically > 100. Hence source locations with arcminute accuracy are possible whenever a source can be detected. This has already been demonstrated by several authors (see the examples in Figures 2 a and b). The possibility of separating two nearby point-sources is demonstrated by Figure 2 (b): Two sources can be separated if their angular distance is $> 3\theta_{68}$. Cherenkov telescopes therefore seem to be the adequate tool for improving the source location of EGRET sources.

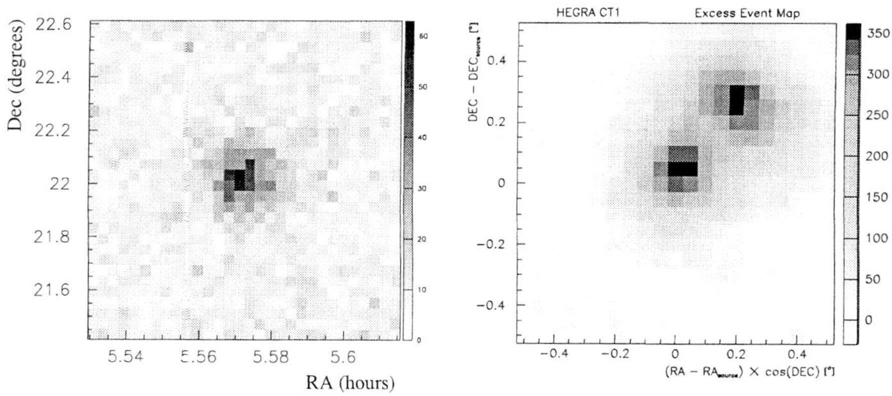

(a) (b)

Figure 2 Point source location and separation with Cherenkov telescopes: (a) Excess event map of the region around the Crab Nebula obtained by Pühlhofer et al. (1997) using the HEGRA System of Cherenkov telescopes. The theoretical accuracy of the location is 0.01°. The reconstructed position deviates from the radio position of the Crab Nebula by 0.01°. (b) Excess event map of a hypothetical double source obtained by superimposing on- and 0.3°-off-axis observations of the Blazar Mkn 501 using the HEGRA CT1 Cherenkov telescope (Petry 1997). The point spread function of CT1 is $\theta_{68} = 0.1°$ resulting in an expected location accuracy of 0.01° degrees. Taking into account a systematic tracking error of 0.1°, the position of Mkn 501 is correctly reconstructed from the on-axis observation. More important, the 0.3° angular distance between the two sources, which should be unaffected by the tracking error, is also correctly reconstructed within the expected 0.01°.

With their high photon statistics due to their large collection areas (typically $10^4 m^2$ at the threshold rising up to $10^5 m^2$ at larger zenith angles and higher energies) IACTs can also find pulsed emission and resolve fast time structures. This may be especially helpful since young pulsars (Kaaret & Cottam 1996) and Geminga-like radio-quiet pulsars (Yadigaroglu & Romani 1995) are among the candidate sources behind EGRET UNIDs.

In this article, I would like to explore what the next generation of IACTs with their improved sensitivity and lower energy threshold can achieve in terms of helping to identify the EGRET UNIDs. The result will mainly be a list of catalog objects for which observations with the new IACTs make sense and for which an identification can be expected. This defines their possible contribution to the solution of the EGRET UNID puzzle. As discussed in the next section, I will limit myself to the projects for IACTs with energy thresholds below 100 GeV. However, the exercise can easily be repeated for other experiments.

2. THE NEXT-GENERATION CHERENKOV TELESCOPES

Since the middle of the past decade, projects are underway to further improve the Cherenkov telescope technology and explore the 10-200 GeV regime with ground-based instruments. At the same time, the new telescopes are aiming to reach much improved sensitivity at higher energies.

In this article, I limit myself to those projects which use *imaging* atmospheric Cherenkov telescopes (IACTs), mostly because it has already been demonstrated that this technology can reach high flux sensitivity and high accuracy in determining source positions. I would like to mention, however, that other technologies are under development which may become competitive at some point, especially the non-imaging experiments using mirror arrays of former solar power plants (CELESTE (e.g. Smith et al. 1998), GRAAL (e.g. Arqueros et al. 1999), STACEE (e.g. Oser et al. 1999), Solar II (e.g. Zweerink et al. 1999) and the large-field-of-view, large-duty-cycle observatories MILAGRO (e.g. McCullough et al. 1999) and ARGO (not yet operational, see e.g. D'Ettore et al. 1999).

Table 1 The four projects for next-generation Imaging Cherenkov Telescopes.

Project name	latitude	longitude	altitude	CTs[a]	$E_{\rm thr}(0°)$[b]
CANGAROO III	31° S	137° E	160 m	4 × 10 m	80 GeV
HESS I	23° S	17° E	1800 m	4 × 13 m	40 GeV
MAGIC I	29° N	17° W	2200 m	1 × 17 m	30 GeV
VERITAS	32° N	111° W	1300 m	7 × 10 m	60 GeV

[a]Number of individual telescopes and diameter of their mirror dish.
[b]Predicted energy threshold for gammas at zenith angle $\vartheta = 0°$.

The next-generation projects for IACTs which will reach thresholds below 100 GeV are CANGAROO III (e.g. Mori et al. 1999), HESS I (e.g. Hofmann et al. 1999), MAGIC I (e.g. Lorenz et al. 1999) and VERITAS (e.g. Krennrich et al. 1999). The locations of the sites chosen for these observatories and their estimated minimum energy thresholds are given in Table 1. The locations are also shown in Figure 3. The roman numbers behind some of the project names denote the project phases.

All four experiments except CANGAROO III have published detailed Monte Carlo studies and presented a sensitivity curve. Figure 4 shows these sensitivity curves in the relevant energy range up to 1 TeV. CANGAROO III will reach a performance roughly similar to VERITAS (Mori

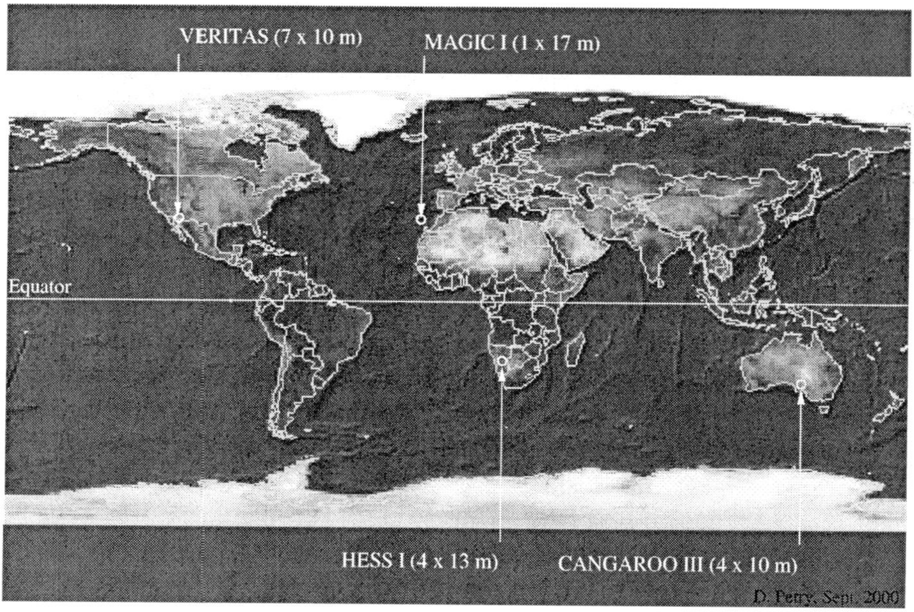

Figure 3 A world map indicating the locations of the four Imaging Atmospheric Cherenkov Telescope projects which may significantly contribute to the identification of EGRET UNIDs. See text for references.

2001). The sensitivity curves show the integral gamma-ray flux which the corresponding instrument can detect above the energy threshold E_0 with 20 % statistical error within an observation time of 50 h at small zenith angles. Also shown is the EGRET flux sensitivity for an observation time of 1 month taking into account the fact that the field of view and the duty cycle of EGRET is larger than that of an IACT.

Furthermore, Figure 4 indicates in which flux regime the bulk part of the EGRET unidentified objects (UNIDs) would be found if their spectrum would continue as a straight power law from the 100 MeV, at which it is measured by EGRET, to 100 GeV. Plotting the distribution of the logarithms of the extrapolated fluxes at 100 GeV gives a nicely symmetric, nearly Gaussian distribution with an average of -10.79 corresponding to 1.6×10^{-11} $\mathrm{cm^{-2}s^{-1}}$ and a standard deviation of 1.2. The dashed lines in Figure 4 outline the flux regime one standard deviation above and below this average.

3. OBSERVABILITY CRITERIA

Figure 4 is misleading as far as the actual observability of EGRET UNIDs by the IACTs under consideration is concerned. There are four

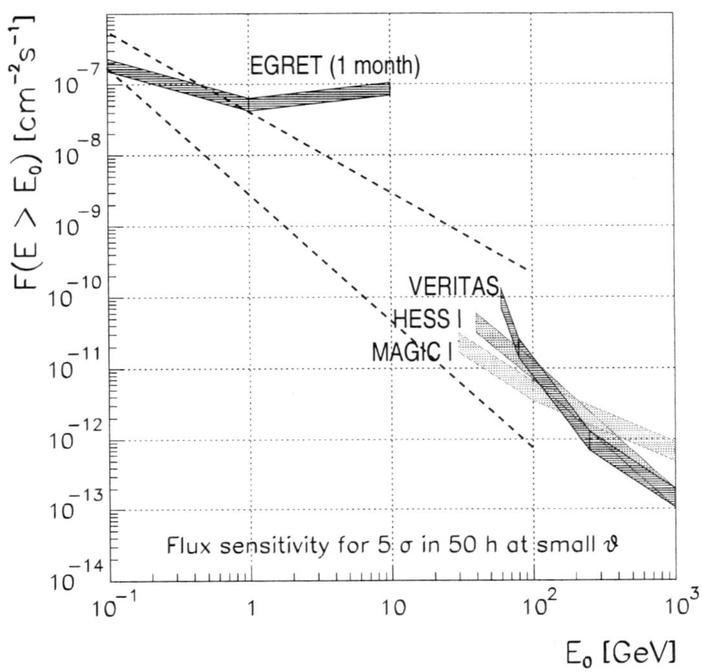

Figure 4 The sensitivity of HESS I, MAGIC I and VERITAS projects as given in Hofmann et al. (1999), Barrio et al. (1998) and Krennrich et al. (1999) respectively. Also shown is the EGRET sensitivity for one month of observing time. The dashed lines indicate the boundaries of the flux regime in which the central 68 % of the EGRET unidentified sources can be found if their 100 MeV fluxes are extrapolated to 100 GeV using the spectrum measured by EGRET.

aspects which have to be taken into account in assembling an observation agenda. These are discussed in the following.

3.1. SPECTRAL STEEPENING

The most obvious problem is that most EGRET sources will show spectral breaks above 100 MeV and the naive extrapolation of the power-law measured at around 100 MeV may be orders of magnitude above the real flux at the IACT threshold.

In order to predict observability more reliably, a certain spectral steepening has to be assumed. At the same time, no bias should be introduced towards any source class. As a first general approach, I will therefore assume a spectral steepening similar to that of the Crab Nebula which is the best studied high energy gamma source. This is also adequate since most of the EGRET UNIDs are expected to be galactic such that cut-offs due to the absorption in the intergalactic infrared background should not be assumed a priori.

The differential spectral index $\gamma_{0.1}$ of the Crab Nebula at 0.1 GeV is given in the Third EGRET Catalog as $\gamma_{0.1} = 2.19 \pm 0.02$ while it is measured by the Whipple telescope at 500 GeV to be $\gamma_{500} = 2.49 \pm 0.06 \pm 0.04$ (Hillas et al. 1998). However, since the EGRET energy range is situated at the transition of the synchrotron component to the Inverse-Compton (IC) component of the Crab spectrum, the index 2.19 is actually an interpolation between the steep synchrotron spectrum and the beginning flat IC spectrum (de Jager et al. 1996). Above 1 GeV, the IC component is dominant. Hillas et al. (1998) show that the steepening of the IC component towards higher energies can be described by an increase in spectral index of 0.15 per decade of energy.

Applied to the extrapolation of the EGRET UNID spectra, this would mean that the integral flux above the energy x [GeV], $10 < x < 1000$ can roughly be expected to be

$$
\begin{aligned}
F(E > x[\mathrm{GeV}]) &= F_0 \cdot 10^{-\alpha} \cdot 10^{-(\alpha+0.15)} \cdot (\tfrac{x}{10})^{-(\alpha+0.30)} \\
&= F_0 \cdot 10^{-(\alpha-0.15)} \cdot x^{-(\alpha+0.30)} \\
&\quad \text{for } \ 10 < x < 100 \\
F(E > x[\mathrm{GeV}]) &= F_0 \cdot 10^{-\alpha} \cdot 10^{-(\alpha+0.15)} \cdot 10^{-(\alpha+0.30)} \cdot (\tfrac{x}{100})^{-(\alpha+0.45)} \\
&= F_0 \cdot 10^{-(\alpha-0.45)} \cdot x^{-(\alpha+0.45)} \\
&\quad \text{for } \ 100 < x < 1000
\end{aligned}
\tag{22.1}
$$

where F_0 is the integral flux above 0.1 GeV of a given source taken from the Third EGRET Catalog and α is the integral spectral index derived by subtracting 1.0 from the differential spectral index stated in the same catalog. E.g. for $\alpha = 1.0$, the steepening leads to an extrapolated integral flux at 100 GeV which is smaller by a factor 2.8 than the extrapolation without steepening.

3.2. SKY ACCESS LIMITS

Due to the decaying optical quality of the atmosphere towards larger zenith angles (water vapor, aerosols, background light from towns etc.), Cherenkov telescopes typically cannot observe at zenith angles much larger than 70°. And even if they can, they will not give priority to observations of objects which can only be observed at zenith angles larger than this.

The zenith angle ϑ at the upper culmination of an astronomical object depends on the the latitude ϕ of the observatory and the declination DEC of the object according to $\vartheta = |\phi - \text{DEC}|$. Hence the condition

$$|\phi - \text{DEC}| < 70^\circ \tag{22.2}$$

has to be imposed in the selection of observable objects. In order to make sure that the objects can be observed for sufficient time at $\vartheta < 70^\circ$, the permitted declination-range has to be narrowed down to $|\phi - \text{DEC}| < 60^\circ$ for objects which are not circumpolar. For the four projects under consideration these conditions mean

$$\begin{array}{lrll} \text{CANGAROO III} & & \text{DEC} & < +29^\circ \\ \text{HESS I} & & \text{DEC} & < +37^\circ \\ \text{MAGIC I} & -31^\circ < & \text{DEC} & \\ \text{VERITAS} & -28^\circ < & \text{DEC} & \end{array} \tag{22.3}$$

Figure 5 shows the DEC distributions for the identified, tentatively identified and unidentified EGRET objects in the third catalog. The figure also indicates the DEC constraints for the four IACT projects. Due to the position of the galactic plane, the projects on the southern hemisphere clearly have better access to EGRET UNIDs. The distributions of identified and tentatively identified objects are nearly symmetrical.

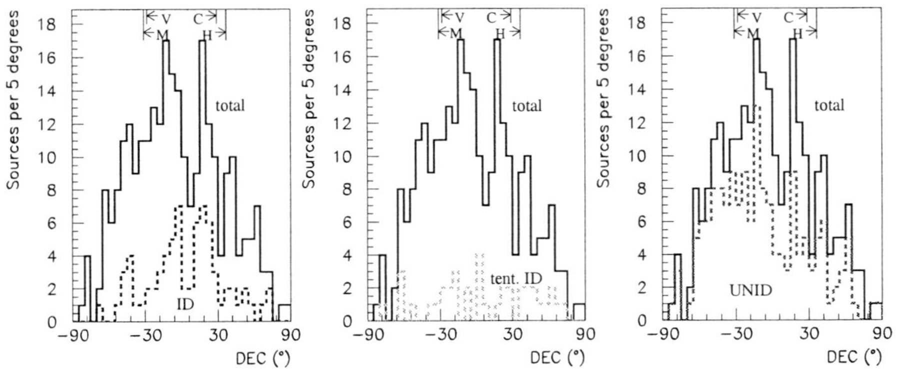

Figure 5 The distributions of the declination (DEC) of the objects in the third EGRET catalog. Left: the identified objects. Middle: the tentatively identified objects. Right: the unidentified objects. On all plots the distribution for the total number of sources is superimposed. At the top of each plot, arrows indicate the sky access limits of CANGAROO III (C), HESS I (H), MAGIC I (M) and VERITAS (V).

3.3. ZENITH-ANGLE DEPENDENCE OF ENERGY THRESHOLD AND FLUX SENSITIVITY

Figure 4 shows the sensitivity of the various IACT projects for small zenith angles ($\vartheta < 20°$) just like it shows the EGRET sensitivity for on-axis observations. Both effective collection area $A_{\rm eff}$ and energy threshold $E_{\rm thr}$ of an IACT are to a good approximation proportional to $\cos^{-2}(\vartheta)$ as explained in the following.

The area perpendicular to the optical axis illuminated by the Cherenkov light at the position of the telescope is proportional to the square of the distance d to the shower maximum. d grows with ϑ as $d \propto 1/\cos(\vartheta)$. The energy threshold $E_{\rm thr}$ of a Cherenkov telescope is, for a given zenith angle, to a good approximation equivalent to a Cherenkov photon density threshold $\rho_{\rm thr}$ at the position of the telescope. $\rho_{\rm thr}$ is an instrumental constant determined by the trigger condition of the data acquisition system. The Cherenkov photon density ρ at the position of the telescope is a function of the primary Energy E and of the zenith angle ϑ. The dependence on the impact parameter can be neglected since we average over it. For a given primary energy, ρ decreases inversely proportional to the illuminated area, i.e. $\rho(E,\vartheta) \propto \cos^2(\vartheta)$. At least for electromagnetic showers, the atmosphere is (again to a good approximation) a linear calorimeter yielding a Cherenkov photon density directly proportional to the primary energy of the incident gamma photon, i.e. for a fixed ϑ we have $\rho(E,\vartheta) \propto E$. Hence, $\rho(E,\vartheta) \propto E \cdot \cos^2(\vartheta)$. It follows that to satisfy the trigger condition $\rho(E,\vartheta) > \rho_{\rm thr}$, E has to increase with ϑ as $\cos^{-2}(\vartheta)$.

$$\begin{aligned} A_{\rm eff}(\vartheta) &= A_{\rm eff}(0°) \cdot \cos^{-2}(\vartheta) \\ E_{\rm thr}(\vartheta) &= E_{\rm thr}(0°) \cdot \cos^{-2}(\vartheta) \end{aligned} \quad (22.4)$$

The values for $E_{\rm thr}(0°)$ are shown in Table 1.

An increase in effective collection area for gammas is accompanied by a proportional increase in hadronic background rate. The gain in flux sensitivity is therefore only the square-root of the gain in area:

$$F_{5\sigma}(E_{\rm thr}(\vartheta),\vartheta) = F_{5\sigma}(E_{\rm thr}(0°),0°) \cdot \cos^{-1}(\vartheta) \quad (22.5)$$

where $F_{5\sigma}(E_{\rm thr})$ is the integral flux above energy threshold $E_{\rm thr}$ which results in a 5σ detection after 50 h of observation time.

3.4. REQUIRED OBSERVATION TIME

Given an integral flux $F(E_{\rm thr})$ above an energy threshold $E_{\rm thr}$ and a flux sensitivity expressed as $F_{5\sigma}(E_{\rm thr})$ as defined in the previous section, and assuming that the flux is so small that the signal is background dominated (the conservative assumption), we can calculate the observation time $T_{5\sigma}$ which is necessary to detect the flux with 5σ significance:

$$T_{5\sigma}(E_{\text{thr}}) = 50\text{h} \left(\frac{F(E_{\text{thr}})}{F_{5\sigma}(E_{\text{thr}})} \right)^{-2} \tag{22.6}$$

Cherenkov telescopes typically have a duty cycle of about 10 % due to the fact that they can only observe during good weather and at night (moon-less night for maximum sensitivity). This results typically in 1000 h of available observation time per year. For an individual source, rarely more than 100 h can be obtained.

4. THE AGENDA

Based on the observability criteria developed above, UNIDs can now be selected for an observation agenda. In the following, those sources are preselected which may be detected in less than 50 h of observation time. These are shown in Tables 2, 3, 4, and 5 for further reference. The prime candidates are then selected requiring that a detection is possible even if the spectrum at 0.1 GeV is steeper than the nominal value by one standard deviation. The prime candidates are summarized in Table 6. A discussion follows in section 5.

4.1. PRESELECTION

Using the formulae derived in section 3, Tables 2, 3, 4, and 5 are compiled. All four tables have the same format. The tables are generated as follows: For each non-identified source of the third catalog and for each of the four IACT projects, the minimum zenith angle of observation ϑ_{min} is calculated using

$$\vartheta_{\text{min}} = |\phi - DEC| \tag{22.7}$$

and the corresponding minimum energy threshold $E_{\text{thr}}(\vartheta_{\text{min}})$ is determined according to equation 22.4. ϑ_{min} and the corresponding threshold are shown in the tables. If a source is not observable according to equation 22.3, it is omitted. Otherwise, the threshold is inserted into equation 22.1 to determine the extrapolated flux $F(E_{\text{thr}})$. From this flux, the necessary observation time $T_{5\sigma}$ is determined using equation 22.6. Flux and observation time together with the spectral index $\alpha(E_{\text{thr}})$ which the extrapolated spectrum would have at the IACT threshold, are also shown in the table. If $T_{5\sigma}$ is more than 50 h, the source is omitted.

In order to take into account the error of the EGRET spectral measurement and to explore a realistic worst case, the procedure is repeated assuming an initial spectral index at 0.1 GeV which is larger by one standard deviation than the nominal value. The resulting modified values for $F(E_{\rm thr})$, $\alpha(E_{\rm thr})$ and $T_{5\sigma}$ are given in brackets after their nominal values. The modified value of $T_{5\sigma}$ is named $T'_{5\sigma}$ in the following.

For the case of VERITAS with its steep dependence of $F_{5\sigma}$ on E below 80 GeV (see Figure 4), it turns out that for all sources the observation time can be reduced if the minimum threshold is raised to 80 GeV before entering the calculation. The table contains the values for this case.

After $T_{5\sigma}$ is calculated, the number of detected primary gamma photons is calculated conservatively assuming an effective collection area of 10^4 m^2. If this number is smaller than 100, the value of $T_{5\sigma}$ is increased such that the number becomes equal to 100. This ensures that the source location can be determined with sufficiently high accuracy to improve the EGRET measurement. The values are marked if this correction was necessary.

4.2. PRIME CANDIDATES

Table 6 shows those sources from Tables 2, 3, 4, and 5 which satisfy the additional restriction that the "modestly worst case" observation time $T'_{5\sigma}$ (see the previous section) is less than 50 h. These sources can be regarded as the prime candidates for allocating the valuable IACT observation time since only if a detection is possible, the EGRET location measurement can be improved. An upper limit on the flux value is only valuable if the observed object has been identified and models can be constrained.

5. DISCUSSION

The observability Tables 2, 3, 4, and 5 and the selection of prime candidates in Table 6 show that the next-generation IACTs may make a major contribution to the identification of EGRET UNIDs. Altogether 78 of them may be in the reach of the IACT sensitivity. Under the modest assumptions of spectral steepening made here, many sources may be detectable within few hours of observation time.

Among the 78 sources are 20 which have a tentative identification (given in the Third EGRET Calalog unless otherwise specified):

0010+7309 (SNR CTA1), 0241+6103 (LS I +61°303), 0617+2238 (SNR IC443), 0808+4844 (QSO 0809+483), 0812-0646 (PKS 0805-077), 0827-4247 (SNR Pup A) 0917+4427 (QSO 0917+449), 1009+4855 (QSO 1011+496), 1323+2200 (EQ 1324+224), 1410-6147 (SNR 312.4-00.4), 1627-2419 (rho Oph), 1800-2338 (SNR W 28), 1854-1514 (LS5039, a microquasar suggested by Paredes et al. (2000)), 1856+0114 (SNR W 44),

1903+0550 (SNR 040.4-00.5), 2020+4017 (SNR γCyg), 2100+6012 (4C 59.33), 2206+6602 (TXS 2206+650), 2255+1943 (QSO 2250+1926), and 2352+3752 (QSO 2346+385).

Some of these have already (unsuccessfully) been observed by present IACTs. They come as no surprise and will be included in observation schedules in any case since the known position of the possible counterpart makes the observation especially promising.

Among the 38 prime candidates shown in Table 6, only 11 of the tentatively identified objects are present, mostly SNR candidates. Observers will face the additional problem here that these sources may be extended in angular size which may increase the necessary observation time. This is also true for all other UNIDs.

Increases in the required observation time will also arise in those few cases where θ_{95}, the radius of the 3EG position probability map, is near the radius of the IACT trigger field-of-view (typically $> 1°$) and the true source position is accidentally on the boundary of the map. But most of the 78 selected UNIDs have θ_{95} well below 1° (Tables 2, 3, 4, and 5).

A decrease in necessary observation time may be possible in case a candidate counterpart-pulsar is known. The IACT sensitivity can be greatly increased by introducing the known pulsar ephemeris into the analysis provided the emission is still pulsed at the IACT's threshold.

A nice opportunity to cross-calibrate all four IACT observatories are the prime candidate objects 1837-0606 and 1856+0114 (SNR W44?) which are observable from all four locations at similar zenith angles (see Table 6).

In summary, the EGRET UNIDs are a promising field for the application of the next-generation IACTs. From the start of operations of the individual observatory it will probably take between 2 and 4 years until the between 16 and 27 prime candidates have been observed and analysed. This estimation takes into account that not the whole observation time will be dedicated to EGRET UNIDs. However, if the observatories cooperate and split the candidate list into four parts of 9 to 10 sources each, a coverage of all 38 prime candidates could be reached within one year since each observatory would only have to invest up to ≈ 500 h of observation time. The attempt to cover the remaining 40 non-prime candidates could equally be shared.

References

Arqueros, F., et al., 1999, in Kieda, D. et al. (eds.) Proc. "26th International Cosmic Ray Conference", 5, 215

Barrio, J.A., et al., 1998, report MPI-PhE/98-5, Max-Planck-Institute for Physics, Munich

de Jager, O.C., Harding, A.K., et al., 1996, ApJ, 457, 253

D'Ettore Piazzoli, B., et al., 1999, in Kieda, D. et al. (eds.) Proc. "26th International Cosmic Ray Conference", 2, 373

Hartman, R.C., et al., 1999, ApJ Suppl. Ser., 123, 79

Hillas, A.M., et al., 1998, ApJ, 503, 744

Hofmann, W., et al., 1999, in Dingus, B.L., et al. (eds.) Proc. "Towards a Major Atmospheric Cherenkov Detector VI", AIP Proc. 515, 500

Kaaret, P. & Cottam, J., 1996, ApJ, 462, L35

Kataoka, J., et al., 1999, ApJ, 514, 138

Krennrich, F., et al., 1999, in Dingus, B.L., et al. (eds.) Proc. "Towards a Major Atmospheric Cherenkov Detector VI", AIP Proc. 515, 515

Lamb, R.C. & Macomb, D.J., 1997, ApJ, 488, 872

Lorenz, E., et al., 1999, in Dingus, B.L., et al. (eds.) Proc. "Towards a Major Atmospheric Cherenkov Detector VI", AIP Proc. 515, 510

McCullough, J.F., et al., 1999, in Kieda, D. et al. (eds.) Proc. "26th International Cosmic Ray Conference", 2, 369

Mori, M., et al., 1999, in Dingus, B.L., et al. (eds.) Proc. "Towards a Major Atmospheric Cherenkov Detector VI", AIP Proc. 515, 485

Mori, M., 2001, private communication

Oser, S., et al., 1999, in Kieda, D. et al. (eds.) Proc. "26th International Cosmic Ray Conference", 3, 464

Paredes, J.M., et al., 2000, Science, 288, 2340

Petry, D., 1997, PhD thesis, report MPI-PhE/97-27, Max-Planck-Institute for Physics, Munich

Pühlhofer, G., et al., 1997, Astropart. Phys., 8, 101

Punch, M., et al., 1992, Nature, 358, 477

Smith, D.A., et al., 1998, in Aubourg, E. et al. (eds.) Proc. "19th Texas Symposium on Relativistic Astrophysics", Nucl. Phys. B Proc. 80

Weekes, T.C., et al., 1989, ApJ, 342, 379

Yadigaroglu, I.-A. & Romani, R.W., 1995, ApJ, 449, 211

Zweerink, J.A., et al., 1999, in Kieda, D. et al. (eds.) Proc. "26th International Cosmic Ray Conference", 5, 223

Table 2 Observability of unidentified and tentatively identified EGRET sources for project CANGAROO III

object name	3EG θ_{95}[b] [°]	α[c]	CANGAROO III observability $\vartheta_{\min}$[d] [°]	E_{thr}[e] [GeV]	$F(E_{\mathrm{thr}})$[f] [$\mathrm{cm^{-2}s^{-1}}$]	$\alpha(E_{\mathrm{thr}})$[g]	$T_{5\sigma}$[h] [h]
0215+1123	1.06	1.03 ± 0.62	42	147	2.95E-11(3.21E-13)	1.48(2.10)	42(> 500)
0616-3310	0.63	1.11 ± 0.24	2	80	2.86E-11(5.74E-12)	1.41(1.65)	25(> 500)
0617+2238[a]	0.13	1.01 ± 0.06	54	228	5.13E-11(3.22E-11)	1.46(1.52)	22(55)
0631+0642	0.46	1.06 ± 0.15	38	128	4.12E-11(1.41E-11)	1.51(1.66)	19(161)
0634+0521	0.67	1.03 ± 0.26	36	123	5.39E-11(8.47E-12)	1.48(1.74)	11(430)
0706-3837	0.90	1.30 ± 0.43	8	81	3.23E-11(1.81E-12)	1.60(2.03)	20(> 500)
0747-3412	0.70	1.22 ± 0.30	3	80	3.09E-11(4.16E-12)	1.52(1.82)	21(> 500)
0824-4610	0.61	1.36 ± 0.07	15	86	2.43E-11(1.51E-11)	1.66(1.73)	36(94)
0827-4247[a]	0.77	1.10 ± 0.12	12	83	9.75E-11(4.35E-11)	1.40(1.52)	2.85‡(11)
0841-4356	0.52	1.15 ± 0.09	13	84	1.14E-10(6.23E-11)	1.45(1.54)	2.43‡(5.43)
0848-4429	0.62	1.05 ± 0.16	14	85	2.31E-10(7.87E-11)	1.35(1.51)	1.20‡(3.53‡)
1014-5705	0.67	1.23 ± 0.20	26	99	4.73E-11(1.19E-11)	1.53(1.73)	11(175)
1027-5817	0.37*	0.94 ± 0.09	27	101	3.48E-10(1.87E-10)	1.39(1.48)	0.80‡(1.49‡)
1048-5840	0.17	0.97 ± 0.09	28	102	2.62E-10(1.41E-10)	1.42(1.51)	1.06‡(1.98‡)
1234-1318	0.76	1.09 ± 0.24	18	88	4.91E-11(9.63E-12)	1.39(1.63)	9.16(237)
1323+2200[a]	0.47	0.86 ± 0.35	53	221	5.98E-11(4.04E-12)	1.31(1.66)	15(> 500)
1410-6147[a]	0.36	1.12 ± 0.14	31	108	8.78E-11(3.30E-11)	1.57(1.71)	3.51(25)
1420-6038	0.32	1.02 ± 0.14	30	106	2.10E-10(7.91E-11)	1.47(1.61)	1.33‡(4.24)
1627-2419[a]	0.65	1.21 ± 0.27	7	81	2.67E-11(4.38E-12)	1.51(1.78)	28(> 500)
1631-4033	0.89	1.25 ± 0.27	10	82	2.24E-11(3.65E-12)	1.55(1.82)	41(> 500)
1655-4554	0.66	1.19 ± 0.24	15	86	4.63E-11(9.16E-12)	1.49(1.73)	9.99(255)
1704-4732	0.66	0.86 ± 0.33	17	87	1.29E-09(1.38E-10)	1.16(1.49)	0.22‡(2.01‡)
1714-3857	0.51	1.30 ± 0.20	8	82	2.70E-11(7.06E-12)	1.60(1.80)	28(409)
1736-2908	0.62	1.18 ± 0.12	2	80	7.32E-11(3.28E-11)	1.48(1.60)	3.79‡(19)
1741-2050	0.63	1.25 ± 0.12	10	83	2.05E-11(9.14E-12)	1.55(1.67)	49(247)
1744-3011	0.32	1.17 ± 0.08	1	80	9.72E-11(5.70E-11)	1.47(1.55)	2.86‡(6.17)
1746-2851	0.13	0.70 ± 0.07	2	80	4.22E-09(2.64E-09)	1.00(1.07)	0.07‡(0.11‡)
1800-2338[a]	0.32	1.10 ± 0.10	7	81	1.46E-10(7.45E-11)	1.40(1.50)	1.91‡(3.73‡)
1809-2328	0.16	1.06 ± 0.08	8	81	1.29E-10(7.57E-11)	1.36(1.44)	2.15‡(3.67‡)
1812-1316	0.39	1.29 ± 0.11	18	88	2.65E-11(1.26E-11)	1.59(1.70)	31(139)
1824-1514	0.52	1.19 ± 0.18	16	86	4.18E-11(1.24E-11)	1.49(1.67)	12(141)
1826-1302	0.46	1.00 ± 0.11	18	88	2.78E-10(1.32E-10)	1.30(1.41)	1.00‡(2.11‡)
1837-0606	0.19	0.82 ± 0.14	25	97	6.30E-10(2.40E-10)	1.12(1.26)	0.44‡(1.16‡)
1850-2652	1.00	1.29 ± 0.45	4	80	6.08E-11(2.99E-12)	1.59(2.04)	5.44(> 500)
1856+0114[a]	0.19	0.93 ± 0.10	32	112	3.33E-10(1.65E-10)	1.38(1.48)	0.83‡(1.68‡)
1928+1733	0.75	1.23 ± 0.32	49	183	4.13E-11(3.74E-12)	1.68(2.00)	27(> 500)

For definitions see the end of table 5.

Table 3 Observability of unidentified and tentatively identified EGRET sources for project HESS I

object name	3EG θ_{95}[b] [°]	α[c]	$\vartheta_{\min}$[d] [°]	$E_{\rm thr}$[e] [GeV]	HESS I observability $F(E_{\rm thr})$[f] [$\rm cm^{-2}s^{-1}$]	$\alpha(E_{\rm thr})$[g]	$T_{5\sigma}$[h] [h]
0215+1123	1.06	1.03 ± 0.62	34	59	1.05E-10(2.02E-12)	1.33(1.95)	13(> 500)
0616-3310	0.63	1.11 ± 0.24	10	41	7.28E-11(1.72E-11)	1.41(1.65)	20(355)
0617+2238[a]	0.13	1.01 ± 0.06	46	82	2.21E-10(1.48E-10)	1.31(1.37)	4.22(9.44)
0631+0642	0.46	1.06 ± 0.15	30	53	1.42E-10(5.53E-11)	1.36(1.51)	6.69(44)
0634+0521	0.67	1.03 ± 0.26	28	52	1.77E-10(3.49E-11)	1.33(1.59)	4.17(107)
0706-3837	0.90	1.30 ± 0.43	16	43	8.92E-11(6.57E-12)	1.60(2.03)	14(> 500)
0747-3412	0.70	1.22 ± 0.30	11	42	8.40E-11(1.38E-11)	1.52(1.82)	15(> 500)
0824-4610	0.61	1.36 ± 0.07	23	47	6.53E-11(4.24E-11)	1.66(1.73)	28(67)
0827-4247[a]	0.77	1.10 ± 0.12	20	45	2.30E-10(1.11E-10)	1.40(1.52)	2.15(9.34)
0841-4356	0.52	1.15 ± 0.09	21	46	2.76E-10(1.59E-10)	1.45(1.54)	1.53(4.61)
0848-4429	0.62	1.05 ± 0.16	22	46	5.24E-10(1.96E-10)	1.35(1.51)	0.53‡(3.04)
0859-4257	0.64	1.32 ± 0.20	20	45	5.09E-11(1.50E-11)	1.62(1.82)	44(> 500)
1014-5705	0.67	1.23 ± 0.20	34	58	1.07E-10(2.98E-11)	1.53(1.73)	13(166)
1027-5817	0.37*	0.94 ± 0.09	35	60	6.67E-10(3.75E-10)	1.24(1.33)	0.42‡(1.08)
1048-5840	0.17	0.97 ± 0.09	36	61	5.09E-10(2.86E-10)	1.27(1.36)	0.59(1.87)
1234-1318	0.76	1.09 ± 0.24	10	41	1.41E-10(3.33E-11)	1.39(1.63)	5.22(94)
1323+2200[a]	0.47	0.86 ± 0.35	45	80	2.19E-10(2.11E-11)	1.16(1.51)	4.23(456)
1410-6147[a]	0.36	1.12 ± 0.14	39	66	1.80E-10(7.28E-11)	1.42(1.56)	5.11(31)
1420-6038	0.32	1.02 ± 0.14	38	64	4.13E-10(1.67E-10)	1.32(1.46)	0.95(5.78)
1627-2419[a]	0.65	1.21 ± 0.27	1	40	7.76E-11(1.54E-11)	1.51(1.78)	17(428)
1631-4033	0.89	1.25 ± 0.27	18	44	5.90E-11(1.14E-11)	1.55(1.82)	32(> 500)
1634-1434	0.49*	1.15 ± 0.23	8	41	5.30E-11(1.33E-11)	1.45(1.68)	37(> 500)
1655-4554	0.66	1.19 ± 0.24	23	47	1.13E-10(2.57E-11)	1.49(1.73)	9.38(180)
1704-4732	0.66	0.86 ± 0.33	25	48	2.55E-09(3.32E-10)	1.16(1.49)	0.11‡(1.11)
1714-3857	0.51	1.30 ± 0.20	16	43	7.44E-11(2.21E-11)	1.60(1.80)	20(224)
1717-2737	0.64	1.23 ± 0.15	5	40	5.19E-11(2.11E-11)	1.53(1.68)	38(229)
1719-0430	0.44	1.20 ± 0.24	18	44	4.87E-11(1.13E-11)	1.50(1.74)	47(> 500)
1736-2908	0.62	1.18 ± 0.12	6	40	2.01E-10(9.78E-11)	1.48(1.60)	2.53(11)
1741-2050	0.63	1.25 ± 0.12	2	40	6.28E-11(3.06E-11)	1.55(1.67)	26(108)
1744-3011	0.32	1.17 ± 0.08	7	41	2.63E-10(1.63E-10)	1.47(1.55)	1.48(3.88)
1746-2851	0.13	0.70 ± 0.07	6	40	8.36E-09(5.49E-09)	1.00(1.07)	0.03‡(0.05‡)
1800-2338[a]	0.32	1.10 ± 0.10	1	40	3.93E-10(2.16E-10)	1.40(1.50)	0.71‡(2.17)
1809-2328	0.16	1.06 ± 0.08	0	40	3.40E-10(2.10E-10)	1.36(1.44)	0.88(2.29)
1810-1032	0.39*	1.29 ± 0.16	12	42	6.17E-11(2.35E-11)	1.59(1.75)	28(193)
1812-1316	0.39	1.29 ± 0.11	10	41	8.91E-11(4.59E-11)	1.59(1.70)	13(49)
1824-1514	0.52	1.19 ± 0.18	8	41	1.28E-10(4.34E-11)	1.49(1.67)	6.28(55)
1826-1302	0.46	1.00 ± 0.11	10	41	7.49E-10(3.86E-10)	1.30(1.41)	0.37‡(0.72‡)

continued on the next page

Table 3 (part 2) Observability of unidentified and tentatively identified EGRET sources for project HESS I

3EG object name	$\theta_{95}{}^{b}$ [°]	α^{c}	$\vartheta_{\min}{}^{d}$ [°]	$E_{\mathrm{thr}}{}^{e}$ [GeV]	HESS I observability $F(E_{\mathrm{thr}})^{f}$ [$\mathrm{cm^{-2}s^{-1}}$]	$\alpha(E_{\mathrm{thr}})^{g}$	$T_{5\sigma}{}^{h}$ [h]
1837-0606	0.19	0.82 ± 0.14	17	44	1.54E-09(6.59E-10)	1.12(1.26)	$0.18^{\ddagger}(0.42^{\ddagger})$
1850-2652	1.00	1.29 ± 0.45	4	40	1.83E-10(1.23E-11)	1.59(2.04)	3.03(> 500)
1856+0114[a]	0.19	0.93 ± 0.10	24	48	9.55E-10(5.15E-10)	1.23(1.33)	$0.29^{\ddagger}(0.54^{\ddagger})$
1928+1733	0.75	1.23 ± 0.32	41	69	1.99E-10(2.46E-11)	1.53(1.85)	4.42(290)
1958+2909	0.57	0.85 ± 0.20	52	106	2.48E-10(6.17E-11)	1.30(1.50)	4.36(71)

For definitions see the end of table 5.

Table 4 Observability of unidentified and tentatively identified EGRET sources for project MAGIC I

3EG object name	$\theta_{95}{}^{b}$ [°]	α^{c}	$\vartheta_{\min}{}^{d}$ [°]	$E_{\mathrm{thr}}{}^{e}$ [GeV]	MAGIC I observability $F(E_{\mathrm{thr}})^{f}$ [$\mathrm{cm^{-2}s^{-1}}$]	$\alpha(E_{\mathrm{thr}})^{g}$	$T_{5\sigma}{}^{h}$ [h]
0010+7309[a]	0.24	0.85 ± 0.10	44	58	7.87E-10(4.16E-10)	1.15(1.25)	$0.35^{\ddagger}(0.67^{\ddagger})$
0215+1123	1.06	1.03 ± 0.62	18	33	2.27E-10(6.22E-12)	1.33(1.95)	$1.23^{\ddagger}$(> 500)
0229+6151	0.49	1.29 ± 0.18	33	43	7.07E-11(2.38E-11)	1.59(1.77)	8.17(72)
0241+6103[a]	0.18	1.21 ± 0.07	32	42	2.15E-10(1.41E-10)	1.51(1.58)	$1.29^{\ddagger}$(2.01)
0323+5122	0.55	1.38 ± 0.41	22	35	3.67E-11(3.32E-12)	1.68(2.09)	25(> 500)
0348+3510	0.74	1.16 ± 0.27	6	30	7.70E-11(1.65E-11)	1.46(1.73)	4.91(108)
0426+1333	0.45*	1.17 ± 0.25	15	32	8.09E-11(1.91E-11)	1.47(1.72)	4.74(85)
0439+1555	0.92	1.27 ± 0.44	13	32	1.44E-10(1.14E-11)	1.57(2.01)	$1.93^{\ddagger}$(233)
0510+5545	0.71	1.19 ± 0.20	27	38	8.72E-11(2.66E-11)	1.49(1.69)	4.75(51)
0613+4201	0.57	0.92 ± 0.26	13	32	2.26E-10(5.07E-11)	1.22(1.48)	$1.23^{\ddagger}$(12)
0617+2238[a]	0.13	1.01 ± 0.06	6	30	8.11E-10(5.75E-10)	1.31(1.37)	$0.34^{\ddagger}(0.48^{\ddagger})$
0628+1847	0.57	1.30 ± 0.10	10	31	6.96E-11(3.92E-11)	1.60(1.70)	6.13(19)
0631+0642	0.46	1.06 ± 0.15	22	35	2.49E-10(1.03E-10)	1.36(1.51)	$1.12^{\ddagger}$(3.15)
0634+0521	0.67	1.03 ± 0.26	24	36	2.89E-10(6.26E-11)	1.33(1.59)	$0.96^{\ddagger}$(8.75)
0808+4844[a]	0.72	1.15 ± 0.45	20	34	6.47E-11(4.71E-12)	1.45(1.90)	7.76(> 500)
0812-0646[a]	0.72	1.34 ± 0.29	36	46	3.12E-11(5.29E-12)	1.64(1.93)	45(> 500)
0910+6556	0.86	1.20 ± 0.26	37	47	5.07E-11(1.02E-11)	1.50(1.76)	18(430)
0917+4427[a]	0.56	1.19 ± 0.14	15	32	7.10E-11(3.16E-11)	1.49(1.63)	6.15(31)
1009+4855[a]	0.75*	0.90 ± 0.37	20	34	1.48E-10(1.71E-11)	1.20(1.57)	$1.88^{\ddagger}$(112)

continued on the next page

Table 4 (part 2) Observability of unidentified and tentatively identified EGRET sources for project MAGIC I

object name	3EG θ_{95} [b] [°]	α [c]	$\vartheta_{\min}$ [d] [°]	$E_{\rm thr}$ [e] [GeV]	MAGIC I observability $F(E_{\rm thr})$ [f] [$\rm cm^{-2}s^{-1}$]	$\alpha(E_{\rm thr})$ [g]	$T_{5\sigma}$ [h] [h]
1234-1318	0.76	1.09 ± 0.24	42	55	9.48E-11(2.09E-11)	1.39(1.63)	5.86(121)
1323+2200[a]	0.47	0.86 ± 0.35	7	30	6.71E-10(9.07E-11)	1.16(1.51)	$0.41^{\ddagger}(3.56)$
1337+5029	0.72	0.83 ± 0.29	21	35	3.50E-10(6.42E-11)	1.13(1.42)	$0.79^{\ddagger}(8.07)$
1631-1018	0.72	1.20 ± 0.27	39	50	3.19E-11(5.96E-12)	1.50(1.77)	47(> 500)
1719-0430	0.44	1.20 ± 0.24	34	43	5.09E-11(1.19E-11)	1.50(1.74)	16(294)
1736-2908	0.62	1.18 ± 0.12	58	108	4.67E-11(2.02E-11)	1.63(1.75)	47(253)
1744-3011	0.32	1.17 ± 0.08	59	114	5.64E-11(3.21E-11)	1.62(1.70)	35(107)
1746-2851	0.13	0.70 ± 0.07	58	106	3.16E-09(1.94E-09)	1.15(1.22)	$0.09^{\ddagger}(0.14^{\ddagger})$
1800-2338[a]	0.32	1.10 ± 0.10	53	82	1.45E-10(7.42E-11)	1.40(1.50)	3.71(14)
1809-2328	0.16	1.06 ± 0.08	52	81	1.31E-10(7.64E-11)	1.36(1.44)	4.55(13)
1810-1032	0.39*	1.29 ± 0.16	40	50	4.60E-11(1.70E-11)	1.59(1.75)	23(167)
1812-1316	0.39	1.29 ± 0.11	42	55	5.66E-11(2.83E-11)	1.59(1.70)	16(66)
1824+3441	0.82	1.03 ± 0.50	6	30	4.05E-10(2.33E-11)	1.33(1.83)	$0.69^{\ddagger}(54)$
1824-1514	0.52	1.19 ± 0.18	44	58	7.48E-11(2.38E-11)	1.49(1.67)	10(99)
1826-1302	0.46	1.00 ± 0.11	42	54	5.22E-10(2.61E-10)	1.30(1.41)	$0.53^{\ddagger}(1.06^{\ddagger})$
1835+5918	0.15	0.69 ± 0.07	30	40	4.50E-09(2.96E-09)	0.99(1.06)	$0.06^{\ddagger}(0.09^{\ddagger})$
1837-0423	0.52	1.71 ± 0.44	33	43	4.44E-11(3.08E-12)	2.01(2.45)	21(> 500)
1837-0606	0.19	0.82 ± 0.14	35	45	1.50E-09(6.38E-10)	1.12(1.26)	$0.19^{\ddagger}(0.44^{\ddagger})$
1850-2652	1.00	1.29 ± 0.45	56	95	4.64E-11(2.12E-12)	1.59(2.04)	43(> 500)
1856+0114[a]	0.19	0.93 ± 0.10	28	38	1.26E-09(6.97E-10)	1.23(1.33)	$0.22^{\ddagger}(0.40^{\ddagger})$
1903+0550[a]	0.64	1.38 ± 0.17	23	35	9.10E-11(3.35E-11)	1.68(1.85)	4.12(30)
1928+1733	0.75	1.23 ± 0.32	11	31	6.75E-10(1.07E-10)	1.53(1.85)	$0.41^{\ddagger}(2.60)$
1958+2909	0.57	0.85 ± 0.20	0	30	1.07E-09(3.43E-10)	1.15(1.35)	$0.26^{\ddagger}(0.81^{\ddagger})$
2016+3657	0.55	1.09 ± 0.11	8	31	3.43E-10(1.83E-10)	1.39(1.50)	$0.81^{\ddagger}(1.52^{\ddagger})$
2020+4017[a]	0.16	1.08 ± 0.04	11	31	1.26E-09(1.00E-09)	1.38(1.42)	$0.22^{\ddagger}(0.28^{\ddagger})$
2021+3716	0.30	0.86 ± 0.10	8	31	2.18E-09(1.23E-09)	1.16(1.26)	$0.13^{\ddagger}(0.23^{\ddagger})$
2022+4317	0.72	1.31 ± 0.19	14	32	6.47E-11(2.16E-11)	1.61(1.80)	7.34(66)
2027+3429	0.77	1.28 ± 0.15	6	30	8.77E-11(3.72E-11)	1.58(1.73)	3.78(21)
2033+4118	0.28	0.96 ± 0.10	12	31	1.47E-09(8.26E-10)	1.26(1.36)	$0.19^{\ddagger}(0.34^{\ddagger})$
2035+4441	0.54	1.08 ± 0.26	16	32	2.83E-10(6.29E-11)	1.38(1.64)	$0.98^{\ddagger}(7.85)$
2046+0933	0.60*	1.22 ± 0.51	19	34	8.42E-11(4.33E-12)	1.52(2.03)	4.57(> 500)
2100+6012[a]	0.48	1.21 ± 0.25	31	41	6.33E-11(1.41E-11)	1.51(1.76)	9.84(199)
2206+6602[a]	0.88	1.29 ± 0.26	37	47	3.87E-11(7.80E-12)	1.59(1.85)	30(> 500)
2227+6122	0.46	1.24 ± 0.14	32	42	1.06E-10(4.55E-11)	1.54(1.68)	3.59(20)
2248+1745	0.94	1.11 ± 0.39	11	31	1.11E-10(1.18E-11)	1.41(1.80)	$2.51^{\ddagger}(215)$
2255+1943[a]	0.87*	1.36 ± 0.61	9	31	1.30E-10(3.93E-12)	1.66(2.27)	$2.14^{\ddagger}(> 500)$
2314+4426	0.78	1.34 ± 0.32	15	32	8.74E-11(1.38E-11)	1.64(1.96)	4.06(164)
2352+3752[a]	0.94	1.47 ± 0.68	9	31	4.18E-11(8.50E-13)	1.77(2.45)	17(> 500)

For definitions see the end of table 5.

Table 5 Observability of unidentified and tentatively identified EGRET sources for project VERITAS

	3EG		VERITAS observability				
object name	θ_{95}[b] [°]	α[c]	$\vartheta_{\min}$[d] [°]	$E_{\rm thr}$[e] [GeV]	$F(E_{\rm thr})$[f] [cm^{-2}s^{-1}]	$\alpha(E_{\rm thr})$[g]	$T_{5\sigma}$[h] [h]
0010+7309[a]	0.24	0.85 ± 0.10	41	141	2.70E-10(1.31E-10)	1.30(1.40)	1.03‡(2.12‡)
0215+1123	1.06	1.03 ± 0.62	21	91	5.86E-11(8.55E-13)	1.33(1.95)	4.74‡(> 500)
0229+6151	0.49	1.29 ± 0.18	30	106	1.63E-11(4.65E-12)	1.74(1.92)	25(308)
0241+6103[a]	0.18	1.21 ± 0.07	29	105	5.34E-11(3.28E-11)	1.66(1.73)	5.20‡(8.47‡)
0348+3510	0.74	1.16 ± 0.27	3	80	1.86E-11(3.06E-12)	1.46(1.73)	15‡(> 500)
0426+1333	0.45*	1.17 ± 0.25	18	89	1.83E-11(3.34E-12)	1.47(1.72)	17(497)
0439+1555	0.92	1.27 ± 0.44	16	87	2.95E-11(1.51E-12)	1.57(2.01)	9.41‡(> 500)
0510+5545	0.71	1.19 ± 0.20	24	96	2.18E-11(5.52E-12)	1.49(1.69)	13‡(196)
0613+4201	0.57	0.92 ± 0.26	10	82	7.02E-11(1.22E-11)	1.22(1.48)	3.96‡(34)
0617+2238[a]	0.13	1.01 ± 0.06	9	82	2.20E-10(1.47E-10)	1.31(1.37)	1.26‡(1.89‡)
0628+1847	0.57	1.30 ± 0.10	13	84	1.40E-11(7.14E-12)	1.60(1.70)	27(104)
0631+0642	0.46	1.06 ± 0.15	25	98	6.16E-11(2.19E-11)	1.36(1.51)	4.51‡(13)
0634+0521	0.67	1.03 ± 0.26	27	100	7.34E-11(1.22E-11)	1.48(1.74)	3.78‡(42)
0808+4844[a]	0.72	1.15 ± 0.45	17	87	1.64E-11(7.80E-13)	1.45(1.90)	20(> 500)
0910+6556	0.86	1.20 ± 0.26	34	116	1.27E-11(2.03E-12)	1.65(1.91)	45(> 500)
0917+4427[a]	0.56	1.19 ± 0.14	12	84	1.71E-11(6.67E-12)	1.49(1.63)	18(118)
1009+4855[a]	0.75*	0.90 ± 0.37	17	87	4.74E-11(3.87E-12)	1.20(1.57)	5.86‡(365)
1234-1318	0.76	1.09 ± 0.24	45	162	1.96E-11(3.33E-12)	1.54(1.78)	26(> 500)
1323+2200[a]	0.47	0.86 ± 0.35	10	82	2.11E-10(2.01E-11)	1.16(1.51)	1.32‡ (14‡)
1337+5029	0.72	0.83 ± 0.29	18	89	1.21E-10(1.68E-11)	1.13(1.42)	2.30‡(20)
1800-2338[a]	0.32	1.10 ± 0.10	56	251	2.61E-11(1.19E-11)	1.55(1.65)	23(110)
1809-2328	0.16	1.06 ± 0.08	55	249	2.47E-11(1.32E-11)	1.51(1.59)	26(89)
1824+3441	0.82	1.03 ± 0.50	3	80	1.11E-10(3.92E-12)	1.33(1.83)	2.50‡(326)
1826-1302	0.46	1.00 ± 0.11	45	160	1.19E-10(5.31E-11)	1.45(1.56)	2.33‡(5.24‡)
1835+5918	0.15	0.69 ± 0.07	27	101	1.80E-09(1.11E-09)	1.14(1.21)	0.15‡(0.25‡)
1837-0606	0.19	0.82 ± 0.14	38	129	4.41E-10(1.62E-10)	1.27(1.41)	0.63‡(1.72‡)
1856+0114[a]	0.19	0.93 ± 0.10	31	108	3.48E-10(1.73E-10)	1.38(1.48)	0.80‡(1.61‡)
1903+0550[a]	0.64	1.38 ± 0.17	26	99	1.62E-11(5.00E-12)	1.68(1.85)	24(249)
1928+1733	0.75	1.23 ± 0.32	14	85	1.45E-10(1.67E-11)	1.53(1.85)	1.92‡(19)
1958+2909	0.57	0.85 ± 0.20	3	80	3.47E-10(9.10E-11)	1.15(1.35)	0.80‡(3.05‡)
2016+3657	0.55	1.09 ± 0.11	5	81	8.92E-11(4.27E-11)	1.39(1.50)	3.11‡(6.50‡)
2020+4017[a]	0.16	1.08 ± 0.04	8	82	3.34E-10(2.55E-10)	1.38(1.42)	0.83‡(1.09‡)
2021+3716	0.30	0.86 ± 0.10	5	81	7.08E-10(3.62E-10)	1.16(1.26)	0.39‡(0.77‡)
2022+4317	0.72	1.31 ± 0.19	11	83	1.38E-11(3.86E-12)	1.61(1.80)	27(349)
2027+3429	0.77	1.28 ± 0.15	2	80	1.88E-11(6.91E-12)	1.58(1.73)	15‡(105)
2033+4118	0.28	0.96 ± 0.10	9	82	4.37E-10(2.24E-10)	1.26(1.36)	0.64‡(1.24‡)
2035+4441	0.54	1.08 ± 0.26	13	84	7.58E-11(1.32E-11)	1.38(1.64)	3.67‡(30)
2046+0933	0.60*	1.22 ± 0.51	22	94	1.78E-11(5.45E-13)	1.52(2.03)	18(> 500)

continued on the next page

Table 5 (part 2) Observability of unidentified and tentatively identified EGRET sources for project VERITAS

object name	3EG θ_{95}[b] [°]	α[c]	VERITAS observability $\vartheta_{\min}$[d] [°]	E_{thr}[e] [GeV]	$F(E_{\mathrm{thr}})$[f] [$\mathrm{cm}^{-2}\mathrm{s}^{-1}$]	$\alpha(E_{\mathrm{thr}})$[g]	$T_{5\sigma}$[h] [h]
2100+6012[a]	0.48	1.21 ± 0.25	28	103	1.57E-11(2.77E-12)	1.66(1.91)	26(> 500)
2227+6122	0.46	1.24 ± 0.14	29	105	2.56E-11(9.65E-12)	1.69(1.83)	11‡(71)
2248+1745	0.94	1.11 ± 0.39	14	85	2.69E-11(1.93E-12)	1.41(1.80)	10‡(> 500)
2255+1943[a]	0.87*	1.36 ± 0.61	12	84	2.46E-11(4.06E-13)	1.66(2.27)	11‡(> 500)
2314+4426	0.78	1.34 ± 0.32	12	84	1.83E-11(2.12E-12)	1.64(1.96)	16(> 500)

[a] The object has already been tentatively identified in 3EG (Hartman et al. 1999).
[b] The radius of the 95% confidence contour in the position probability map in 3EG.
[c] The integral spectral index at 0.1 GeV (see equ. 22.1) with statistical error from 3EG.
[d] The minimum zenith angle, see equ. 22.7.
[e] The IACTs minimum energy threshold for this source, see equ. 22.4.
[f] The expected flux at the minimum energy threshold, see equ. 22.1; in brackets the flux which is obtained from an extrapolation using an initial spectrum at 0.1 GeV steeper by one standard deviation.
[g] The integral spectral index at the IACTs minimum energy threshold for this source, see equ. 22.1; in brackets the index obtained if the spectrum at 0.1 GeV was steeper by one standard deviation.
[h] The observation time to obstain a detection with 5 σ significance; in brackets the observation time necessary if the spectrum at 0.1 GeV was steeper by one standard deviation.
* The shape of the position probability map is irregular (see 3EG).
‡ The detection is photon flux limited: the observation time was increased such that 100 gammas are detected.

Table 6 The prime candidates for EGRET-UNID observations by the next-generation Imaging Cherenkov Telescopes

3EG name	tentative ID	$\vartheta_{\min}(C)^a$ [°]	$\vartheta_{\min}(H)^b$ [°]	$\vartheta_{\min}(M)^c$ [°]	$\vartheta_{\min}(V)^d$ [°]
0010+7309	SNR CTA 1	-	-	44	41
0241+6103	LS I +61°303	-	-	32	29
0613+4201	-	-	-	13	10
0617+2238	SNR IC443	-	46	6	9
0628+1847	-	-	-	10	-
0631+0642	-	-	30	22	25
0634+0521	-	-	-	24	27
0827-4247	SNR Pup A	12	20	-	-
0841-4356	-	13	21	-	-
0848-4429	-	14	22	-	-
0917+4427	QSO 0917+449	-	-	15	-
1027-5817	-	27	35	-	-
1048-5840	-	28	36	-	-
1323+2200	EQ 1324+224	-	-	7	10
1337+5029	-	-	-	21	18
1410-6147	SNR 312.4-00.4	31	39	-	-
1420-6038	-	30	38	-	-
1704-4732	-	17	25	-	-
1736-2908	-	2	6	-	-
1744-3011	-	1	7	-	-
1746-2851	-	2	6	58	-
1800-2338	SNR W 28	7	1	53	-
1809-2328	-	8	0	52	-
1812-1316	-	-	10	-	-
1826-1302	-	18	10	42	45
1835+5918	-	-	-	30	27
1837-0606	-	25	17	35	38
1856+0114	SNR W 44	32	24	28	31
1903+0550	SNR 040.5-00.5	-	-	23	-
1928+1733	-	-	-	11	14
1958+2909	-	-	-	0	3
2016+3657	-	-	-	8	5
2020+4017	SNR γCyg	-	-	11	8
2021+3716	-	-	-	8	5
2027+3429	-	-	-	6	-
2033+4118	-	-	-	12	9
2035+4441	-	-	-	16	13
2227+6122	-	-	-	32	-

[a] the minimum zenith angle for observations with CANGAROO III
[b] the minimum zenith angle for observations with HESS I
[c] the minimum zenith angle for observations with MAGIC I
[d] the minimum zenith angle for observations with VERITAS

GAMMA RAY ASTRONOMY IN THE ERA OF *GLAST*

Neil Gehrels
NASA/Goddard Space Flight Center
Mail Code 661
Greenbelt, MD 20771, USA
gehrels@lheapop.gsfc.nasa.gov

on behalf of the *GLAST* Science Team

Keywords: Gamma-ray astronomy; Space astrophysics; Future missions

Abstract The Gamma Ray Large Area Space Telescope (*GLAST*) is a high energy (20 MeV to 300 GeV) gamma-ray astronomy mission scheduled for launch in 2005. The underlying theme of the *GLAST* mission is to explore sites of particle acceleration in the universe. *GLAST* will build on discoveries made by EGRET/*CGRO*, examining blazars, gamma-ray bursts, supernova remnants, pulsars, dark matter and solar flares. In addition, *GLAST* observations may begin to uncover the mystery of the $\sim$ 170 unidentified high-energy gamma-ray. The sensitivity will be $< 6 \times 10^{-9}$ cm^{-2} s^{-1} for a one year survey at high galactic latitude, a factor of $\sim$ 15 better than EGRET.

1. INTRODUCTION

The *GLAST* mission's closest predecessor is the Energetic Gamma-Ray Experiment Telescope (EGRET) instrument on the *Compton Gamma-Ray Observatory* (*CGRO*). The first sensitive all-sky gamma-ray survey was performed by EGRET between 1991 April and 1992 November. After that time EGRET observed objects based on peer reviewed proposals, working under a full duty cycle until 1996 and under a reduced duty cycle until *CGRO* was deorbited in 2000. EGRET made many significant findings during its lifetime, including the detection of $\sim$ 70 AGN with gamma-ray emission (Hartman et al. 1999) and the discovery of long lasting gamma-ray emission from gamma-ray bursts (GRBs) (Hurley et al. 1994). *GLAST* will build on these findings, as

A. Carramiñana et al. (eds.),
The Nature of Unidentified Galactic High-energy Gamma-ray Sources, 321–330.

well as forge a new discovery space. (See http://glast.gsfc.nasa.gov for more information on the *GLAST* mission.)

2. *GLAST* OVERVIEW

The Gamma Ray Large Area Space Telescope (*GLAST*) is scheduled for launch in 2005. *GLAST* will carry a wide-field imaging telescope which will observe gamma-rays in the energy range 20 MeV to 300 GeV. The *GLAST* mission is a collaboration between NASA, the Department of Energy, and institutions in France, Italy, Japan, Sweden and Germany.

A comparison of the mission parameters for EGRET and *GLAST* is listed in Table 1. Some of *GLAST*'s advanced capabilities include:

- *Huge FOV:* the field of view of the *GLAST* mission covers $> 20\%$ of the sky.
- *Broadband:* the *GLAST* instruments cover four decades in energy, including the unexplored region of > 10 GeV.
- *Excellent PSF:* the point spread function of *GLAST* is unprecedented for gamma rays and is a factor of > 3 better than EGRET.
- *No Expendables:* without the expendables of other missions (i.e. the neon/argon gas used in EGRET's spark-chambers), *GLAST* will enjoy a long mission without degradation.
- *Large Area:* *GLAST* will have an area of a factor > 7 larger than EGRET.
- *Improved Sensitivity:* the combination of *GLAST*'s field of view, larger effective area and higher angular resolution will result in a factor of > 15 improvement in sensitivity over EGRET.

3. *GLAST* SCIENCE

Among EGRET's notable detections are gamma-ray emission from blazars, prolonged gamma-ray emission from gamma-ray bursts, and ~ 170 unidentified sources. The following sections describe how *GLAST* will build on these findings to begin to solve some of the mysteries left open by EGRET.

3.1. UNIDENTIFIED SOURCES

The third EGRET catalog includes 172 unidentified sources (Hartman et al. 1999), more than half of the 271 total sources. Proposed

Table 1 *GLAST* LAT Performance Compared with EGRET.

Quantity	EGRET	*GLAST* (LAT)
Energy Range	20 MeV - 30 GeV	20 MeV - 300 GeV
Energy Resolution ($\Delta E/E$)	0.1	<0.1
Effective Area (peak)	1500 cm^2	>8000 cm^2
Field of View	0.5 sr	>2 sr
Angular Resolution	5.8° @ 100 MeV 0.5° @ 10 GeV	< 3.5° @ 100 MeV < 0.15° @ < 10 GeV
Sensitivity (> 100 MeV)*	$\sim 10^{-7}$ cm^{-2} s^{-1}	$< 6 \times 10^{-9}$ cm^{-2} s^{-1}
Mass	1810 kg	3000 kg
Lifetime	1991 - 1997	2005 - 2010

* One year survey at high galactic latitude.

phenomena behind these unidentified sources include supernova remnants, radio-quiet pulsars, massive stars, and molecular clouds. Of course, the unidentified sources could also represent new phenomena altogether. Several different sub-populations are evident in the EGRET data (Grenier & Perrot 1999; Gehrels et al. 2000). The EGRET sources have positions that are $\sim 0.5°$ or greater in size, which is far too large to search for counterparts in other wavebands. *GLAST*, on the other hand, will have positions that are more accurate, between 30 arcsec and 5 arcmin, so multiwavelength correlated observations will be possible and will provide essential information on identifying these sources.

3.2. GAMMA-RAY BLAZARS

The third EGRET catalog lists 66 high confidence observations of gamma-ray blazars (Hartman et al. 1999). Most of these sources were only visible to EGRET while they were flaring, so only a few blazars were detectable during their quiescent emission. This means that for most detected blazars, EGRET was only able to observe them on one or two occasions. *GLAST*, with its large field of view, will be able to monitor most of these blazars in the sky at all times. Also, with its higher angular resolution, it will be able to localize the blazars more accurately than EGRET. In addition, because of its higher sensitivity, *GLAST* will be able to view many of these sources during their quiescent state. Besides monitoring those AGN already discovered by EGRET, *GLAST* will hugely increase the number of AGN detected. Based on the log N-log S plot shown in Figure 1, *GLAST* will detect $\sim$ 3000 AGN at 5σ. Finally, *GLAST* observations combined with TeV observatories will be able to improve spectral studies of blazars, filling in the important band from 20 MeV to 10 TeV with high significance data for hundreds of AGN.

Armed with this additional information from *GLAST*, astronomers will begin to answer the following questions about gamma-ray blazars left open by EGRET:

- What is the population of high-energy blazars?
- What is the nature of the quiescent emission?
- What is the relation to ratio luminosity and variability?
- What are the high-latitude unidentified variable sources?

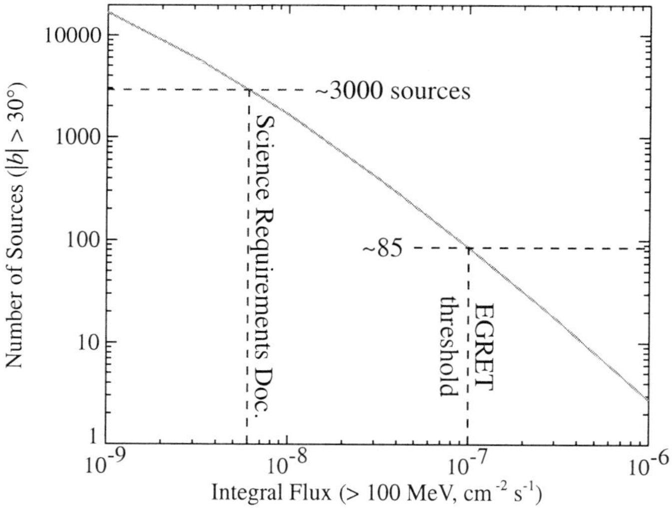

Figure 1 log N-log S plot for estimating the detection rate of AGN for *GLAST*. The curved line is calculated under the assumption that the gamma-ray luminosity function of blazars is proportional to that of radio blazars.

3.3. SUPERNOVA REMNANT ORIGIN OF COSMIC RAYS

Supernova shock acceleration models for cosmic rays (CRs) predict the correct spectra and energetics, and evidence of TeV electrons in supernova remnants (SNRs) has been found in both non-thermal X-ray emission (Koyama et al. 1995) and ground-based gamma-ray observations (Tanimori et al. 1998). However, definitive proof that SNR are indeed the source of CRs is still elusive. Detection of an extended gamma-ray source, such as an interstellar cloud, located next to a SNR (which may be the case for SNR γ-Cygni) would be a smoking gun for the SNR origin of CRs. The left hand panel of Figure 2 shows the EGRET observation of γ-Cygni; the right hand panel shows a simulation of a *GLAST* observation. If the extended gamma-ray source shown in the *GLAST* simulation were, indeed, to be detected, it would provide proof that nuclei as well as electrons are accelerated as predicted. *GLAST* will be able to search for SNRs among the unidentified EGRET sources, as well as resolve some of the SNR sources already detected. These observations may finally lock down the SNR-CR acceleration connection.

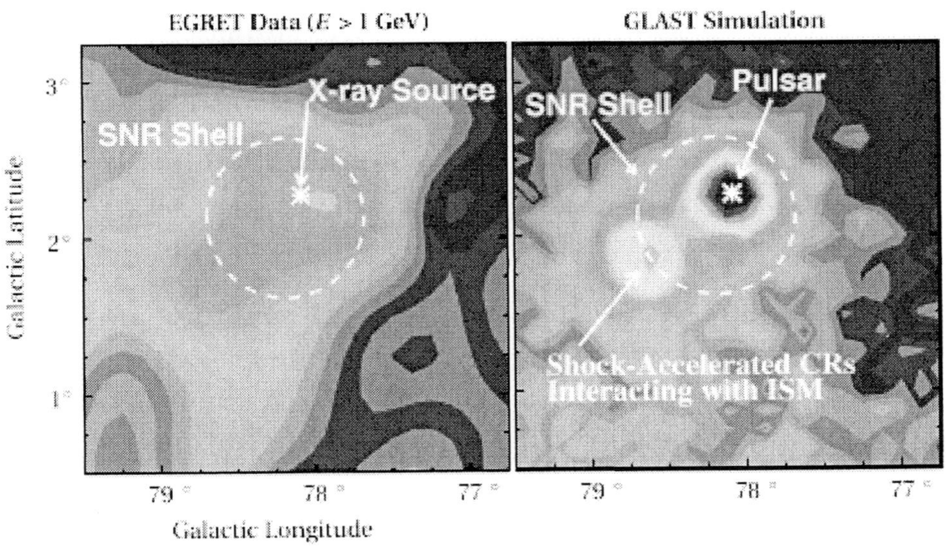

Figure 2 The left-hand panel shows the EGRET data for E > 1 GeV for supernova remnant γ-Cygni. The right-hand panel depicts a simulation of a *GLAST* observation of the same region. The extended gamma-ray source, if present, would provide proof that cosmic rays are accelerated in supernova shocks.

3.4. GAMMA RAY BURSTS

EGRET observed that GRBs can have high energy (> 50 MeV) emission during the burst and extending for about an hour afterwards (e.g. Hurley et al. 1994). The presence of these high energy photons before and after a burst are key to understanding the acceleration mechanisms during the blast and as the blastwave interacts with its surroundings. The results from EGRET, however, provide poor statistics with only 4 bursts detected during the main event and one detected in the long-lasting afterglow. *GLAST* will greatly improve these statistics by detecting $\sim$ 100 bursts/year. It will not be deadtime limited like EGRET was for high count rates. In addition, *GLAST* will be able to observe the main burst for many events and will improve the localization of the events to enable follow-up observations in other wavebands.

4. *GLAST* INSTRUMENT AND INTERDISCIPLINARY SCIENTIST SELECTION

The instruments selected in February and March 2000 to be flown on the *GLAST* mission are the Large Area Telescope (LAT) and the *GLAST* Burst Monitor (GBM). In addition to instrument selection, interdisciplinary scientists (IDSs) have been chosen to participate in the mission. The following sections describe the basic components of each instrument and identify the selected IDSs.

4.1. LARGE AREA TELESCOPE

The LAT (PI: P. Michelson, Stanford) design consists of four elements: Si tracker, CsI calorimeter, anticoincidence shield and data acquisition system. Figure 3 shows a schematic of the LAT system. (See http://glast.stanford.edu/index.html for further information on the LAT)

- *Si Tracker:* the Si tracker will be built as a collaboration between Stanford, University of California Santa Cruz (UCSC) and institutions in Italy and Japan. The design consists of sixteen tower modules, with each module containing interleaved planes of silicon-strip detectors and lead converter sheets.

- *CsI Calorimeter:* the calorimeter will be built by a collaboration from Navy Research Laboratory (NRL) and institutions in France and Sweden. The calorimeter consists of CsI bars arranged in a seg-

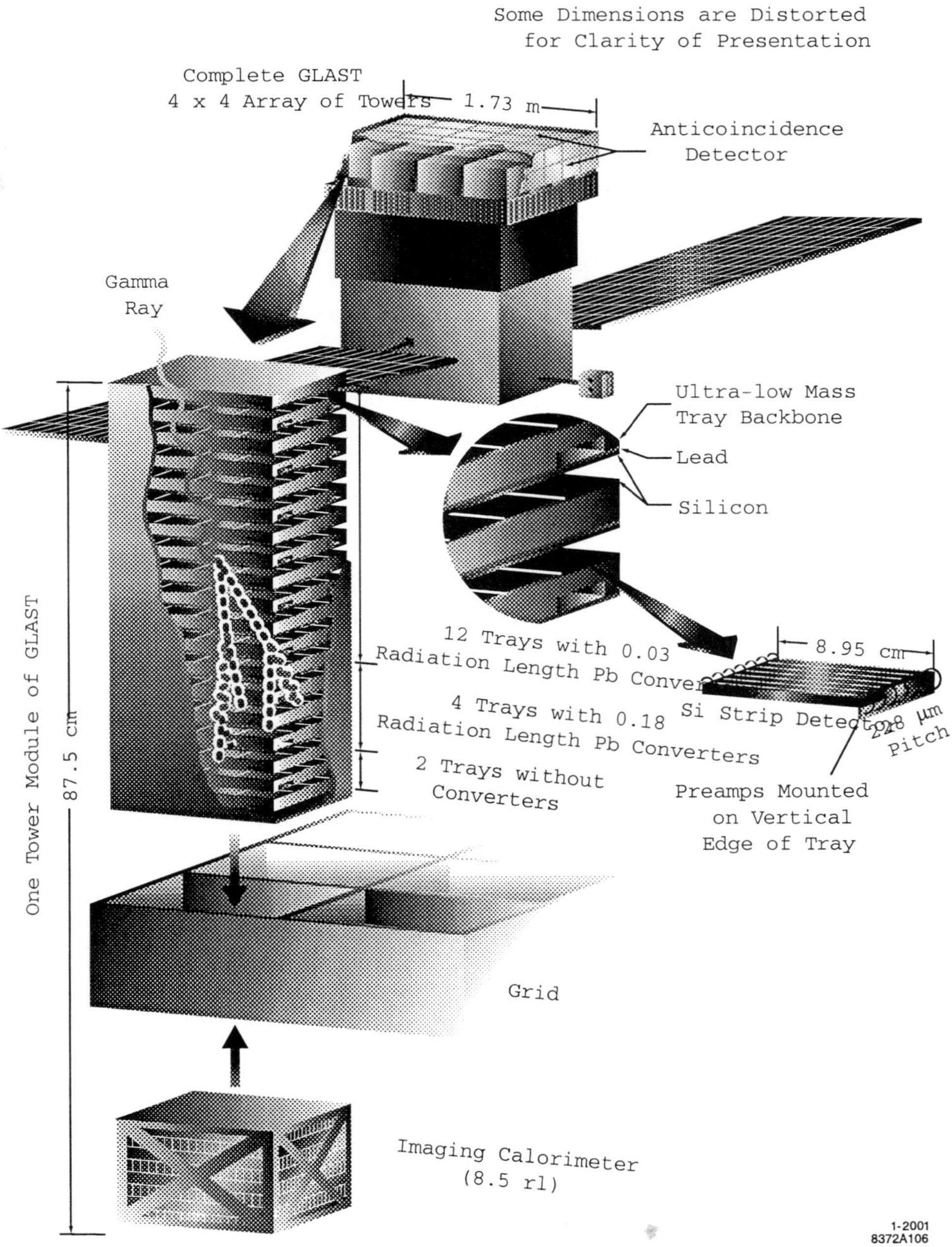

Figure 3 *GLAST* LAT in detail.

Table 2 *GLAST* GBM Performance Compared with EGRET.

Quantity	BASTE	GBM
Energy Range	25 keV - 1.9 MeV (LAD) 7 keV - 10 MeV (SD)	5 keV - 1 MeV (LED) 150 keV - 30 MeV (HED)
Detectors	NaI	NaI/BGO
Field of View	All Sky Not Occulted by Earth	All Sky Not Occulted by Earth
GRB Final Location*	1.7°	< 1.5°
Burst Sensitivity (50-300 keV)	0.2 cm^{-2} s^{-1}	<0.5 cm^{-2} s^{-1}

* Calculated onboard, 1-σ radius for a burst with brightness 10 cm^{-2} s^{-1} in 50-300 keV band and duration 1 s.

mented fashion. These bars will give both longitudinal and transverse information about the energy deposited from the shower.

- *Anticoincidence Shield:* the anticoincidence shield will be constructed at Goddard Space Flight Center and consists of segmented plastic scintillator tiles read out by wave-shifting fibers and photomultiplier tubes.

- *Data Acquisition System:* the data acquisition system will be developed by a a collaboration of Stanford and NRL.

4.2. *GLAST* BURST MONITOR

The GBM (PI: C. Meegan, Marshall Space Flight Center) consists of two types of scintillation detectors: sodium iodide (NaI) and bismuth germanate (BGO). The NaI scintillators will cover the energy range from 5 keV to 1 MeV and will provide burst triggers and locations. The BGO detectors will detect the energy range from 150 keV to 30 Mev, overlapping the NaI scintillators at the lower end and the LAT at the higher end. A comparison of the GBM capabilities to BATSE is given in Table 2. Figure 4 shows a schematic of both

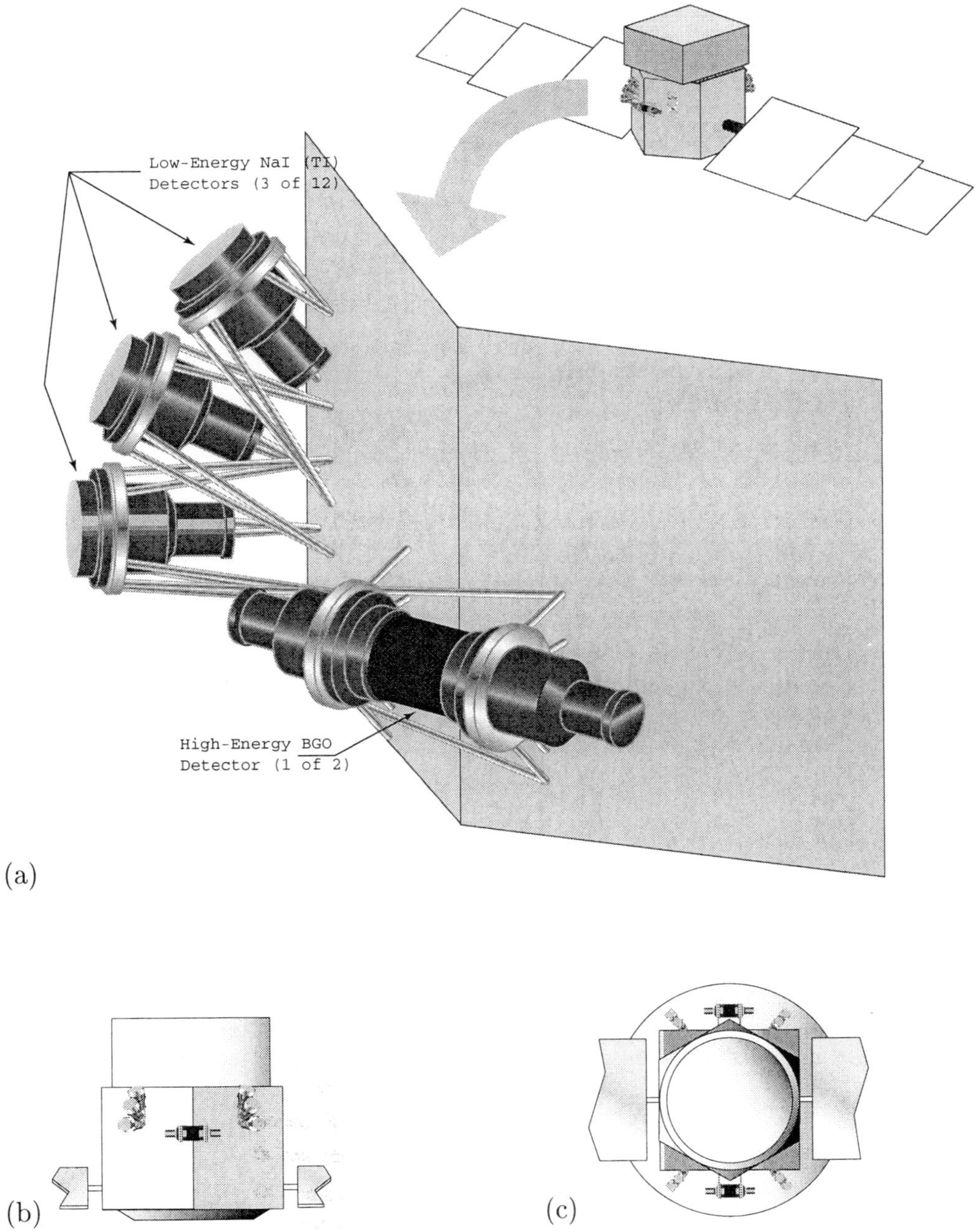

Figure 4 *GLAST* GBM in detail. Panel (a) shows a detail of the placement of the GBM detectors on the side of the *GLAST* spacecraft. The placement of the GBM detectors about the spacecraft from a top and side is shown in panels (b) and (c), respectively.

types of GBM detectors and their possible placement on the spacecraft. (See http://gammaray.msfc.nasa.gov/gbm/ for more information on the GBM).

4.3. IDS SELECTION

The following interdisciplinary scientists (IDSs) have been chosen for the project (each listed with area of expertise): C. Dermer from NRL (non-thermal universe); B. Dingus from Wisconsin (transients); M. Pohl from Ruhr University (diffuse galactic emission); and S. Thorsett from UCSC (pulsars).

References

Gehrels, N., Macomb, D. J., Bertsch, D. L., Thompson, D. J., & Hartman, R. C. (2000). *Nature*, 404:363-365.

Grenier, I. A., & Perrot, C. (1999). *Proc. XXVI Int. Cosmic Ray Conf. Salt Lake City*, 3:476-479.

Hartman, R.C., et al. (1999). *The Astrophysical Journal Supplement Series*, 123:79-202.

Hurley, K., et al. (1994). *Nature* 372, 652.

Koyama, K., et al. (1995). *Nature*, 378:255.

Tanimori, T., et al. (1998). *Astrophysical Journal Letters*, 497:L25.

THE *AGILE* GAMMA-RAY ASTRONOMY SATELLITE

S. Mereghetti, M. Tavani, A. Argan*, A. Chen*, P. Caraveo, F. Perotti, S. Vercellone
Istituto di Fisica Cosmica G.Occhialini, CNR via E.Bassini 15, I-20133 Milano, Italy

G. Barbiellini, M. Prest*, E. Vallazza
Università di Trieste and INFN, via A.Valerio 2, I-34147 Trieste, Italy

G. Di Cocco, C. Labanti, M. Trifoglio
TESRE - CNR, via Gobetti 101, I-40129 Bologna, Italy

E. Costa, M. Feroci, I. Lapshov, A. Rubini, P. Soffitta
IAS - CNR, via Fosso del Cavaliere, I-00133 Roma, Italy

P. Picozza, A. Morselli, V. Cocco*, C. Pittori*
Univ. "Tor Vergata" and INFN, via della Ricerca Scientifica 1, I-00133 Roma, Italy

D. Zanello, P. Lipari
Università "La Sapienza" and INFN, P.le A.Moro 2, I-00185 Roma, Italy

F. Longo
Università di Ferrara, via Paradiso 12, I-44100 Ferrara, Italy
(*) Affiliated to CIFS

Keywords: High energy instrumentation, gamma-ray astronomy satellites

Abstract The *AGILE* satellite will be a unique tool for high-energy astrophysics in the 30 MeV - 50 GeV range before *GLAST*. The scientific performances of *AGILE* are comparable to those of EGRET, despite the much smaller weight and dimensions. The *AGILE* mission will be optimized for the imaging capabilities above 30 MeV and for the study of transient astrophysical sources.

1. Introduction

The *AGILE* satellite will be the next gamma-ray satellite mission operating in the 30 MeV–50 GeV energy range. It will fill the time gap between the end of the EGRET era and the advent of *GLAST*.

A. Carramiñana et al. (eds.),
The Nature of Unidentified Galactic High-energy Gamma-ray Sources, 331–338.

AGILE (*Astro-rivelatore Gamma a Immagini LEggero*, Tavani et al. 1999, Mereghetti et al. 2000) is the first satellite of the Program for Small Scientific Missions of the Italian Space Agency (ASI). It is expected to be operative in the years 2003-2006.

The progress achieved in the technology of solid state silicon detectors makes it possible now to develop within a very small ($\sim$60$\times$60$\times$54 cm^3) and light payload ($\sim$ 65 kg) an instrument as powerful as EGRET (actually more powerful than EGRET in a few respects, as described below).

AGILE, being optimized from the point of view of its imaging performances, will give a significant contribution to solve the mystery of the unidentified gamma-ray sources. The *AGILE* Gamma Ray Imaging Detector (GRID) will cover the 30 MeV - 50 GeV range with a very good angular resolution (allowing a source location accuracy $\sim 5' - 20'$ for intense sources), a very large field of view ($\sim$3 sr), and a sensitivity comparable to that of EGRET for on-axis, and substantially better for off-axis, point sources. *AGILE* will also carry a hard X–ray detector (Super-AGILE) using an imaging coded mask system to simultaneously monitor in the 10-40 keV range the sources observed in the central part of the GRID field of view. The possibility of simultaneously observing the same part of the sky in the hard X–ray and gamma-ray bands, with a combination of detectors on the same satellite, is an unique capability of *AGILE* that will also be very useful in the study of unidentified high energy sources.

2. OVERVIEW OF THE *AGILE* INSTRUMENT

The *AGILE* scientific payload (Figure 1) can be regarded as a single instrument consisting of three detectors (GRID, MCAL and Super-AGILE) plus two auxiliary systems (Anticoincidence and Data Handling).

2.1. GRID

The heart of the GRID (Gamma Ray Imaging Detector) is a silicon Tracker, consisting of 14 planes of Si strip detectors. This will provide the unambiguous identification of incident gamma-rays by recording the characteristic track signature of the e$^-$-e$^+$ that result from pair creation in thin layers of converting material. A CsI Mini-calorimeter placed below the tracker will provide energy information with a resolution of the order of ΔE/E$\sim$1 at 200 MeV. A better energy resolution can be

Figure 1 The *AGILE* payload. Starting from the top, the following subsystems are visible: top Anticoincidence panel, SuperAGILE collimator (the SuperAGILE coded mask is not shown), 14 Silicon Tracker trays, Mini-calorimeter. The whole payload is surrounded by the lateral anticoincidence panels (not shown on two sides for clarity).

achieved at lower energies by the combined use of the Mini-calorimeter and the multiple scattering measured in the Tracker.

Each plane of the Tracker (with the exception of the two at the bottom) is made of three layers: a 245 μm thick photon pair converter in tungsten (0.07 X_0) is followed by two planes of silicon strip detectors (thickness 410 μm) with the strips arranged in orthogonal directions to provide the X and Y coordinates of the particle tracks. The Si detectors have a pitch (distance between strips) of 121 μm, but only half of the strips will be read out, with a floating strip every two strips, yielding a total of ~43,000 channels for the whole Tracker. If a particle crosses a floating strip, charges are induced on the two adjacent read-out strips through capacitive coupling. In this way, more than one strip has signal, thus enabling to use an interpolation algorithm to recover the original spatial resolution.

The Mini-calorimeter consists of two planes of CsI(Tl) bars. Each bar has dimensions of 2.3×1.5×40 cm^3. The two planes have the bars arranged in orthogonal directions to provide the X and Y location of the showers. The scintillation light is collected by photodiodes placed at both ends of each bar.

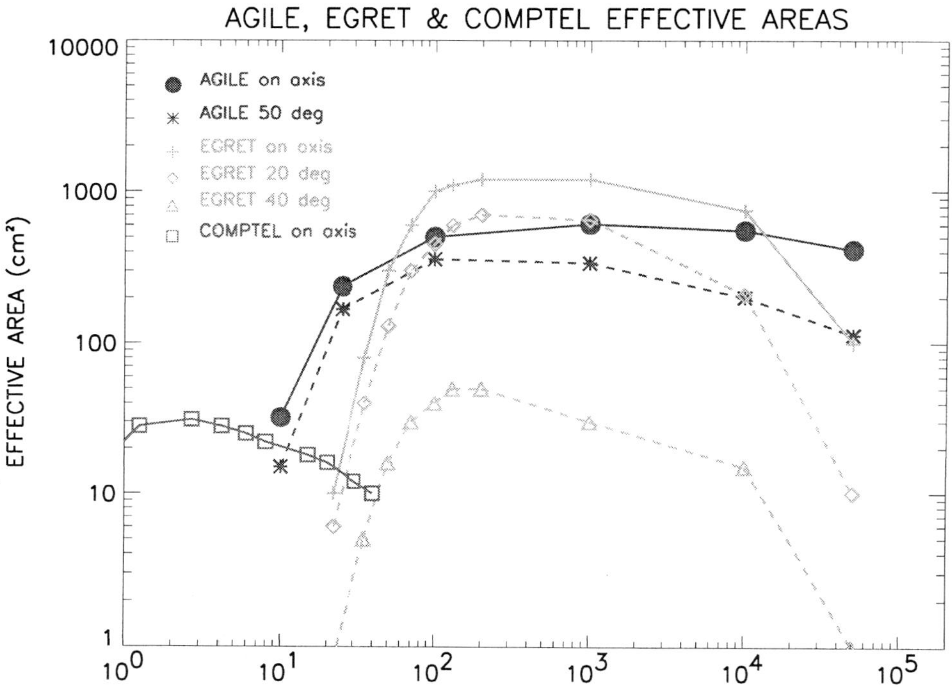

Figure 2 Comparison of the effective area of *AGILE* with those of previous instruments on *CGRO*. The curves refer to different off-axis angles. Note the small decrease at 50° off-axis for the *AGILE* effective area.

The Mini-calorimeter, with a thickness of 1.5 radiation lengths, will allow to determine the energy of the incident photons recorded by the Tracker. In addition, the Mini-calorimeter will be used to study gamma-ray bursts and other transient phenomena in the ~250 keV - 100 MeV energy range (without directional information).

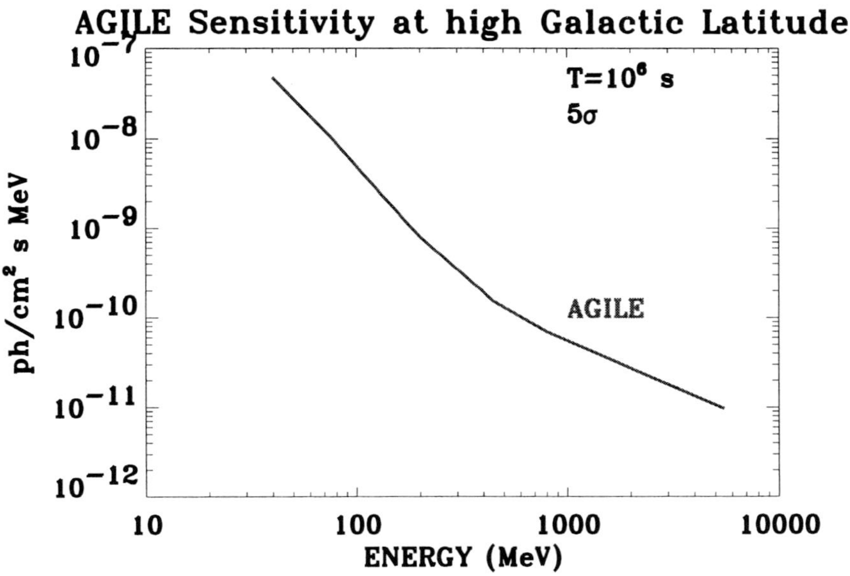

Figure 3 Expected *AGILE* sensitivity for a gamma-ray point source at high galactic latitude.

2.2. SUPER-AGILE

The hard X-ray detector (Super-AGILE) is based on the same building blocks (9.5×9.5 cm^2 silicon tiles) that are used for the Tracker planes. These will be placed on the top of the Tracker (above the first tungsten layer) to form an additional detection plane sensitive in the 10-40 keV range and used in conjunction with a coded mask to obtain imaging capabilities. The mask will be placed at a distance of ~14 cm from the detectors (below the top anticoincidence). Since the silicon microstrips provide pixels only along one dimension, Super-AGILE will consist of four equal modules arranged in two pairs, giving mono-dimensional sky images along two orthogonal directions. The mask will be supported by an ultra-light carbon fiber structure that will also serve as a collimator to reduce the background, which, at these energies, is dominated by the cosmic diffuse radiation.

3. SCIENTIFIC PERFORMANCES

The expected scientific performances of *AGILE* are summarized in Table 1. One of the main characteristics of the *AGILE* GRID is its very large field of view (see Figure 4). As shown in Figure 2 the *AGILE*

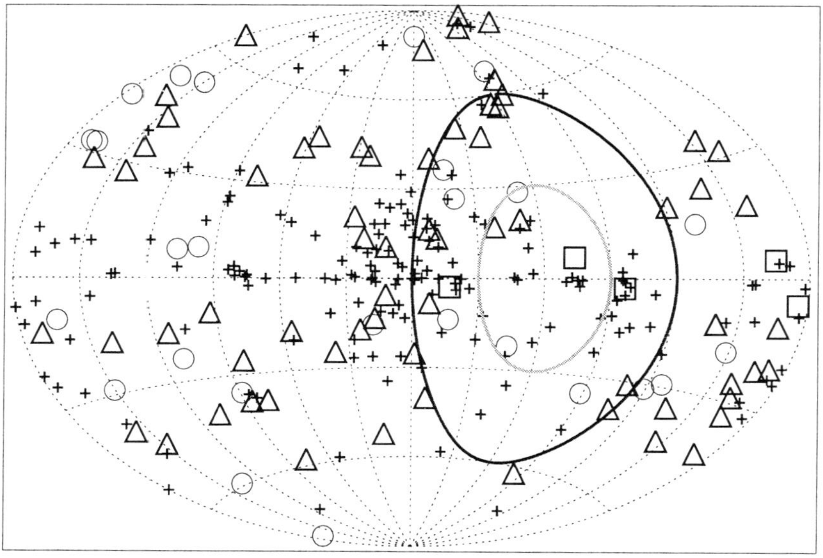

Figure 4 Comparison of the *AGILE* GRID and EGRET fields of view. The symbols indicate the sources of the 3EG catalog (AGNs: triangles, possible AGNs: circles, Pulsars: squares, Unidentified: crosses)

effective area remains almost constant for large off-axis angles. This will allow to simultaneously monitor a large number of sources in a single pointing and will also result, at the end of the mission, in a large exposure factor for each region of the sky (Mereghetti et al. 1999). The great sky exposure factor, coupled with the good angular resolution, will allow a detailed study of the diffuse Galactic and Extra-Galactic emission and to better locate the unidentified EGRET sources. The expected sensitivity for regions outside the Galactic Plane is shown in Figure 3.

Another important characteristic of the *AGILE* Tracker is the very small dead time, ~100 μsec (i.e. three orders of magnitude smaller than for EGRET). This will be crucial in the study of the high energy emission of gamma-ray bursts, which are expected in the field of view at a rate of ~5-10 per year. Furthermore, an on-board GPS system will allow to time tag the photons with an absolute accuracy of only a few μsec.

The extension in the hard X–ray range of a gamma-ray mission, made possible by the Super-AGILE detector, is an innovative concept that will allow the study of correlated variability for sources of different classes, ranging from active galactic nuclei and blazars to unidentified galactic transients.

4. THE MISSION

The *AGILE* payload will be carried in an equatorial (inclination $<5°$), circular orbit (altitude $\sim$550 km), by a spacecraft of the MITA class, developed by Gavazzi Space as prime contractor. The total mass of the satellite will be of the order of 180-200 kg. The satellite will point with a 3-axis attitude stabilization with an accuracy of the order of 1°. A post-facto attitude reconstruction at the arcmin level will be obtained by means of star sensors.

A typical *AGILE* pointing will last 2-3 weeks. *AGILE* will also have the possibility to quickly repoint (within 1 day) in order to react to interesting targets of opportunity.

An Operations and Control Center will have the primary responsibility of satellite operations and communications using the ASI ground base at Malindi (Kenya). The equatorial orbit will allow a single contact per orbit at a downlink rate of 500 kbit s^{-1}.

The *AGILE* Data Center will be devoted to the monitoring of the instrument scientific performance, to the quick look analysis, and to the processing, distribution and archival of scientific data products and the associated calibration data.

Part of the *AGILE* data will be available to the international astrophysical community through a Guest Observers Program managed by ASI, while a number of long term scientific projects involving the systematic analysis of large amount of data and requiring a deep knowledge of the instrument will be carried out by the *AGILE* Team. The Super-AGILE detector will also act as a wide field monitor to study the long term variability of bright sources in the 10-40 keV range and to discover new transients. The Super-AGILE data will be routinely analyzed at the *AGILE* Data Center and the results will be made publicly available as a service to the community, in a way similar to what is currently done with the ASM data on board RossiXTE.

Acknowledgments

We wish to thank all the members of the *AGILE* System Team. In particular M.Conte, L.Soli, E.Rotta, G.De Paris, A.Bernabeo and C.Mangili.

References

Mereghetti S., et al. (1999), Mem.S.A.It., 70, p. 229.

Mereghetti S., et al. (2000), *Proc. GeV-TeV Gamma-ray Astrophysics Workshop*, AIP 515, p. 467.

Tavani M., et al. (1999), Mem.S.A.It., 70, p. 201.

Table 1 *AGILE* Scientific Performances

Gamma-ray Imaging Detector (GRID)	
Energy Range	30 MeV – 50 GeV
Field of view	~3 sr
Sensitivity at 100 MeV	6×10^{-9} ph cm^{-2} s^{-1} MeV^{-1} (5σ in 10^6 s)
Sensitivity at 1 GeV	4×10^{-11} ph cm^{-2} s^{-1} MeV^{-1} (5σ in 10^6 s)
Angular Resolution at 1 GeV	36 arcmin (68% cont. radius)
Source Location Accuracy	~5-20 arcmin S/N>10
Energy Resolution	$\Delta E/E \sim 1$ at 300 MeV
Absolute Time Resolution	~1 μs
Dead time	~100 μs
Hard X–ray Detector (Super-AGILE)	
Energy Range	10-40 keV
Field of view	$107^\circ\times68^\circ$ FW at Zero Sensitivity
Sensitivity	~5 milliCrabs (5σ in 1 day)
Angular Resolution	6 arcmin
Source Location Accuracy	~2-3 arcmin S/N~10
Energy Resolution	$\Delta E<4$ keV
Absolute Time Resolution	~4 μs
Dead time (for each of the 16 readout units)	~4 μs
Mini-Calorimeter	
Energy Range	250 keV - 100 MeV
Energy Resolution	~1 MeV above 1 MeV
Absolute Time Resolution	~3 μs
Dead time (for each of the 32 bars)	~20 μs

PRESENTATIONS OF THE MULTIWAVELENGTH ROUND-TABLE DISCUSSION

Dirk Petry
Dept. of Physics and Astronomy
Iowa State University, Ames, IA 50011
petry@iastate.edu

Sandro Mereghetti
Istituto di Fisica Cosmica G.Occhialini, CNR
via Bassini 15, Milano, I-20133, Italy
sandro@ifctr.mi.cnr.it

Alberto Carramiñana
Instituto Nacional de Astrofísica, Óptica y Electrónica
Luis Enrique Erro 1, Tonantzintla, Puebla 72840, MEXICO
alberto@inaoep.mx

Gustavo E. Romero
Instituto Argentino de Radioastronomía (IAR)
Casilla de Correos No. 5, 1894 Villa Elisa
Buenos Aires, Argentina
romero@irma.iar.unlp.edu.ar

Abstract The round-table discussion was held at the end of the workshop, with short presentations on what can be done in the short future to study unidentified γ-ray sources. The presentations were followed by an intense discussion which -unfortunately- cannot be repoduced here. Brief summaries of four of the six presentations are included here.

A. Carramiñana et al. (eds.),
The Nature of Unidentified Galactic High-energy Gamma-ray Sources, 339–346.

1. THE ROLE OF CHERENKOV TELESCOPES IN SOLVING THE EGRET-UNID PUZZLE

by Dirk Petry

Several new ground-based high-energy gamma-ray observatories are presently under construction or in their startup phase: the Imaging Atmospheric Cherenkov Telescope (IACT) observatories CANGAROO III (e.g. Mori et al. 1999), HESS I (e.g. Hofmann et al. 1999), MAGIC I (e.g. Lorenz et al. 1999) and VERITAS (e.g. Krennrich et al. 1999), observatories using mirror arrays of former solar power plants, CELESTE (e.g. Smith et al. 1998), GRAAL (e.g. Arqueros et al. 1999), STACEE (e.g. Oser et al. 1999), Solar II (e.g. Zweerink et al. 1999), and the large-field-of-view, large-duty-cycle observatories MILAGRO (e.g. McCullough et al. 1999) and ARGO (not yet operational, see e.g. D'Ettore et al. 1999).

The physics program of these new observatories will concentrate on the following source classes (in order of priority) and related aspects for which I only give keywords:

- Blazars (source models, cosmology, diffuse IR-background)
- SNRs (source models, cosmic ray origin)
- Pulsars (source models)
- GRBs (source models, cosmology, quantum gravity)
- Galactic Center (search for WIMP decay line, dark matter)
- Microquasars, CVs, XRBs (source models)

Out of these, all classes except the GRBs are present among the possible candidates for counterparts for EGRET UNIDs. It can therefore be expected that EGRET UNIDs will receive a considerable fraction of the observation time for pointed observations.

For IACTs, which will reach the highest point source sensitivities and lowest energy thresholds in the first phase of the projects, the annual observation time depends on the weather quality and varies between 900 and 1100 hours. If 30 % of this time, i.e. 300 h, can be dedicated to EGRET UNID observations, the four IACT projects combined can observe $300/50 \times 4 = 24$ UNID objects per year down to the nominal sensitivity which is reached after 50 h observation time. At this pace, it will take about two years to cover the most promising targets among the UNIDs (see my other contribution to these proceedings).

For those EGRET UNIDs which can be detected, Cherenkov telescopes will provide flux measurements, spectra, high photon statistics for detecting a possible pulsed component and most of all improved source locations with accuracies of 1' or better which will lead to clear identifications in most cases.

Due to the favourable arrangement of the observatory locations around the earth, always at least two observatories will be able to observe any given object. In this way mutual confirmations and cross-calibrations will easily be possible. The absolute flux calibration of the observatories will later very much be improved by the overlap in the observable energy range with satellite experiments such as GLAST.

2. X–RAY OBSERVATIONS

by Sandro Mereghetti

During this workshop we have seen a number of multi-wavelength investigations of the EGRET error regions. It seems that the great majority of these studies resulted in the discovery of "peculiar" objects that can be considered as likely counterparts of the high-energy sources. In several cases it happens that more than one possible counterpart is found. This is probably an unavoidable consequence of the very good sensitivity, angular resolution, and energy resolution achieved at radio, IR, optical and X–ray wavelengths. Thus, despite the large number of interesting discoveries involved in these multi-wavelength approach, the problem of unanbiguosly associating specific γ-ray sources with other objects is not easily solved. Only the signature of a γ-ray periodicity is left as a sure mean of source identification, in the lack of more accurate localizations in the γ-ray range.

Therefore, looking for pulsars has always been, and will be in the future, an obvious objective, and X–ray observations have in this respect the following advantages:

- young, energetic pulsars are often surrounded by bright synchrotron nebulae. Besides being easily recognizable (by the current X–ray instruments) thanks to their spatial morphology and characteristic spectral shape, their visibility is not affected by beaming.

- X–ray emission from middle aged pulsars is dominated by thermal radiation from the neutron star surface. Again this has the advantage of being emitted over a 4π solid angle in a nearly isotropic way.

The other side of the coin is that interstellar absorption at energy below a few keV limits the volume that can be efficiently sampled in the soft X–ray range, especially at low galactic latitudes.

The two major X–ray satellites currently available, Chandra and XMM-Newton have excellent imaging capabilities up to ~10 keV. They should therefore allow us to discover also highly absorbed, distant galactic objects. The challenge here will be to recognize potentially interesting counterparts among the large number of detected sources, that for typical limiting fluxes of the order of a few 10^{-15} erg cm^{-2} s^{-1} will number hundreds per square degree.

3. OPTICAL OBSERVATIONS

by Alberto Carramiñana

Present attempts to optically identify EGRET sources must follow a multiwavelength approach with an intermediate observation to constrain the source position. Unless an obvious counterpart is present, optical observers need error boxes of radii below 10 to 20 arcsec for a useful search. X-rays provide the best opportunity to properly locate (within 10") optical targets; radio observations can also be useful, although we know some EGRET sources do not show radio emission. Even when *GLAST* comes into orbit, X-ray observations are likely to be needed to procure arc-second locations for follow-up optical observations. Only in exceptional cases will a *GLAST* position directly lead to an optical identification.

The first step to search for optical counterparts is to check archival data, like the Palomar Observatory Sky Survey. If an optical object is present in the X-ray error box spectroscopic identification is the next step; if no object is present one still needs optical observations in search of fainter candidates. The type of objects known or believed to be Galactic gamma-ray sources are among the worst possible targets for an optical observer: isolated neutron stars, black holes and extended diffuse sources (which might even be isolated!). As a consequence, most of the observations made so far have been aimed at discarding objects present in the X-ray error boxes. Empty boxes are suggestive of compact objects, and to bound alternatives upper limits are useful results, as they constrain the nature of the source. In that respect 2m class telescopes have helped but observations in larger telescopes ($\geq 4-6$ m) are needed to set limits on sources fainter than 25-th magnitude. In examples like that of the CTA-1 source, 8-10m telescopes are required if one wants to go from the know colours (6m observations) to a spectroscopic identification.

Three types of observations can be considered:

1 spectroscopic identification: this has been used to discard sources in the cases of γ-Cygni and 3EG J1835+5918. In more than one case there is an object too faint for identification and future spectroscopic observations are likely to require large telescopes.

2 colours: a neutron star is faint and blue. In that respect HST observations can be very useful, as in the example of CTA-1. HST has the power to detect isolated neutron stars, and to measure their colours, as proven by the detection of PSR B1055–52.

3 period searches: although X-rays are the preferred band to search for periodicity in Geminga-like objects, the optical region should not be underestimated. Several high quantum efficiency devices with rapid response are now becoming available. Two examples represented in this workshop are OPTIMA (MPE Garching; Straubmeier & Kanbach 2000) and SIS (Stanford Univ.; Romani et al. 1999). These might be able to find pulsations of unknown frequency for sources just a magnitude or two brighter than the detection limit. It is important to seek the use of these instruments in large telescopes. Confirmation of periodicity of known frequency can be made cheaply using a stroboscopic wheel and a standard CCD (like the one used for Crab and pulsar studies at the Cananea Observatory - Galičič 1999)

Finally, studies of sources in the Galactic plane must not overlook the effects of extinction. Infrared observations should be in the agenda, in order to penetrate deeper into the Galactic plane. These observations should first measure (infrared) colours and, whenever feasible, infrared spectra.

4. RADIO OBSERVATIONS

by Gustavo Romero

4.1. WHY RADIO OBSERVATIONS?

The same population of relativistic electrons that generates gamma-rays through IC scattering of ambient photons in many astrophysical situations should also produce radio emission by synchrotron mechanism if a magnetic field is present at the sources. Even if the observable gamma rays are the result of hadronic interactions, the same mechanism that accelerates the hadrons should operate on the leptons, producing relativistic electrons and, once again, radio emission. Thus, radio emission should almost always accompany the high-energy gamma-ray radiation. The problem, of course, is whether the radio flux is significant enough as to be detected with the current technology.

Basically, we have two types of radio observations: single dish and interferometric observations. Observations of the first kind are useful to detect large and weak sources in the galaxy, like supernova remnants. If the source is relatively close to the galactic plane, then filtering techniques are required to remove the background contaminating emission from the Galaxy and emphasize the fine structure of the sources. Some regions are so confusing that is is not always possible to obtain a satisfactory map, despite our recent progress in image processing. The main problem of single dish observations in the radio continuum is that the probability of chance association with an unidentified gamma-ray source (UNID) is not too low if the detected radio source is extended. In order to get these probabilities to a lower level, simultaneous line observations (HI, CO) should be performed. If an atomic or molecular cloud is detected, for instance, at the same position where a new supernova remnant is overlapped with an UNID, the a priori probability of a physical relation could be not negligible (see, for instance, the case studied by Combi et al. 1998).

Interferometric observations, on the contrary, are useful for observing compact sources. The main problem here is that the location error boxes of the UNIDs are so large that we have many potential radio counterparts. A typical VLA or ATCA mosaic image of the inner gamma-ray confidence contours can contain 10-20 point-like sources. How can we recognize the best candidates? When we should use more powerful imaging capabilities like the VLBA or VLBI networks?. The answer seems to be in cross correlation and variability studies. Frequent VLA or ATCA observations can allow us to isolate those non-thermal radio sources which are variable on timescales from month to years. Cross correlated studies with optical and X-ray observations can lead then to the identification of the most promising candidates with yet unnoticed compact objects like early-type stars or peculiar stellar binary systems. High-resolution observations of these objects can reveal features like jets or evidence of colliding stellar winds that could be indicative of gamma-ray emission. This methodological approach has been recently adopted by several researchers with very promising results (e.g. Paredes et al. 2000, Benaglia et al. 2000).

4.2. WHAT CAN WE DO IN THE NEXT YEARS?

We must wait at least five years till GLAST and its improved capabilities. However, in the meanwhile, a lot of work should be done from the point of view of a radio astronomer. First at all, the radio fields of the UNIDs should be explored and candidates (non-thermal point-like sources) should be clearly pointed out. These sources will be used in cross correlation studies in the next years. At the same time, some particular sources, which are within the contours of clearly variable UNIDs, should be followed up in order to establish their time behaviour. The variable ones should be observed with improved angular resolution when possible.

On the other hand, we expect that many nearby early-type stars will be detected by GLAST. A detailed knowledge of the radio spectra of these sources is desirable in order to constrain the models of gamma-ray production. Since the IC contribution from these stars is expected to peak at a few MeV (Benaglia et al. 2000), a few of them could be even detected with IBIS imager of the forthcoming INTEGRAL satellite. The radio spectrum of these stars could be used to estimate the IC flux and then we could differentiate at higher energies the fraction of radiation coming from hadronic interactions.

Single dish observations of the radio fields around UNIDs will continue in the next years, in the hope of finding new supernova remnants that could be producing gamma-rays through interactions with interstellar clouds or that could be associated with the creation of a potentially gamma-ray emitting compact object. In the latter case, the gaseous remnants could be very useful to estimated the age of the compact relic. A combination of both single dish and interferometric observations at several radio bands, as recently implemented by Punsly et al. (2000), could lead to interesting discoveries and new challenges. Even if our search fails in leading us to a clear identification of the counterparts of the UNIDs in the Third EGRET catalog, we will be preparing the scenario to optimize the early analysis of GLAST's results. Many of the things that we could find will be surely worthy of our effort, independently of whether they are, or not, associated with high-energy gamma-ray sources.

Acknowledgments

Thanks Alberto for a wonderful workshop (*ger*).

References

Arqueros, F., et al., 1999, in Kieda, D. et al. (eds.) Proc. "26th International Cosmic Ray Conference", 5, 215

Benaglia P., Romero G.E., Stevens I.R., Torres D.F. 2000, A&A, in press [astro-ph/0010605]

Combi J.A., Romero G.E., Benaglia P. 1998, A&A 333, L91

D'Ettore Piazzoli, B., et al., 1999, in Kieda, D. et al. (eds.) Proc. "26th International Cosmic Ray Conference", 2, 373

Galičič, M., 1999, PhD Thesis, Univ. of Ljubljana

Hofmann, W., et al., 1999, in Dingus, B.L., et al. (eds.) Proc. "Towards a Major Atmospheric Cherenkov Detector VI", AIP Proc. 515, 500

Krennrich, F., et al., 1999, in Dingus, B.L., et al. (eds.) Proc. "Towards a Major Atmospheric Cherenkov Detector VI", AIP Proc. 515, 515

Lorenz, E., et al., 1999, in Dingus, B.L., et al. (eds.) Proc. "Towards a Major Atmospheric Cherenkov Detector VI", AIP Proc. 515, 510

McCullough, J.F., et al., 1999, in Kieda, D. et al. (eds.) Proc. "26th International Cosmic Ray Conference", 2, 369

Mori, M., et al., 1999, in Dingus, B.L., et al. (eds.) Proc. "Towards a Major Atmospheric Cherenkov Detector VI", AIP Proc. 515, 485

Oser, S., et al., 1999, in Kieda, D. et al. (eds.) Proc. "26th International Cosmic Ray Conference", 3, 464

Paredes J.M., Martí J., Ribó M., Massi M. 2000, Science 288, 2340

Punsly B., Romero G.E., Torres D.F., Combi J.A. 2000, A&A, in press [astro-ph/0007465]

Romani, R.W., et al. 1999, ApJ 521, L153

Smith, D.A., et al., 1998, in Aubourg, E. et al. (eds.) Proc. "19th Texas Symposium on Relativistic Astrophysics", Nucl. Phys. B Proc. 80

Straubmeier, C., Kanbach, G., 2000, ASP Conf Ser 202, 311

Zweerink, J.A., et al., 1999, in Kieda, D. et al. (eds.) Proc. "26th International Cosmic Ray Conference", 5, 223

VII

SUMMARY

UNIDENTIFIED GAMMA-RAY SOURCES: SUMMARY AND CONCLUSION

David J. Thompson
Laboratory for High Energy Astrophysics
NASA Goddard Space Flight Center
Greenbelt, Maryland
USA
djt@egret.gsfc.nasa.gov

Keywords: Gamma-ray sources, unidentified

Abstract As the era of the Compton Gamma Ray Observatory draws to a close, the problem of the unidentified Galactic sources remains one of the great unsolved mysteries. Participants in this meeting have made substantial progress in defining the problem, suggesting possible candidate objects, and anticipating the future steps that will lead to a resolution.

1. INTRODUCTION

The goal of this meeting was to examine current work and develop better ways to approach the problem of the unidentified sources in the future. In this brief summary, I organize the conference material in terms of the present, the near future, and the more distant future. All references are to papers in this volume or material presented at the conference.

2. THE PRESENT

Although systematic uncertainties remain a concern (Reimer), population studies (Gehrels, Grenier, Mereghetti), individual source spectra (Mukherjee), time variation (Gehrels, Punsly), spatial distribution (Romero, Torres, Parades), multiwavelength comparisons (Mukherjee, Roberts), and correlation analyses (Romero, Carramiñana) all point to one important conclusion: There are new populations of high-energy

A. Carramiñana et al. (eds.),
The Nature of Unidentified Galactic High-energy Gamma-ray Sources, 349–352.

gamma-ray sources among the presently unidentified sources of the Third EGRET Catalog! This fact strongly motivates new work.

Among the candidate objects considered at this meeting were:

- Unidentified Blazars (Mukherjee)
- Microquasars (Romero, Rodríguez, Martí, Parades)
- Diffuse Sources (Digel)
- Magnetized Black Holes (Punsley)
- Early Stars with Strong Winds (Romero)
- SNR (Torres, Ellison, Slane, Fegan)
- Variable Pulsars (Romero)
- Older Pulsars (Mukherjee)
- Young Pulsars in Binaries (Kaaret, Romani)
- Young Isolated Pulsars (Kanbach)
- Radio-Quiet Pulsars (Romani)
- Pulsar Wind Nebulae (Roberts)
- Off-beam Pulsars (Baring)
- Millisecond Pulsars (Grenier, Romani)

The consensus was that no single class of objects is sufficient to explain the variety of gamma-ray sources already seen.

3. THE NEAR FUTURE

In the next few years, progress can be made despite the absence of new gamma-ray observations. This work will fall into two broad areas of research:

- Multi-Wavelength Observations, Correlations, and Spectral Energy Distributions

  This powerful technique, several examples of which were presented by Mukherjee, Roberts, Kaaret, and Parades, works downward in photon energy from gamma rays to X-rays, optical, and radio. The goal is to find unusual objects in the relatively large gamma-ray error boxes. This approach has produced some good candidates, but not all gamma-ray sources have yet been studied. Of particular interest will be the new generation of X-ray telescopes (Slane).

- Theory and Modeling

  Just as important as observing candiddate sources is the study of what the various candidate source classes expect to produce at other wavelengths. Ways to explore implications of suggested correlations and specific predictions for the future were included in talks by Romero, Romani, Baring, Kaaret, Ellison, Rodríguez, Parades, and Punsly.

4. THE MORE DISTANT FUTURE

Ultimately, new gamma-ray observations will be required in order to localize sources, measure source spectra and time variability, and test the models that have been developed based on existing observations. This work will encompass two broad areas:

- New Gamma-Ray Telescopes

  The new generation of Very High Energy (VHE) telescopes will include CELESTE, STACEE, HESS, CANGAROO-2, MAGIC, and VERITAS, all of which will have higher sensitivity and lower energy thresholds than the present generation of ground-based gamma-ray telescopes (Petry, Fegan). These telescopes will give better source locations, photon statistics, and time resolution.

  The small Italian successor to EGRET, AGILE, will be launched in 2002 (Mereghetti). With a source sensitivity comparable to EGRET but a larger field of view and better angular resolution, AGILE will confirm and extend EGRET results.

  The next major advance in high-energy gamma-ray astronomy will be GLAST (Gehrels). GLAST will detect thousands of sources and will provide arcmin positioning for many brighter sources.

  At lower energies, INTEGRAL will give good source locations and high spectral resolution (Grenier). Another possible medium-energy gamma-ray telescope that would have much higher sensitivity than COMPTEL is MEGA, currently under study in Germany (Kanbach).

- Correlated Observations across the Spectrum

  Even the new gamma-ray telescopes will not come close to matching the spatial and spectral resolution possible at longer wavelengths. Multiwavelength astronomy has proven to be essential for many astrophysical topics, particularly those non-thermal sources that are seen strongly in gamma rays. It will be important for gamma-ray astrophysics to cooperate with other wavelengths to study variability on all time scales as well as simultaneous, broad-band spectral measurements. In order to achieve the maximum astrophysical information, coordinated efforts will be required.

5. CONCLUSION

The study of unidentified gamma-ray sources promises to be a dynamic, exciting, ongoing field of research.

Topic Index

Published in
ASTROPHYSICS AND SPACE SCIENCE LIBRARY

- **Volume 262: Whistler Phenomena Short Impulse Propagation**
  Csaba Ferencz, Orsolya E. Ferencz, Dániel Hamar, János Lichtenberger (Eds.)
  Hardbound, ISBN 0-7923-6995-5, June 2001
- **Volume 260: Solar Cosmic Rays**
  Leonty I. Miroshnichenko
  Hardbound, ISBN 0-7923-6928-9, May 2001
- **Volume 259: The Dynamic Sun**
  Arnold Hanslmeier, Mauro Messerotti, Astrid Veronig
  Hardbound, ISBN 0-7923-6915-7, May 2001
- **Volume 258: Electrohydrodynamics in Dusty and Dirty Plasmas Gravito-Electrodynamics and EHD**
  Hiroshi Kikuchi
  Hardbound, ISBN 0-7923-6822-3, June 2001
- **Volume 257: Stellar Pulsation - Nonlinear Studies**
  Mine Takeuti, Dimitar D. Sasselov
  Hardbound, ISBN 0-7923-6818-5, March 2001
- **Volume 256: Organizations and Strategies in Astronomy**
  André Heck
  Hardbound, ISBN 0-7923-6671-9, November 2000
- **Volume 255: The Evolution of the Milky Way Stars versus Clusters**
  Francesca Matteucci, Franco Giovannelli
  Hardbound, ISBN 0-7923-6679-4, January 2001
- **Volume 254: Stellar Astrophysics**
  K.S. Cheng, Hoi Fung Chau, Kwing Lam Chan, Kam Ching Leung
  Hardbound, ISBN 0-7923-6659-X, November 2000
- **Volume 253: The Chemical Evolution of the Galaxy**
  Francesca Matteucci
  Hardbound, ISBN 0-7923-6552-6, May 2001
- **Volume 252: Optical Detectors for Astronomy II State-of-the-art at the Turn of the Millennium**
  Paola Amico, James W. Beletic
  Hardbound, ISBN 0-7923-6536-4, December 2000

- **Volume 251: Cosmic Plasma Physics**
  Boris V. Somov
  Hardbound, ISBN 0-7923-6512-7, September 2000
- **Volume 250: Information Handling in Astronomy**
  André Heck
  Hardbound, ISBN 0-7923-6494-5, October 2000
- **Volume 249: The Neutral Upper Atmosphere**
  S.N. Ghosh
  Hardbound, ISBN 0-7923-6434-1, July 2000
- **Volume 247: Large Scale Structure Formation**
  Reza Mansouri, Robert Brandenberger
  Hardbound, ISBN 0-7923-6411-2, August 2000
- **Volume 246: The Legacy of J.C. Kapteyn**
  Studies on Kapteyn and the Development of Modern Astronomy
  Piet C. van der Kruit, Klaas van Berkel
  Hardbound, ISBN 0-7923-6393-0, August 2000
- **Volume 245: Waves in Dusty Space Plasmas**
  Frank Verheest
  Hardbound, ISBN 0-7923-6232-2, April 2000
- **Volume 244: The Universe**
  Visions and Perspectives
  Naresh Dadhich, Ajit Kembhavi
  Hardbound, ISBN 0-7923-6210-1, August 2000
- **Volume 243: Solar Polarization**
  K.N. Nagendra, Jan Olof Stenflo
  Hardbound, ISBN 0-7923-5814-7, July 1999
- **Volume 242: Cosmic Perspectives in Space Physics**
  Sukumar Biswas
  Hardbound, ISBN 0-7923-5813-9, June 2000
- **Volume 241: Millimeter-Wave Astronomy: Molecular Chemistry & Physics in Space**
  W.F. Wall, A. Carramiñana, Luis Carrasco, P.F. Goldsmith